MW00837839

THE AUDIOPHILE'S PROJECT SOURCEBOOK

THE AUDIOPHILE'S PROJECT SOURCEBOOK

G. Randy Slone

McGraw-Hill
New York Chicago San Francisco Lisbon London Madrid
Mexico City Milan New Delhi San Juan Seoul
Singapore Sydney Toronto

Library of Congress Cataloging-in-Publication Data

Slone, G. Randy.
The audiophile's project sourcebook / G. Randy Slone.
p. cm.
ISBN 0-07-137929-0
1. Sound—Recording and reproducing—Equipment and supplies—Amateurs' manuals.
I. Title.

TK9968.S56 2002
621.389'3—dc21 2001044227

McGraw-Hill

A Division of The ***McGraw-Hill*** *Companies*

6 7 8 9 0 DOC/DOC 0 9

ISBN 978-0-07-183263-2

The sponsoring editor for this book was Scott Grillo, the editing supervisor was Daina Penikas, and the production supervisor was Pamela A. Pelton. It was set in Melior per the CMS design specs by Joanne Morbit of McGraw-Hill Professional's Hightstown composition unit.

CONTENTS

PREFACE

The Audiophile's Project Sourcebook is intended for electronics hobbyists and audiophiles who enjoy constructing their own "high-end" audio equipment, and who also desire a more complete understanding of the fundamentals involved with audio design. In addition, the information contained in this book will equip audio enthusiasts with the practical knowledge for experimentation and modifications to existing audio equipment, as well as clarifying issues of application compatibility.

A well-rounded variety of domestic and professional audio projects is provided, including preamplifiers, filter circuits, headphone amplifiers, power amplifiers, effects circuits, power supplies, protection circuits, and display circuits. For the most part, these projects represent the state of the art in performance capabilities, and will exceed the performance parameters of most commercially available equipment. In addition to the complete schematics and technical descriptions, I have provided the PC board artwork and layout illustrations for many of the more complex designs.

I have assumed that the reader is experienced in the fundamentals of electronics and electronic construction, but I have made few assumptions regarding the reader's expertise relating specifically to audio electronics. Appendix A provides a reasonably exhaustive clarification of audio electronics terms and modern "buzz" words as a further barrier against confusion.

The Audiophile's Project Sourcebook is for hobbyists and audiophiles who have been discouraged by the apparent schism between audio professionals and extravagant esoterics. It is designed to help anyone experience the ultimate in sonic quality at a reasonable expense. The concepts and principles presented are not subjective, mythical, or traditional; they are scientific and extensively supported with provable analysis. Many readers will be pleasantly surprised at the low cost involved with even the best audio systems.

Above all, I have directed this book toward people who want to experience the fun, satisfaction, and fulfillment of superb and practical electronic projects (not to mention the significant cost savings). This text is not infested with watered-down demonstration or educational projects—it is a sourcebook of impressive, high-performance designs that lend legitimacy to the reader's construction efforts. I sincerely hope that many will enjoy it and benefit from it.

ACKNOWLEDGMENTS

As a Christian, my first and foremost expression of appreciation belongs to my Lord and Savior, Jesus Christ, who keeps me and assists me in all honorable and productive goals. Also along the personal vein, a special debt of gratitude goes out to my wife, Mary Ann, who somehow manages to provide me with great measures of understanding and encouragement, even though she seldom understands what the heck I'm doing.

From the professional perspective, I am greatly indebted to my friend and associate at ZUS Audio, Russell Torlage, for his continual feedback and support of this project. A heartfelt thanks also goes out to Joe Koenig, Luis Alves, and the exemplary team at Electronics Workbench for their assistance in many problematic technical areas.

The heart of this book has been beating in audiophiles for the past four decades, so the majority of credit for this work really belongs to many pioneers, both past and present. Progress in the audio fields is effectively a group effort, with many excellent audio engineers and dedicated hobbyists deserving the majority of credit.

And finally, all of my efforts and support would be meaningless without the professionalism and expertise provided by the incomparable group at McGraw-Hill. I am greatly indebted to their progressive leading and encouragement.

G. Randy Slone

THE AUDIOPHILE'S PROJECT SOURCEBOOK

CHAPTER 1

ESTABLISHING THE BASICS

It is prudent to establish a basic foundation of concepts and goals as an opening prelude to this textbook. It is only fair to begin with a few brief comments that lend legitimacy to the serious time and money invested in the audio fields by many dedicated hobbyists and professionals, followed by a brief discussion of the *common denominators* inherent to all audio circuit designs.

The Psychological Tie Between Sound and the Human Mind

For a variety of reasons, sound is often placed on the back burner of human senses—we have the tendency to take it for granted. It is common to associate the importance of a sense with our ability to function without it. From this perspective, most every person will acknowledge our sense of sight as being of paramount importance, which is undoubtedly true. But have you ever taken a few moments to seriously consider how tightly your sense of hearing is integrated into the deepest recesses of your very being?

For example, when was the last time that you had an experience of such beauty that you felt the tendency to cry? Was it due to something you saw or something you heard? Are the majority of the most profound events in your life associated with things you heard or things you saw? Does a beautiful painting incite the same levels of emotional swell inside of you as a beautiful song? How many musical CDs do you own compared to the number of artistic paintings in your home? There is no question that the same emotional heights reached at sporting events can also be seen at musical concerts. Fan clubs established for famous musicians have equal enthusiasm with fan clubs devoted to movie stars. From the perspective of the *effect* generated in our lives, it would be difficult to assign a greater degree of importance to either sound or sight.

Hundreds of years before the birth of Christ, Aristotle (BC 384–322) offered a somewhat enigmatic statement, "Why do rhythms and melodies, which are composed of sound, resemble the feelings; while this is not the case for tastes, colors, or smells?" Those professionals who have undertaken serious research into the emotional, psychological, and spiritual connections between sound and human response (called *psychoacousticians*) have discovered profound relationships that are not fully documented or adequately understood. Such underfunded research continues as I write this text. Even with the shortcomings of understanding associated with psychoacoustics, it is a well-documented fact that for thousands of years (presumably since the beginning of humankind), music and song have been the fundamental basis for celebration, entertainment, religious worship, and almost every facet of recreation in general. Patriotic song has always been a tool for unifying a large mass of people in nationalistic fervor. Drums have been used by civilizations all over the world, throughout history, to trigger aggressive tendencies as preparation for war, and trumpets have traditionally voiced the battle cries. Every significant cause has an associative song. Angels are associated with song and music throughout the Bible, and Christmas music is the heart of the yearly holiday season. We listen to music in our homes, cars, and businesses; we even wear portable tape players while jogging. There is always a multitude of boom boxes at the beach, stereo systems at parties, background music in commercial businesses, and even elevator music in elevators!

The physics relevant to sound is a fascinating study to undertake. The human ear and eye are both incredible at their sensitivity and dynamic range. For example, the eye is capable of perceiving a single photon, while the ear is capable of hearing a sound created by air moving over a distance equal to the radius of a hydrogen atom! Comparing both of these sensitivity levels to the brightest flash of light or the loudest boom of sound, the dynamic range of our audio/video senses is approximately 1000 million times. However, when we compare the bandwidth of light to audio, the difference is quite dramatic. The entire visible light spectrum is concentrated into approximately one octave of bandwidth, whereas the audio spectrum spans 10 octaves. Here is another interesting thought. We see things in three dimensions—height, width, and depth. How many dimensions do you listen in? I'm not sure that anyone has accurately pinned down that answer yet.

Sound is a survival sense, a tool of education and communication, and an artform. Threatening sounds warn us of impending danger, we listen to our educators and verbally converse with our family and peers, and we appreciate beautiful music as we appreciate a classic painting, a fine glass of wine, or a gourmet meal. It is difficult to place a relative importance on the effect of sound in our lives, and it is abstract, to say the least, to attempt to understand why a variety of frequencies, harmonic qualities, sound pressure levels, and periodical rhythms can be summed together to stir our deepest emotional nature. I'm confident that an improved understanding of the human psychology relative to sound will emerge as more research is continued in the field of *virtual reality* (VR), since aural simulation will play a dominant role in the overall VR effect. The future of audio in our lives should be very, very interesting.

Why Audio Electronics?

Most individuals who are deeply interested in achieving high-quality performance from their audio equipment (commonly called *audiophiles*) have a natural and logical interest in the mechanics of sound. In the beginning, such interest may manifest itself in simpler, less technically oriented tasks, such as experimenting with speaker cables, adding commercial subwoofer units, or auditioning differing program formats. However, if the audiophile's interest is sufficiently fueled by "messing around" with the easier stuff, the natural progression is to become involved with the deeper, more complex areas of audio electronics. To aid the audiophile in this endeavor, there is a consistent outpouring of hobbyist material, in the form of books and periodicals, which serves to keep the interest level sustained while simultaneously training the audiophile in both practical experience and technical prowess. I am repeatedly amazed at the technical knowledge of many audio electronic hobbyists who have never received any formal education in the field of electronics. Oftentimes, they appear to have a better practical knowledge of the field than many formally trained engineers.

As a logical extension of becoming highly proficient in the audio electronics field, there are a variety of professional directions to pursue. Most notable audio engineers started out as weekend hobbyists, and later discovered that they could convert their beloved hobby into a lucrative career. Many audio equipment manufacturers have started from the same humble beginnings. Fortunately, it isn't necessary to develop a complete line of audio products to achieve such a goal. Rather, it is usually more practical to devote your primary attention to one small facet of the industry that you have the talents to excel in. For example, I have a friend who began by devoting his hobby energies toward developing high-performance passive crossover networks for speaker systems. After investing the time and money to develop a good reputation for himself, he developed a full-time career in manufacturing a variety of crossover networks (and associated components) for the audio marketplace. If you have a great love for audio electronics, the goal of turning your favorite pastime into a profitable career is certainly a strong motivating force to continue and grow in your capabilities.

Most individuals who purchase this book will probably not harbor any aspirations of developing a career for themselves in the field of audio electronics. But from the hobbyist perspective, there are many other motivational factors that come into play as time progresses. One factor is the extraordinary monetary savings that can be realized by building your own equipment.

It is no secret within the electronics manufacturing environment that the materials cost of a product may only represent 10% (or less) of its final selling price. The reasons for this situation are simple. First, the R&D (i.e., research and development) costs of the product must be reclaimed in the selling profit. Second, the labor costs in the manufacture and assembly of the item must be added. Third, the wholesale and retail merchants must receive their profit, along with the manufacturer. And finally, there are marketing, packaging, shipping, and technical support costs that must be added. If the typical audio electronics hobbyist has the necessary technical information, coupled with a sufficient degree of assembly experience, it is possible for the hobbyist to realize a cost savings of 100% to 1000% in comparison to equivalent commercial equipment. Recognizing that an elaborate "high-end" domestic audio system may cost up to $5000, or more, the savings incurred by constructing much of the equipment in one's spare time can add up to a substantial amount.

Another motivational factor for being an audio electronics hobbyist is the capability of constructing many items that may not be commercially available (or exorbitantly expensive and/or difficult to obtain). Many audio electronics books or periodicals contain construction and technical information detailing unique amplifiers, preamplifiers, speaker systems, and many other hardware items, which simply may not be available on the commercial market. In many cases, these items may be superior to commercial units, or they may fix a unique problem that is highly application dependent. In any event, it is enjoyable, interesting, and gratifying to construct a variety of these projects. And it never hurts to bolster your ego a little by being able to say to your friends, "I built this myself." (Be careful with such harmless boasts—they often backfire! It isn't unusual to have a dozen friends come back at you with the request, "Will you build one for me, considering the fact that we're such good friends?" There goes your spare time for a month!)

One of the strongest motivational factors among audio electronics hobbyists is the fact that they can actually construct audio equipment that is audibly superior to commercial equipment. Manufacturers must remain as competitive as possible, and this results in a large quantity of compromises in the end products. Also, for the most part, the larger manufacturers have not invested very much money into research and development of superior audio products. The result is that the big manufacturers are still producing audio equipment representing designs of two decades ago, even though significant advances have been made since then. I should mention here that I am not including the high-end audio equipment manufacturers in the same category as the larger, more diversified electronic manufacturers. Many of the high-end manufacturers are investing bundles of money in research, and this is clearly evident in the price tag. Within this same context, I should also mention that not all so-called high-end manufacturers are producing high-performance equipment. The actual performance of some high-end audio equipment in the marketplace today would have been considered poor as far back as the 1960s, so the purchase of such equipment is really a *buyer beware* situation.

Another motivational factor for getting involved with audio electronics is that it is one of the few remaining electronic fields open to the do-it-yourselfer. In the good ole days of being an electronics hobbyist, it was possible to build your own color television sets, radios, shortwave sets, or almost anything based on electronics. (If you happen to be a younger reader, you may find it incredible to believe that anyone could build his or her own TV set. However, back in the 1970s, the Heathkit Company offered some excellent-quality TV kits that were successfully constructed by many enthusiasts.) With the advent of our modern mass production techniques, coupled with the desire for higher performance in a smaller size, it has become impossible for the weekend hobbyist to construct or work on most electronic equipment. Few home labs contain elaborate surface-mount reworking stations, and the typical multilayer PC board is a "throw-away" item. The traditional home-town TV and radio repair shops have all but disappeared, because modern electronic equipment is not really manufactured to be repaired—it is meant to be replaced. For the most part, the enthusiasm of the modern electronics hobbyist is diminished due to the inability to build anything impressive or practical in the home arena. The major exception to this rule is the field of audio electronics.

Believe it or not, there is another field of electronics besides the digital and communications fields—it's called *analog*! (Excuse my sarcasm.) It is common, at present, to encode much of our audio program material into digital formats, such as used in conjunction with CD, DAT, and MP3 programs. However, the accurate high-level reproduction of such material requires analog processing, so the vast majority of all domestic and professional sound systems will continue to rely on linear techniques. We will go into this subject in more detail later on in this text, but for now, allow me to simply state that there are certain laws of physics that will continue to force the field of audio electronics to remain in the analog domain. Past attempts at integrating, downsizing, and miniaturizing high-performance audio equipment have not met with a great deal of success. Consequently, these factors add up to a situation that allows the home-brew hobbyist to actually compete against the large industrialized manufacturers, and win!

It should also be noted that a considerable number of advancements in the design of superior audio equipment have been made by the at-home hobbyist. This illustrates another unique facet of the audio electronics field. It is doubtful that you would have the capabilities of making a "breakthrough discovery" in the field of particle physics or high-density RAM memory development within the confines of a home-based electronics lab, but it is very possible for the at-home hobbyist to refine and perfect a better audio electronics design. This is due to the fact that audio equipment, for the most part, is dependent upon *discrete component construction,* since the requirements of high voltages, high slew rates, and the comparatively high-power dissipation factors require such construction. Also, I can state with a great deal of confidence that this situation is not likely to change in the near future (again, due to certain insurmountable limitations of physics). Such a situation results in mandatory construction techniques that allow the audio electronics hobbyist to utilize components and materials that he or she has the capability of working with, and all the while, the definite (although elusive) possibility exists that a design advancement could be accomplished. In the case of many audio electronics hobbyists, the motivation to construct a better piece of equipment is the driving force behind their efforts. (There are few pioneers left in the world today, but they are rampant in the field of audio electronics!) We should not conclude, however, that such an individual will suffer the pangs of failure if such a goal is never realized—it is the excitement of the *possibility* that adds spice to the hobby. The audio electronics hobbyist looks at a handful of electronic components in the same manner that a professional writer looks at a blank sheet of paper—the potential is there, and it's just a matter of getting it down right!

We live in an age wherein it is increasingly difficult for the do-it-yourselfer to achieve the pride and self-gratification of an impressive homemade project. Time is at a premium, and many fields have grown too complex for the typical hobbyist. I often receive feedback from audiophiles who spend hundreds of dollars in having customized control knobs machined for their amplifiers. It isn't uncommon for an enthusiastic audiophile to spend countless hours hand finishing a custom walnut cabinet to house a new project. From my perspective, such behavior isn't any more extreme than the woodworking hobbyist who spends a month's worth of spare time making a new coffee table for the living room (considering that an automated machine could probably turn out one in less than 5 minutes). I suppose the creative spirit within all of us is the driving force to build, construct, and accomplish a goal that we can appreciate and take pride in. I hope this book will incite you to daydream at times, visualizing the new "master" audio project that will embody a little of yourself, or a new concept with the potential of revolutionizing the audio industry. Dreams are adrenaline for the human spirit, and such dreams are, in themselves, an excellent motivation.

I have come to appreciate the fact that there are many good reasons to become an audio electronics hobbyist, some professional, some economical, some educational, some creative, or a combination of all of the aforementioned. Audio electronics is a hobbyist field that is continually progressive, adaptable to any budget, personally gratifying, and I can guarantee that it will still be around long after this textbook has become obsolete.

The Goal of High-Performance Audio Equipment

"The trouble with most people is not what they don't know, but what they know for certain that isn't true."

—Mark Twain

Upon first consideration, the typical person would probably assume that the field of audio electronics is a highly scientific one, since we normally consider the broad field of electronics, in general, to be the crowning glory of our present state of advanced scientific technology. That assumption isn't necessarily true, although it certainly should be. In reality, the field of audio electronics is inundated with myth, sonic folklore, and a pseudoreligious type of mainstream indoctrination. Audio electronics is the only electronic field that I am aware of in which a subjective opinion can be provided that contradicts up to five branches of the physical sciences, and still be accepted as fact by a large percentage of the more esoteric audiophile community.

From my experience, I believe the major underlying problem generating this confused state of affairs is the fact that almost all audio program material (consisting primarily of music) is subjective. You may like your music with a lot of bass; I may like mine with a lot of treble. I may prefer country music, and you may prefer classical. Some like their music very loud, while others like it soft. Virtually everything associated with music appreciation is subjective. Therefore, it seems natural and reasonable to consider the musical reproduction system as subjective also. Audiophiles tend to search for an audio system that sounds *nice*. Is *niceness* the performance goal of a high-quality audio system?

Suppose we compared the audio field to another subjective field—video. Green is a very pleasing color to the human eye, so why not adjust the color control on your TV set so that all the peoples' faces are green? I doubt that there are very many people who would enjoy watching a TV that displayed green faces on everyone. Why? Because it is not accurate with reality. There are still individualistic adjustments that are viable for a TV set, such as more or less contrast or more or less brightness, but generally speaking, we desire a sense of accuracy (or reality) associated with our viewing.

The same comparison holds true for audio. I may prefer one style of music to another, or I may prefer differing tonal settings, but I want a flute to sound like a flute, and I don't want a guitar to sound like a harp. *The primary goal of a high-performance audio system is accuracy, not niceness. Accuracy is not subjective; it is analytical.* The goal of accurate sonic reproduction is not an attempt to imprison any individual preferences, but rather, it is a baseline upon which to build a high-quality system that can meet any demand placed on it by the user.

For example, modern recording studios are routinely turning out program material on CD and DAT formats that achieve THD (total harmonic distortion) performance of 0.004% or better. If you play this material through an audio power amplifier exhibiting a THD performance of 1%, you will never hear an accurate reproduction of the program. If the distortion content of the audio power amplifier is primarily even-order, the program may sound nice, but it won't be an accurate reproduction. If you could hear it undistorted, it may sound even nicer, but you'll never know this unless your audio reproduction system is capable of very low-distortion performance. Along this same line of discussion, we must give a little credit to the high-paid studio engineers who tailor the program material to sound as good as possible. If 1% even-order harmonic distortion was going to improve the sound of the recording, it would probably be added before the recording ever left the studio.

In general, the goal of any high-quality audio reproduction system is the highest possible level of sonic accuracy and the lowest possible level of self-generated noise artifacts. The *holy grail* of the audio industry is the proverbial *straight wire with gain.* Of course, there are other performance characteristics that must be evaluated in a high-quality system, such as bandwidth, slew rate, damping factor, power rating, adjustment capabilities, input/output requirements, etc., but *distortion performance* and *signal-to-noise ratio* are by far the most important, and they are the general performance parameters of almost any piece of audio equipment.

Harmonic Distortion

All audio systems fall short of perfect linearity, so they will always inject a certain amount of nonlinearity, or distortion, into the original program material. Such distortion will generate *harmonics* of the original signal content, and these harmonics will appear at the output, combined with the original input signal. If such *harmonic distortion* is significant, it can be audible and possibly very distasteful.

The most common specification for measuring the linearity of an audio system is called *total harmonic distortion,* abbreviated THD. This is the ratio, usually expressed in percent, of the summation of the root mean square (RMS) voltage values for all harmonics present in the output of an audio system, as compared to the RMS voltage at the output for a pure sinewave test signal that is applied to the input of the audio system. If this definition sounds a little "eggheaded," allow me to clarify. Conventional THD tests are performed by applying a *pure* (i.e., absent of any significant harmonic content) sinewave test signal to the input of the audio system under test. This pure sinewave test signal is referred to as the *fundamental* signal. The audio system is set to amplify the fundamental signal by some amount (usually to its rated 0-dB level) and the output of the audio system is applied to a dummy load (i.e., an artificial load, usually chosen to imitate the anticipated real-life loading of the audio system). The fundamental signal is then subtracted from the output signal, leaving only the internally generated harmonics in the output signal. The RMS values of these harmonics are then summed, and this summation is compared to the amplified output of the fundamental signal. For example, assume the output amplitude of the fundamental signal happened to be 100 V RMS, and the sum of the RMS values of the generated harmonics came out to 0.1 V. This represents a ratio of 0.1 part in 100, or 0.1% THD. Occasionally, THD is specified in terms of decibels. Using the previous example, 0.1% THD = 0.1 part in 100, or 1 part in 1000 = −60 dB.

Technically speaking, the output signal of the audio system under test in the previous example would also contain a certain amount of internally generated *noise.* Therefore, a more accurate way of stating the previous 0.1% THD specification is *total harmonic distortion plus noise* (abbreviated THD+N). Most modern specifications take it for granted that the reader knows that the THD specification will include the noise content, so the "+N" suffix is often left off.

Unfortunately, the method used by many manufacturers for measuring THD (often called the *conventional method* or the *single-point method*) is grossly inadequate at providing the *full picture* of the overall sonic performance of an audio system. It consists of applying a 1-kHz fundamental test frequency to the device being tested, adjusting the device for a specified output level and loading condition, and then summing the distortion components for the final calculation. Such conventional testing will only indicate the distortion performance under the exact, static conditions of the test, but the real-life distortion performance under varying frequency and level conditions may be dramatically different. For example, I have bench-tested many commercial amplifiers specified at less than 0.1% THD, but due to a variety of design shortcomings, the high-frequency, low-volume distortion performance exceeded well over 1% (such levels of higher-frequency distortion can easily be heard by experienced listeners).

Figure 1-1 illustrates a conventional audio amplifier topology that can be used to illustrate a few of the technical points made thus far. (Figure 1-1 is a very high-performance, *audiophile-quality* audio power amplifier design that can be considered a "cookbook project" if you so desire. Don't be concerned if it looks somewhat complicated at present—the technicalities of audio power amplifiers will be discussed at length within Chap. 5.) The broadband −3-dB distortion performance of this topology is illustrated graphically in Fig. 1-2.

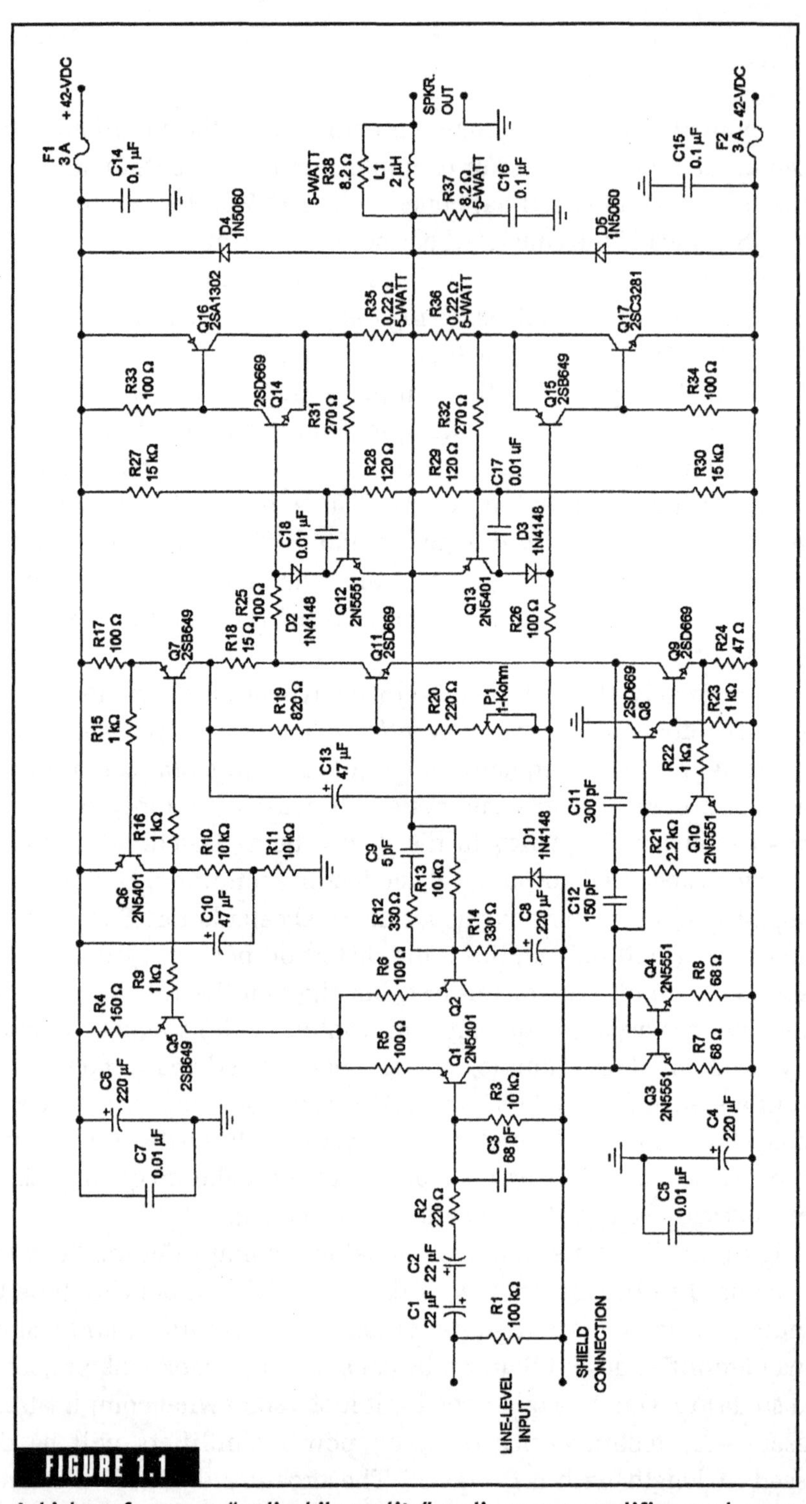

FIGURE 1.1

A high-performance "audiophile-quality" audio power amplifier used as an example for fundamental discussions.

A slightly modified version of the Fig. 1.1 audio power amplifier.

Referring to Fig. 1-2, note the relatively flat performance region occurring between the frequencies of approximately 100 Hz and 2 kHz. Based on the test conditions of the amplifier, the harmonic summations (equaling about 150 μV) calculate out to a distortion performance of about 0.0013% THD. If we move further up in frequency, to about the 20-kHz position, note that the harmonic summations have increased to approximately 400 μV, which calculates out to about 0.0036% THD. Going up further in frequency, the THD performance is about 0.0091% at 50 kHz. In a conventional topology such as Fig. 1-1, if the output amplitude is reduced down to minimum listening levels, the THD performance will typically increase by a factor of about four or five times (such an increase will be almost entirely composed of

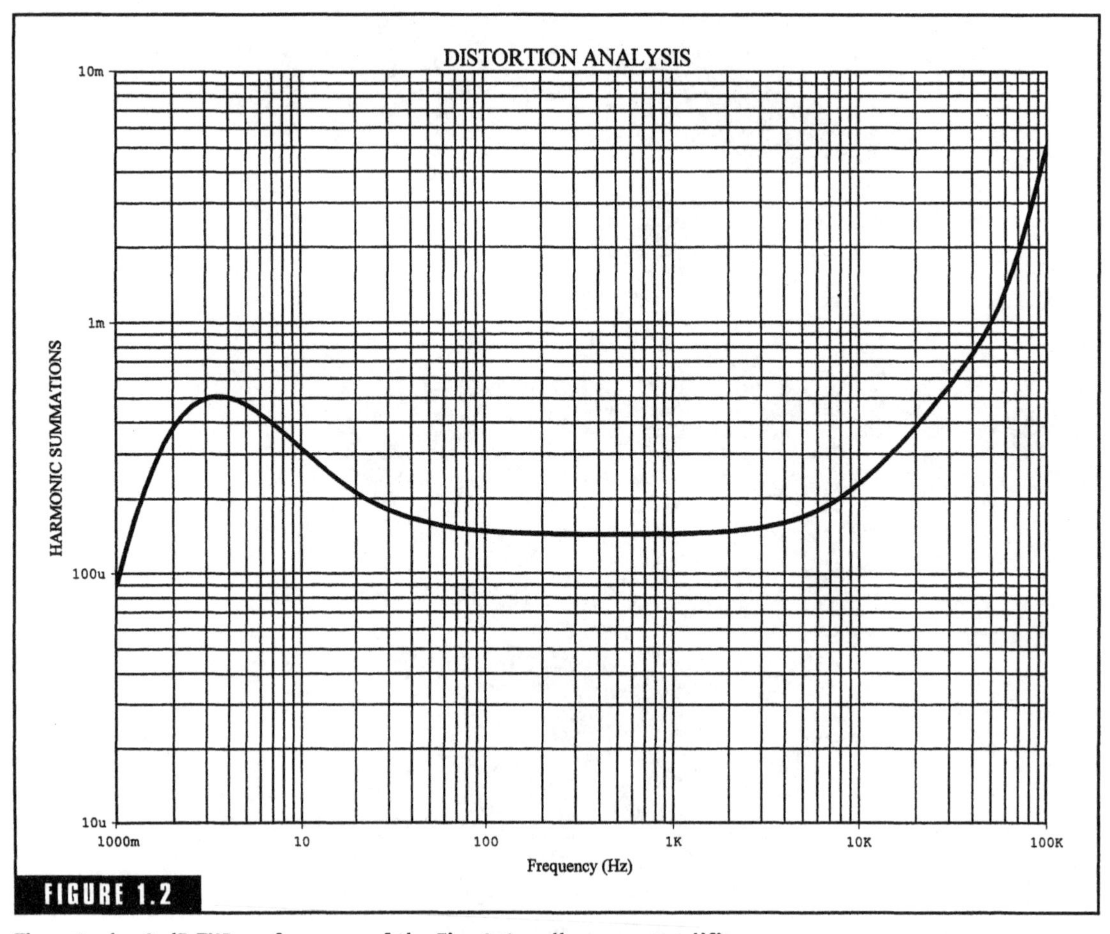

FIGURE 1.2

The actual −3-dB THD performance of the Fig. 1-1 audio power amplifier.

crossover distortion, which is one of the worst types of distortion). Therefore, the previously mentioned distortion levels could rise to about 0.0065, 0.018, and 0.0455%, respectively.

If the amplifier design of Fig. 1-1 were available in the consumer market, it would probably be advertised as providing a THD performance of 0.001% (i.e., the near-optimum 1-kHz test conditions). Note, however, that at higher-frequency, lower-volume conditions, the actual THD performance would probably be about 18 times higher (i.e., about 0.018% @ 20 kHz). Also keep in mind that the Fig. 1-1 amplifier is a very high-performance design. The optimum distortion performance of many consumer-priced audio power amplifiers is specified at 0.01 to 0.1%, which means their higher-frequency, lower-volume distortion is actually going to come out to about 0.18% to 1.8%, or worse (in some cases, much worse).

The previous discussion should be considered in light of the fact that any type of musical program will present a very wide *dynamic range* of varying amplitudes to the audio reproduction system. Regardless of how you adjust the master volume control, you will still be listening to a considerable amount of softer, low-volume passages throughout the program. Consequently, it is quite impossible (not to mention impractical) to raise the volume of the program material above the various types of low-volume, high-frequency distortion.

Another shortcoming of conventional THD testing is a total disregard for the exact *type* of distortion being measured. In other words, the % THD specification tells you the *amount* of total harmonic distortion being produced by the audio device under test, but it doesn't provide any information regarding the *frequency* and *relative levels* of the distortion residuals. In reference to the overall quality of the sonic reproduction, the *type* of distortion is actually of greater importance than the literal *amount* of distortion.

As a general rule, lower-frequency harmonics cannot easily be heard, nor do they sound objectionable in relatively high doses. This is especially true of even-order harmonics (i.e., 2d, 4th, 6th, etc.). Practical testing throughout the past few decades indicates that the lowest humanly perceivable level of low-frequency even-order harmonic distortion is about 1%. Some audiophiles consider higher levels of this type of distortion (i.e., 2 to 4%) desirable, which explains the modern trend toward high-distortion vacuum tube equipment (vacuum tube-based audio equipment tends to produce higher levels of lower-frequency even-order distortion in comparison to their solid-state counterparts). In contrast, higher-frequency odd-order harmonics are easily perceived and highly discordant to human senses. Such types of distortion can be heard at levels as low as 0.3%, or possibly lower, depending on the hearing capabilities of the listener and the specific frequencies and amplitudes involved. Consequently, it is very possible for an audio device producing 3% THD to sound superior to one producing 0.3% THD, depending on the type of distortion being generated. If we add the variable sensitivity and hearing capabilities of each listener into this picture, it becomes readily apparent that the issue of THD performance is a complex one.

Due to the many shortcomings associated with conventional THD specifications, there are some audiophiles who argue that THD specifications are totally irrelevant and should be discontinued. As illustrated in the previous discussion, there is some merit to this argument, but to totally discount THD performance is an extremist position. Providing that the THD measurement is accomplished properly, and in accordance to the applicable range of circumstances, it can be one of the most valuable specifications associated with any piece of audio equipment. For example, many *high-end* audio equipment manufacturers will specify THD over the full range of frequency and amplitude conditions applicable to the equipment. Such specifications are usually provided in the form of charts and graphs, such as the distortion analysis graph of Fig. 1-2. Other manufacturers may provide a "never to exceed" THD specification, also applicable to the full range of amplitude, frequency, and loading conditions anticipated for the equipment. Whenever such complete THD specifications are provided, they are extremely valuable in assessing the quality and performance of the audio equipment.

Although the particulars of THD analysis can be quite complex, its significance to the typical consumer, hobbyist, or experimenter can be simplified. Remember that the quality guideline of audio equipment is *sonic accuracy,* not niceness. It is true that the sonic niceness of a high-distortion vacuum tube amplifier will probably be superior to a mediocre-quality solid-state amplifier (if I had to choose between the two, I would probably choose the vacuum tube amplifier). But the substitution of an objectionable distortion type in preference for a more pleasing distortion type is analogous to *sweeping the dirt under the carpet*—the dirt may not be visible any longer, but you have to hide the floor with the carpet to cover the dirt! In other words, the distortion can be blanketed, but at the expense of coloring the program material. The reasonable and realistic goal is to reduce the THD residuals (whatever type they may be) to such low levels that they cannot be perceived at all throughout the entire range of normal operational conditions.

The precise specification for the lowest level of humanly perceivable distortion is presently a debatable issue within the audiophile community. In regard to the worst-case types of distortion, it seems that about 0.3% THD is the most common opinion. However, it is argued by some that levels as low as 0.1% are perceivable on an almost subliminal level (possibly resulting in *ear fatigue* to the listener). It should be noted that human sensitivity to distortion perception seems to have increased over the years as the quality of audio reproduction improved. Therefore, it is difficult to pin down a precise low-level limit and adamantly state that any lower levels of distortion are totally negligible. My personal opinion is that any piece of audio equipment generating less than 0.1% THD under *worst-case* conditions can be said to be audibly transparent. Such an assumption should not be interpreted to mean that we should stop striving for better distortion performance, however. There is an *additive effect* of distortion within a complete audio chain, so it is desirable to keep the distortion levels of each individual piece of audio equipment as low as possible.

It is unfortunate that many audiophiles have arrived at *across-the-board* judgments of various types of audio equipment based on personal subjective testing of *incomparable equipment.* For example, it is totally unreasonable to subjectively compare a 20-year-old solid-state amplifier with a high-quality vacuum tube amplifier and proclaim that all vacuum tube amplifiers sound better. Likewise, a high-performance metal oxide silicon field-effect transistor (MOSFET) amplifier shouldn't be compared with a mediocre-quality bipolar amplifier in the attempt to prove the superiority of MOSFET designs. *If the overall distortion performance of any piece of audio equipment is maintained well below human perception levels, there will not be any subjective differences in sonic quality regardless of the component types used in the construction.* Of course, such a statement assumes that the other performance specifications relating to bandwidth and noise characteristics are also equivalent. The important point is that a unique *sound characteristic* produced by a specific piece of audio equipment can be traced to a coinciding characteristic in the THD analysis. Consequently, accurate and thorough THD specifications are very important to the serious audiophile, but conventional single-point THD specifications are almost worthless.

Most of the larger, highly diversified commercial manufacturers have the tendency to focus on the advertised conventional THD specifications (marketing people refer to this situation as *the numbers game*). As a result, the motivation to improve the *overall* THD performance may be almost nonexistent. Also, it must be remembered that the majority of the larger manufacturers are not attempting to satisfy the audiophile community. Their marketing efforts are directed toward the general populace, and the majority of consumers will not pay any attention to subtle inadequacies that might be extremely irritating to an experienced audiophile.

There is a final issue relating to harmonic distortion that I feel obligated to mention, even though it is somewhat distasteful. Bluntly stated, not all manufacturers are totally honest relative to their published distortion specifications. This is a gray area with some manufacturers, but with others it is totally blatant and intentional. Consumers are often surprised to discover that large manufacturers can get away with publishing misinformation about their product specifications, but it does happen. The various federal and international organizations that monitor the integrity of manufactured products for consumer protection are primarily focused on the areas of *safety* and *fraud*. Generally speaking, if a manufacturer lies about a product specification that does not relate to consumer safety, and if the manufacturer offers a money-back guarantee if the consumer is not satisfied, it is unlikely that any consumer protection organization will take any type of action against the dishonest manufacturer. The wise approach is to examine an unbiased test report on a specific piece of audio equipment before deciding to purchase it.

What About Intermodulation Distortion?

Intermodulation distortion (abbreviated IM or IMD) is a very *nasty* type of distortion, created by the *nonlinear* mixing of two or more frequencies within the circuitry of an audio processing system. Referring back to electronic fundamentals, when two signals are applied simultaneously to a totally linear circuit, the result is simply a *mixing* (linear summation) of the original two signals. If the same two signals are applied to a nonlinear circuit, a *modulation* of the two signals will result at the output, containing the original two signals, plus the *sum* and *difference* signals. The extent of the modulation effect will depend on the degree of nonlinearity inherent to the mixing circuit, and it should also be understood that the original two signals will be distorted proportional to the degree of modulation. We can state that IM distortion is a function of the linearity of an audio circuit, and the linearity of an audio circuit is defined by the THD specification. Consequently, as we improve the THD performance, the IM distortion performance improves right along with it. This general relationship can be accepted as a rule of thumb for almost all practical audio systems.

During the earlier days of audio equipment development, IM distortion specifications were quite important in detailing the overall performance, since IM distortion tended to be quite high. This situation was a result of older design methodologies, poorer component quality, and less-critical performance demands required by the consumer marketplace. Modern well-designed solid-state amplifiers incorporate superior components and are primarily direct-coupled. Consequently, their linearity *can be* excellent, reducing IM distortion to very low and insignificant levels. The same situation does not always hold true for modern vacuum tube amplifiers, however. Many single-ended monoblock vacuum tube amplifiers will exhibit up to 3% THD performance. In such cases, IM distortion could be significantly high.

IM distortion can be considered a *nonissue* relative to modern amplifier designs exhibiting very low overall THD specifications. It is true that there are certain types of component failure or circuitry design that can promote high levels of IM distortion while exhibiting low conventional THD specifications, but such anomalies will manifest themselves in other performance areas, or in overall THD analysis. Simply stated, any audio equipment design capable of good bandwidth and excellent linearity throughout the entirety of the bandwidth will promote very little IM distortion. Therefore, the design goal is still extremely low distortion, which will automatically eliminate any IM distortion concerns.

Noise

As the term *noise* is defined relative to professional audio and the context of this book, it is the random generation of complex electrical signals resultant from the physics of electronic component operation. Noise in audio systems is primarily due to *recombinational noise* (noise generated in the junctions of solid-state devices) and *Johnson noise* (random noise generated from passive devices). Although *hum* is not generally thought of in terms of being randomly generated or complex, it is still included as a component of the noise spectrum. Hum originates from interference signals introduced into the signal pathways resulting from AC "mains" frequency components, the power supply rectification process, electromagnetic interference (EMI), radio-frequency interference (RFI), and a variety of undesirable feedback and ground loop problems.

In the general sense, noise is the descriptive term applied to almost any audible sound that isn't supposed to be there, encompassing a host of intermittent pops, crackles, buzzes, thumps, and fizzles. Any unwanted sound emanating from a speaker system is deemed noise. In some cases, certain types of harmonic distortion, intermodulation distortion, or loudspeaker defects may be erroneously classified as noise.

The common term defining noise characteristics of audio systems is the *signal-to-noise ratio* (abbreviated SNR or S/N), expressed in RMS value averaged across the audio spectrum and designated in negative decibel levels relative to 0 dBr (i.e., the reference full-output level of the audio system). Normally, the inputs of the audio system under test are shorted for noise measurements. If the noise is measured and averaged without the aid of any specialized audio filtering, the noise specification is said to be *unweighted.* If filters are incorporated to limit the noise bandwidth during evaluation testing, the noise specification is said to be *weighted,* with a designating letter to define the type of filter used. The rationale for weighting a noise specification is relevant to the spectra of the noise being generated—the argument being that it is irrelevant to measure noise signals that are beyond human hearing capabilities.

During the early days of audio electronic development, SNR specifications of −60 or −70 dB were considered quite good. Most of this early noise was actually hum, resulting from a variety of design limitations. Modern audio systems are typically capable of −90- to −120-dB SNR performance, which means that your normal breathing is probably going to sound louder to you than the noise generated from a high-quality audio system.

It should be noted that SNR specifications apply only to internally generated electrical noise inherent to the electrical or electronic circuitry. It does not apply to "physical" noise. Consequently, in the case of audio equipment incorporating cooling fans, the fan noise can be much louder than the electronic "hiss." Fan noise can be quite irritating in very quiet environments, so the audiophile should take any fan noise into consideration as a noise component apart from the specified SNR specifications.

Frequency Response (Bandwidth)

It should come as no surprise that the frequency response, or bandwidth, of an audio system is designed to accommodate the frequency response of the human ear. Normally, this is rather loosely considered to be from 20 to 20,000 Hz. In reality, human hearing is not flat across this frequency spectrum (i.e., the ear is more sensitive to certain frequencies than others), and few people can actually hear frequencies to the designated extremes. Fortunately, such matters are of no concern to the audio processing system. Its only job is to reproduce the program material *exactly* as it was originally produced.

The complexities of music and other audio program sources are not limited to the audio bandwidth. Typical musical programs contain myriad high-frequency and low-frequency harmonics, many of which are well beyond human hearing capabilities. (Frequencies above human hearing are called *ultrasonic frequencies,* and frequencies below human hearing are called *subsonic frequencies,* or sometimes *infrabass material.*) Upon first thought, it would seem logical that an audio processing system could simply cut off, or attenuate, all frequencies below 20 Hz and above 20 kHz, since they cannot be heard. In reality, this is not the case.

The problems associated with high-frequency response are complex, multiple, and interactive (this is one of those issues that could easily fill several chapters of discussion). However, to put it simply, high-frequency, high-amplitude program transients have the tendency to cause the linearizing negative feedback loop (inherent to most audio processing systems) to temporarily *lose control* of the output signal, if the audio system is somewhat sluggish in high-frequency response. Such a condition injects a type of distortion, or *coloration,* into the output signal that was not in the original program material. This type of distortion is commonly referred to as *slew-induced distortion,* or SID, because it relates to an audio system's ability to respond to rapid level changes (defined by the *slew response* parameter of the system). In order to eliminate the effects of SID, the response of the negative feedback loop must be very rapid (i.e., high-frequency). Since the negative feedback signal is conventionally derived from the output signal of the audio system, it can never respond faster than the output signal response, which automatically necessitates a high-frequency response at the output of the audio system. This all boils down to a high-frequency response requirement at the output of the audio system as a method of keeping the negative feedback loop under control during the processing of high-level high-frequency transients.

Figure 1-3 illustrates the actual AC analysis graphs for the Fig. 1-1 audio power amplifier. The upper graph shows that the frequency response of the amplifier begins to roll off at about 200 kHz and reaches the unity gain point (shown by the lower darkened horizontal line) at about 5.5 MHz. In truth, such a high-frequency response is a little too high for optimum stability insurance (in my opinion). In most of my designs, I like the rolloff point to start at between 80 and 100 kHz.

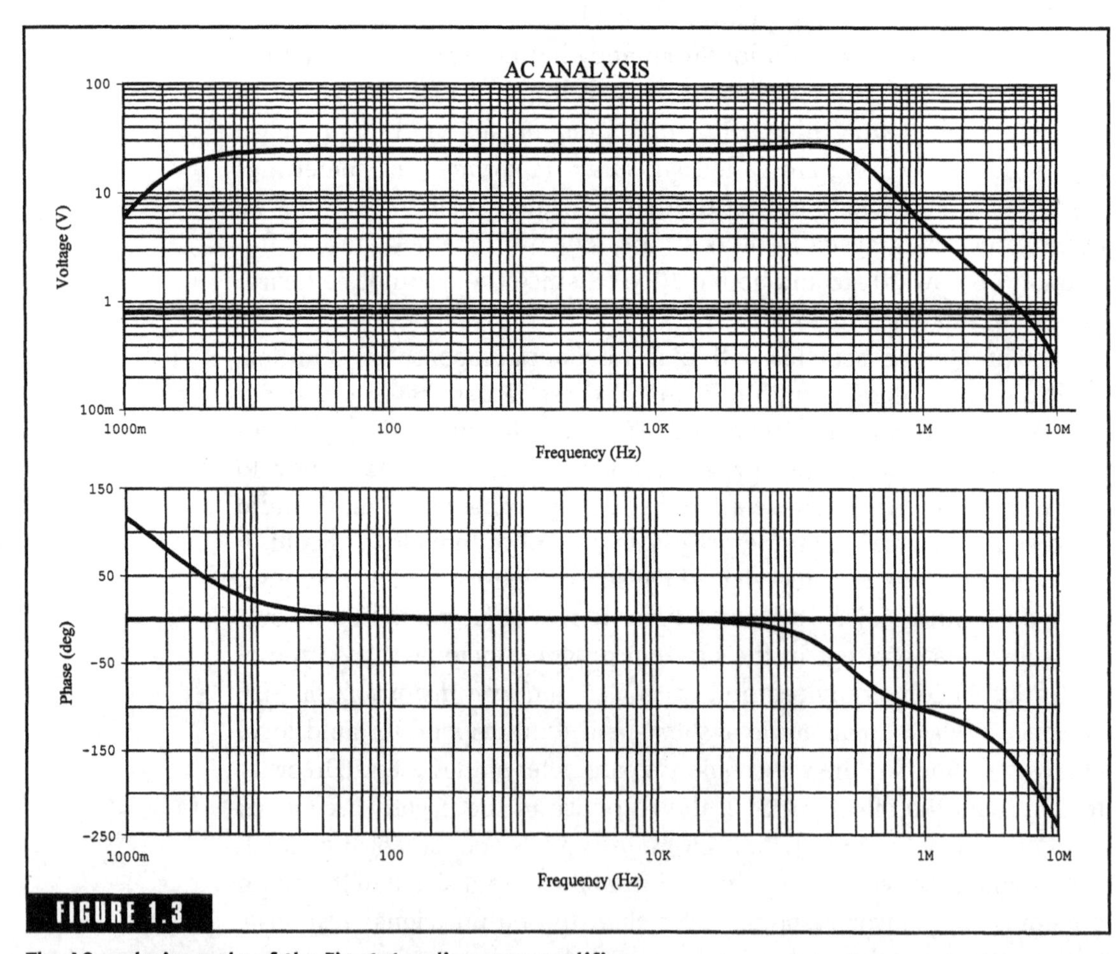

FIGURE 1.3

The AC analysis graphs of the Fig. 1-1 audio power amplifier.

(Such a response would be achieved by replacing C12 with a 220-pF capacitor and C11 with a 560-pF capacitor.)

Another desirable attribute of designing an audio system with an extended high-frequency response relates to high-frequency distortion performance. The effectiveness of any negative feedback loop at linearizing an amplification system is dependent upon the *open-loop gain* of the amplification system. Extending the high-frequency response of the system will also extend the high-frequency response of the open-loop gain, which will result in improved high-frequency distortion performance. For example, an audio amplifier that has a high-frequency bandwidth up to 100 kHz will exhibit much less distortion at 20 kHz than the same amplifier with a bandwidth up to 40 kHz.

As a means of illustrating the previous point, refer to the distortion analysis graph of Fig. 1-2. Note how dramatically and rapidly the distortion begins to rise above 10 kHz. Such a rise in THD levels is due to the rolloff of the open-loop gain, which is a necessary evil to keep the Fig. 1-1 amplifier stable. Don't worry if this topic is confusing to you at present. The concepts of *open-loop gain, closed-loop gain, compensation,* and *stability* will be discussed in more detail within Chap. 6. The only important point to remember is that increasing an audio amplifier's *ultrasonic response* will result in improving the distortion performance in the upper frequencies of the *audible range.*

The situation with subsonic frequencies is entirely different. The real reason for extending the low-frequency response of a high-quality audio system is to eliminate the effects of *capacitor distortion,* resulting from the multitude of signal coupling capacitors that will inevitably be placed in the signal path. Most modern high-quality audio systems are direct-coupled by nature, meaning that it would be very easy to convert them to a pure DC response (colloquially referred to as *direct-coupled operation*). Direct-coupled operation is not desirable in practical audio systems for a number of reasons. First, no domestic speaker system is going to be capable of reproducing frequencies much below 20 Hz to any practical perception levels, so the capability is useless. Secondly, very small DC levels existent at the input of the audio system could be manifested into significant DC levels at the speaker systems, causing activation of DC speaker protection circuits and/or the destruction of the speaker voice coils. Thirdly, subsonic frequencies applied to a speaker system can waste enormous levels of power being supplied by the power amplifier, which can result in overheating the speaker coils, overheating the power amplifier, or distortion resulting from the excessive power drain.

The elimination of destructive DC levels throughout the audio chain is accomplished by incorporating *coupling capacitors,* typically at the input and output of line level devices, and on the input of the power amplifier stages. Coupling capacitors block DC levels while effectively allowing the AC program signals to pass. Unfortunately, if the capacitance value of the coupling capacitors is chosen so that frequencies below 20 Hz begin to be attenuated (i.e., by the capacitor action), the frequencies in the 40-Hz range and below will probably suffer from a type of distortion caused by the nonlinear charge-discharge curve of the coupling capacitors. Logically enough, this type of distortion is called *capacitor distortion.* To remedy this problem, the capacitance value of the coupling capacitors is chosen so that the low-frequency response is in the 10-Hz range, thereby ensuring that all significant capacitor distortion artifacts are eliminated in the audible realm. Therefore, the goal of the low-frequency response inherent to most high-quality audio systems is not to reproduce extreme low-frequency signals. It is to force the capacitor distortion effects down into the subsonic region so that the beginning of the audible frequency region will be free of capacitor distortion effects.

Again, this point can be illustrated by referring to the Fig. 1-2 distortion analysis graph. Note how the distortion begins to rise below 20 Hz. This is mostly due to the effect of the input coupling capacitors, C1 and C2 in Fig. 1-1, with a small portion resulting from the effect of the negative feedback capacitor, C8. If the capacitance values of C1 and C2 were reduced, the resulting capacitor distortion would begin at higher audible frequencies, and the level of the distortion would also be increased. With the values shown, the majority of the distortion remains in the subsonic region and the level stays insignificant until the signal attenuation begins to drop dramatically between 1 and 2 Hz.

In summation, we can say that we desire good ultrasonic frequency response from a high-quality audio system to eliminate any coloration caused by SID and to improve the high-frequency distortion performance. We also desire good subsonic frequency response to guard against any significant coupling capacitor distortion effects. So how far is it necessary to push these extremes?

In my experience, there are no limits to the extremes that some enthusiastic audiophiles will take. I have talked with many audiophiles who are constantly teetering on the brink of catastrophic instability failures because they demand their audio system to be capable of DC to 400-kHz operation. I don't criticize such eccentricities, because they can be part of the enjoyment that many audiophiles receive from their hobby. After all, diamond rings and two-seater sports cars aren't practical either—the fun you receive from a hobby doesn't have to be practical. But the majority of hobbyists and professionals are concerned with the acceptable bandwidth limits that will provide the best possible "no-compromise" performance without getting into risky extremes.

It is generally agreed that a 10-Hz to 80-kHz bandwidth will provide the ultimate performance from any audio system. I have the tendency to push these limits a little in the effort to remove all doubt of performance issues relative to bandwidth, so the majority of the applicable designs in this book will provide a frequency response of approximately 5 Hz to 100 kHz. Of course, you can choose to stretch these extremes further. The disadvantages of going to a lower subsonic frequency response are the additional expense and increased physical size of the coupling capacitors. The primary disadvantage of attempting to extend the high-frequency response is the increased risk of stability problems.

Slew Rate

Slew rate is the colloquial term used to define the *maximum output transition speed* of audio processing equipment, as well as operational amplifiers and other types of signal-processing electronic devices. Simply stated, the slew rate defines how *fast* an electronic system can respond. Some professionals argue that slew rate is a poor term, because it doesn't actually specify that a particular *rate* happens to be the *maximum rate.* Perhaps it would have been better to standardize on a term such as *slew limit* or *slew response,* but I'm not about to buck the established convention of decades of op-amp development, so I will continue to use the term *slew rate* as it is typically applied.

Many hobbyists confuse the meanings of slew rate and frequency response, so I will try to clarify the difference. As a function of *time,* a 1-kHz frequency represents a 1-millisecond (ms) time period. An amplifier capable of reproducing a 1-kHz signal will be required to reproduce each cycle of the signal in a 1-ms time period. If the amplifier can accomplish this, it is said to have a 1-kHz frequency response. In contrast, the required *transitional speed* at the output of the amplifier relates to the *amplitude* of the 1-kHz tone, rather than the time period. Crudely illustrated, if the peak-to-peak output amplitude of the 1-kHz signal happened to be 2 V, the amplifier's output would have to vary in amplitude throughout the entire 2-V range in a period of 1 ms. However, if the peak-to-peak output amplitude of the signal happened to be 20 V, the amplifier's output would have to vary in amplitude throughout the entire 20-V range in the same period of time. As can be readily perceived, the second set of circumstances required the amplifier's output *speed of transition* to be 10 times greater than in the first set of conditions.

Slew rate is typically defined in terms of *volts per microsecond* (V/μs) when applied to audio equipment and operational amplifier performance parameters. Slew rate can be used to define *rise time* (or *positive slewing*) and *fall time* (or *negative slewing*) independently, but as a general rule (without any qualifying statements), it is most commonly used as a singular expression to define the slowest of the two transition types. In most cases, slew rate considerations are going to be more important in analyzing the performance of audio power amplifiers, since low-amplitude line-level signals inherent to preamplifers and other types of low-level audio processing equipment will not require high slew rate performance. Audio processing equipment with poor or inadequate slew rate performance will produce the same type of SID as detailed in the previous bandwidth discussion.

Although slew rate and bandwidth are not directly comparable, they are interrelated. For example, if you need to know the slew rate required for reproduction of a sinusoidal waveshape at a specific frequency and amplitude level, the equation is

$$\text{Sinusoidal slew rate} = \frac{6.28 \times \text{frequency} \times E_{\text{SINE (PK)}}}{1{,}000{,}000}$$

For example, if you wanted to calculate the slew rate required for a 95-kHz sine wave at a 35.3-V RMS amplitude, you would begin by converting the RMS voltage to its corresponding peak value. 35.3 V multiplied by 1.414 comes out to 50 V peak (I made this problem easy on myself). 6.28 multiplied by 95,000 Hz multiplied by 50 V comes out to 29,830,000. Dividing 29,830,000 by 1 million comes out to a slew rate of 29.8 V/μs, or approximately 30 V/μs.

If you happen to know the slew rate performance of a piece of equipment and you want to know the maximum sinewave frequency it can reproduce at a specific output amplitude, the equation is

$$\text{Maximum sine wave frequency} = \frac{\text{Slew rate} \times 1{,}000{,}000}{6.28 \times E_{\text{SINE (PK)}}}$$

For example, working the previous problem in reverse, a slew rate of 30 V/μs multiplied by 1 million comes out to 30,000,000. The amplitude you are calculating for is 35.3 V RMS, which equates to a 50-V peak output amplitude. Multiplying 50 V by 6.28 results in 314. Dividing 30,000,000 by 314 provides the maximum sinusoidal frequency, which comes out to 95,541 Hz, or approximately 95 kHz.

I should mention in this context that the requirement of very high slew rates for high-power audio power amplifiers is one of the reasons that such amplifiers continue to rely on *discrete component construction*. For example, a typical 500-W audio power amplifier capable of driving an 8-ohm load up to an 80-kHz bandwidth will require a slew rate of 45 V/μs. Such high slew rates (especially under relatively high-voltage conditions) are typically difficult to obtain without resorting to discrete components.

Phase Considerations

Phase considerations are important when evaluating a high-quality audio system because the complex phase relationships inherent to musical programs provide the listener with a quality of *spatial perception*. In the same manner that human vision senses rely on two eyes to perceive *depth* (i.e., the third dimension of vision), human hearing perception relies on two ears to provide spatial perception (pertaining to where a sound originates in space). The characteristics of both amplitude and phase are a part of the overall quality of spatial perception. There are a variety of terms used in the audio industry relating to varying qualities of spatial perception, such as *presence, soundstage, radiation pattern, surround-sound,* and *three-dimensional sound.* In addition, there are a host of other more nebulous descriptions, such as *narrow, lifeless, artificial, full-bodied,* etc.

Absolute phase is a term used to describe the in-phase relationship between the input of an audio system and its output at the speaker systems. In other words, if the output of an audio system is in perfect phase with the input program material, it is said to be in absolute phase. As long as all channels of an audio system are in-phase with each other, the phase relationship between the program material and the system output is of no importance at all. Therefore, the concern over absolute phase is a *nonissue*.

Phase distortion is a term that can generally be applied to any type of phase problem that can result in audible or spatial differences between the original program material and the reproduced sound of the audio system. For example, if you ever accidentally reverse the speaker wires on "one" speaker of a stereo system, you will probably notice a peculiar type of unreality in the reproduced music, and it will seem to vary dramatically as you walk around the listening area. Certain frequencies may seem to be canceling each other at specific listening points, and at times the effect may seem to mimic harmonic distortion or intermodulation distortion. Such an effect is created by the two speaker systems being 180-degrees out of phase with each other. It is possible for similar, but less dramatic, phase distortions to occur due to a variety of problems within a typical audio reproduction system.

Beginning with the program source (CD player, FM receiver, phonograph, tape player, etc.), the concerns over any problems relating to phase distortion are almost nonexistent, providing the equipment is in good working order. For example, it is highly unlikely that any CD player will be designed so that the phase relationship between the right and left channels (i.e., the *line-level outputs*) is significantly different—even in the case of very inexpensive units. We are not concerned if all of the line-level outputs are 180-degrees out of phase with the original program on the CD itself (another highly unlikely situation), or if they are in absolute phase with the program material. The important point is that they are in-phase with *each other.* Barring any weird or unusual circumstances, the concerns over phase relationships at the program source should be laid to rest.

The next step in a domestic hi-fi system is typically the preamplifier stage. The most common functions of a preamplifier are to provide (1) voltage amplification to low-level input signals, (2) attenuation to line-level input signals, (3) RIAA equalization and signal amplification for phono inputs, (3) level and balance adjustments, and (4) tone control adjustments for the compensation of speaker and environmental shortcomings. If the preamplifier is of conventional design and the tone controls are adjusted to their "flat" position, there shouldn't be any adverse effects relating to phase relationships. However, tone controls are simply active or passive filters, and electronic filters have a profound effect on phasing.

Throughout the past few decades, there has been a tremendous amount of controversy relating to the so-called *sonic corruption* of tone controls. Tone controls, in general, are accused of causing a variety of distortion problems as well as defiling the spatial purity of the program material with phase anomalies. In truth, well-designed tone control circuitry should not create any significant levels of distortion, and while it is true that tone controls will affect the phase relationships of the original program material, such deviation is not always bad. If it were possible to construct perfect speaker systems, operating into a perfect listening environment, at an optimum listening level, then we might flatten all of the tone settings (or flick the tone bypass switch). In reality, most speaker systems are far from perfect, the listening area of most domestic hi-fi systems is going to corrupt phasing nuances far more than the tone controls, and the human sensitivity level to a variety of frequencies is highly dependent on the volume level (this is the reason that you probably have a *loudness* control built into your preamplifier). Tone controls are simply a method of compensating for a variety of uncontrollable factors involving the overall listening experience.

It should be remembered that the program material has already passed through dozens of filter stages before ever leaving the recording studio. It if was true that tone controls corrupted the program material to unacceptable levels, then you would have to accept the fact that virtually all original program material is unacceptable—what are you going to listen to? My advice is to forget about the majority of the controversy over tone controls. Set your tone controls according to your personal preferences and enjoy your music. I can assure you that almost every professional studio engineer in the world will appreciate (and utilize) tone controls in their own home hi-fi systems.

The next step in a typical domestic hi-fi could be an *equalizer* (i.e., *graphic equalizer*). A graphic equalizer is a larger number of more selective tone control filters, with the *slide potentiometers* (i.e., filter adjustments) arranged on the front control panel in such a manner to provide a visual "graph" of the final, broadband settings. Graphic equalizers are normally placed between the preamplifier outputs and the audio power amplifier inputs. Their function is to correct for frequency-related shortcomings inherent to the audio power amplifier, the associated speaker system, and the listening environment. Regarding phase distortion problems, the user has to be a little more cautious in going to extremes with the adjustment capabilities of most graphic equalizers, since the filter adjustments apply to *individual channels*. [The tone control circuitry in most preamplifiers is designed to adjust "both" channels (i.e., of a stereo system) by the same amount, thereby keeping the phase changes equal between the two channels.] If the individual channels of a graphic equalizer are adjusted drastically different from each other, it is possible to experience some types of phase distortion similar to phase distortion caused by wiring the speaker systems out of phase with each other. With a little common sense and moderation in keeping both channel adjustments similar to each other, the phase concerns are about the same as for the general types of tone controls.

The audio power amplifier is probably the next piece of equipment in the audio chain. Happily, this will be a short discussion, because a well-designed audio power amplifier should be in almost perfect, absolute phase throughout the entire audio bandwidth. Refer to the phase versus frequency graph of Fig. 1-3. Note that the output of the amplifier remains in almost perfect phase from about 20 Hz all the way up to over 50 kHz. Such a response relates to the physics of operation, and almost every solid-state audio power amplifier will exhibit similar phase characteristics. Consequently, the question of phase problems inherent to a well-designed solid-state audio power amplifier should never be a concern. (More will be discussed on the physics behind this phenomenon in a later chapter.)

Before leaving the topic of phase relative to audio power amplifiers, I should mention that vacuum tube audio power amplifiers may exhibit some pretty significant phase shifts at the opposing extremes of the audio bandwidth, due to the output transformers that are incorporated into most vacuum tube designs. Again, this shouldn't be a concern if you are using multiple vacuum tube amplifiers of the same fundamental type, since any phase shifts will happen in all of them by about the same amount.

In my experience (and speaking from a general perspective), the serious audiophile shouldn't be overly concerned with phase problems originating from all of the sources I have mentioned thus far. The high levels of negative feedback incorporated into most preamplifiers and power amplifiers will force an absolute phase condition without even trying, and the phasing problems associated with tone controls and graphic equalizers are a necessary evil and typically benign. However, in the interest of pursuing the ultimate level of sonic experience, there are two areas relating to phasing within the audio chain that could be significantly improved (depending on your personal tendencies toward perfection and the limits of your budget). These are (1) the phase relationship of any added subwoofer systems and (2) the correction of phase shortcomings in speaker crossover networks.

Phase distortion is often a shortcoming with external subwoofer systems. This is due to the significant phase shifts resulting from the low-pass filters incorporated into most subwoofer amplifiers. Some less-expensive subwoofer systems include a phasing switch, which simply sets the subwoofer amplifier to either inverting or noninverting mode. Other subwoofer systems include a continuously variable phasing control, which equips the user with a much more accurate method of setting the phase for the optimum listening performance.

The passive *crossover networks* inherent to virtually all multielement speaker systems exhibit a definite, discernable phase distortion shortcoming. Unfortunately, the problem is expensive and complicated to remedy. Due to the physics involved with all types of passive crossover networks, varying levels of phase distortion will be introduced into the program that are especially prominent around the crossover frequencies, resulting in a frequency-dependent nature to the speaker system's radiation pattern. To remedy this condition, the audiophile must drive the individual speaker elements directly with a multiple power amplifier configuration, and the frequency bands pertinent to the individual speaker elements must be separated with an active *phase-linear crossover filter.* Such a technique is commonly called *bi-amping* or *tri-amping,* depending on the number of speaker elements to be driven. More about phase-linear filters and bi-amping/tri-amping techniques will be discussed in upcoming chapters.

Phase concerns relating to *home theater* and *surround-sound* systems are essentially the same as for typical stereo systems, but the spatial benefits of surround-sound systems are more critically dependent upon adequate phasing techniques. The simple rule-of-thumb guideline to follow with such systems is to try to keep all of the amplification and speaker system equipment reasonably comparable. Obviously, it wouldn't be wise to mix up a variety of vacuum tube and solid-state power amps within a single surround-sound system, nor would it be advisable to incorporate an extreme variety of speaker system types. Home theater and surround-sound systems are highly dependent on (1) amplitude levels, (2) phase relationships, and (3) the precise listening position. Therefore, some good old-fashioned common sense should be applied to setting up such a system for optimum performance.

On a final note regarding home theater and surround-sound systems, I would like to point out an important consideration. Personally, I am neither for nor against surround-sound systems—they simply represent a different (and extended) facet of the overall listening experience. However, it should be noted that the sonic quality of an audio reproduction system will only be as good as its poorest-performing link in the audio chain. The implementation of a dozen poor-quality speaker systems throughout a room is just going to spread the distortion throughout the room. The sonic quality is not improved by increasing the number of sound sources. Likewise, increasing the number of cheap audio power amplifiers will increase the *sound pressure level* (SPL), but the sonic quality will remain poor. I have talked to many individuals who purchased a relatively expensive surround-sound system, and later became very disappointed with their equipment when they heard an excellent-performing stereo system for the first time. My advice to the beginning audiophile is to start with a simple system of high-quality equipment, and then build on the system as time, experience, and money will permit (remember to place "experience" before "money" in that last statement).

Additive Effects in Audio Systems

When considering any type of audio equipment, you will always want the noise and distortion characteristics to be as low as possible. This presumes that you don't want singers to sound like chipmunks and you would rather not have your stereo system hissing like a large cobra. Noise and distortion performance become even more critical when you consider the length of your *audio chain,* because these effects will be somewhat additive within a complete audio system.

The additive effect of noise, distortion, and other sonic inadequacies is one of the main reasons that audiophiles are constantly trying to achieve better performance characteristics. A single piece of equipment may be capable of performance that will produce inaudible levels of noise or distortion, but what if you have the need to daisy-chain three such units together? It is not unusual for an audio signal to have to pass through a dozen individual processing stages in a professional audio recording studio, and a modern domestic audiovisual home theater system can easily surpass a dozen interactive pieces of audio equipment. Therefore, the effort in professional and high-end audio engineering is to achieve the best possible performance within reasonable limits of complexity and cost.

A Final Word on Performance

The field of audio electronics is difficult to pin down to a rigid standardization because it exists in a constant atmosphere of change, coupled with a somewhat enigmatic and artistic nature. Listener sophistication continues to improve along with technical and material advancements. Typical methods of measuring specifications are not always adequate, and many advertising methods are erroneous or unethical. Audio electronics is a science, but it is a science that promotes an artform. Where does the science end and the artform begin?

I don't associate any *magical* or *analytically incomprehensible* attributes to the field of audio electronics—it is simply a progressive field, like so many that we are routinely immersed in. The broad spectrum of controversy inherent to the audiophile community has produced beneficial as well as detrimental results. On the one hand, the esoteric audiophile community has forced the more stuffed-shirt scientific community to reexamine and reprove many fundamentals of audio technology, which has led to a more complete understanding (and appreciation) of many subtleties that would have probably been otherwise overlooked. Unfortunately, such broad-based criticism has developed a sense of rejection and mistrust of accurate scientific methods, with such attitudes fueling all sorts of erroneous mythology.

From my personal perspective, I believe that many audiophiles have gotten so involved with the performance of the equipment that they have forgotten how to enjoy the program. Controversy is good—it keeps our sensitivity and intellect sharp. But don't forget how to enjoy your involvement in this field. It's difficult to appreciate beautiful music when your ulcers are acting up!

Safety Considerations

This book is designed for audiophiles and electronics hobbyists who are already experienced in the electrical/electronic fields. This is not the place to learn basic electronics. For this reason, I have assumed the reader to be experienced in electrical safety procedures as well. I am not going to attempt to do justice to the broad realm of electrical safety, but in keeping with the general theme of this book, there are a few relevant safety issues that I would like to touch on.

I am frequently involved with electronics technicians and engineers who are attempting to construct their first audio project. It is not unusual for highly competent electronics personnel to get a little "rusty" with electrical safety if they are not exposed to hazardous electrical conditions on a routine basis. Therefore, if you happen to fall into this category, please take the time to dig up a good textbook detailing the general procedures for electrical safety and invest some well-spent time in reviewing them.

The most hazardous projects contained within this book are the ones involving power supplies, audio power amplifiers, and bench-testing power supplies. Some power supply projects in this book will call for power transformers rated at 750 VA, or higher, with primary *and* secondary voltage/current levels that are very lethal. Such transformers will typically be used in conjunction with reservoir capacitors that are about the size of large coffee cups. High-capacity, high-voltage capacitors can blow up with considerable force if improperly wired, and their levels of stored charge can literally vaporize, or blow up, conductive items that might be inadvertently dropped onto their connection posts. They are also capable of holding high charge levels for many days, even if the power supply is not connected to the AC mains power. Therefore, when working with high-power projects such as these, abide by the following safety tips:

1. Always wear eye protection when testing or applying operational power to large power supplies or power amplifiers.

2. Install bleeder resistors in your power supply projects—they could save you as well as someone else from a serious stored-charge accident.

3. Before making or breaking connections to power supplies, always make sure the AC mains power is off and ensure that all reservoir capacitors are discharged.

4. If you have the need to discharge a high-capacity capacitor, never use a clip lead, piece of wire, or anything else to short-circuit the terminals. Use an appropriate resistor of sufficient resistance and power rating to provide a controlled discharge.

5. Always install appropriate fusing in transformer primaries, power supply outputs, and/or audio amplifier power supply rails. Do this before attempting any type of "quick test" to see how the unit will perform.

6. Take the time and obtain the proper materials to make quality wiring connections, and make sure all exposed high-voltage connections are well insulated. Avoid using crimp-type terminals. I recommend soldering wires together and insulating with heatshrink tubing for all "butt-splice" connections, and the use of screw-down terminal blocks, solder posts, or solder lugs for other connection requirements. (A large number of "hum" problems associated with audio power amplifiers result from crimp-type connectors used in ground wiring.)

7. When working with AC mains power, make sure the AC power is turned off and be double-safe by unplugging the AC line cord.

8. Whenever you are taking circuit measurements with operational power applied, get in the habit of using only one hand to manipulate the oscilloscope or meter leads. Use a clip lead to make ground or reference connections before applying operational power.

9. Make sure that you never work in an environment wherein any part of your body is in routine contact with anything electrically conductive, such as wet floors, metal chairs, metal table legs, etc.

10. Always use an isolation transformer if you are using any kind of AC line-powered test equipment to take any kind of electrical measurements involving the AC mains power.

11. Don't attempt to catch falling objects, or reconnect a connection that falls off, while operational power is applied. Make sure your test

connections are secure, and if one does happen to come loose during testing, turn off the operational power and discharge any high-capacity capacitors before attempting to reconnect.

12. Be patient, and take the time to analyze whatever you are about to do methodically. The majority of stupid mistakes I've made regarding electrical safety have been due to my overanxious nature or my tendency to get in a hurry. Besides being dangerous, stupid mistakes can be very expensive as well.

As a final safety comment, please don't take your ears for granted. It is a pity that so many people each year must revert to hearing aids due to nerve deafness as a result of listening to chronically loud music for too long. If you often hear a "ringing" in your ears after listening to your audio system, your listening levels are too high and you are damaging your ears! Why not turn it down a little? After all, being an audiophile is a pretty dull hobby if you can't hear the music.

CHAPTER 2

BEGINNING AT THE BEGINNING

In the introduction of this book, I explained that its goal is to provide the necessary technical information, construction techniques, and cookbook designs for the most common audio building blocks in a typical audio reproduction chain. I have intentionally left out anything pertaining to program sources (program players, such as CD players, MP3 players, tape decks, FM receivers, etc.), because even a casual examination of all of these devices would fill a significantly large book. Likewise, I have excluded speaker systems for the same reason (also, there are already many good books on the market focusing on speaker systems, so I feel that my inputs would be somewhat redundant). So our "starting point" in this book will be the first link of the audio chain that the audio signal sees as soon as it leaves the program source—the front end (i.e., the input section) of the preamplifier.

At this point, I would also like to mention that I do not intend on going into very much basic operational amplifier theory throughout this book. If you feel a little weak in this area, or you think you could use some review, I highly recommend the book *IC Op-Amp Cookbook* by Walter G. Jung. This is an industry classic, and should be available from almost any bookstore.

To Balun or Not to Balun? That Is the Question!

What the heck is a *balun?* Balun is actually a colloquial term formed from the expression "*bal*anced to *un*balanced." It originally referred to a type of transformer that would accept a *balanced input signal,* and through transformer action, convert the signal into an *unbalanced,* or *single-ended* signal. Presently, balun is used to describe any device (transformer or solid-state circuit) capable of *receiving a balanced signal* or *driving a balanced signal.* In other words, a balun is a device that converts a balanced signal to an unbalanced signal, or a device that converts an unbalanced signal to a balanced signal. Baluns designed to accept a balanced input signal and convert it to a single-ended signal are called *balanced receivers.* Baluns designed to accept a single-ended input signal and convert it to a balanced output signal are called *balanced drivers.* In some cases, the term balun is used to describe the entire unbalanced to balanced to unbalanced system in an audio chain.

Before proceeding any further into balun discussions, let us examine the need for balanced signal lines. Balanced signal lines contain three wires: (1) signal common, (2) the noninverted audio program signal, and (3) the inverted audio program signal. Signal cables of any type are susceptible to undesirable noise pickup from numerous types of radiated fields. Regardless of what type of stray noise may be picked up by a connection cable, the noise will be *common mode,* meaning that the electrical field cutting the cable wire will induce the same level and phase of noise contamination into all of the wires simultaneously. If such common-mode noise is injected into a balanced signal line, the *balanced receiver* will ignore the noise content and only amplify the "difference" signals of the program material. In other words, the *common mode rejection characteristic* of an op-amp or balanced transformer can be used to reject the common-mode noise, while cleanly amplifying the audio signal. In addition, since the balanced receiver is looking at the difference signal contained in the two audio signal lines, the circuit common reference is not actually used in the signal extraction process, so a host of hum problems (i.e., from ground loops and various grounding problems) can be isolated from one piece of audio equipment to another.

In the not-too-distant past, baluns were used almost exclusively for professional audio systems, due to the need to stretch long lengths of microphone and signal cable across the stage area to a variety of soundboards and amplifiers. In recent years, however, audiophiles have begun to incorporate baluns into domestic audio systems. As with any new innovation into the audio fields, there have been fantastic claims made in detailing the benefits derived from incorporating balanced lines throughout a domestic audio system. Are any of these claims true?

There is an old proverb that states, "If it ain't broke, don't fix it." Is your domestic audio system quiet and relatively free of excessive noise or hum? If your answer is yes, then you don't need to go with balanced signal lines. If your answer is no, then consider a few other things first. Are you using a good grade of shielded signal cable throughout your system? Are all of your system components grounded properly? Have you attempted to cure such problems in more conventional manners? The point is, the majority of problems resulting in excessive noise and hum in a domestic hi-fi system can be traced to easily remedied faults that should be corrected regardless of whether or not you incorporate balanced signal lines.

Balanced lines are a good choice if you have long runs of signal wire. For example, you might want to use a balun system if you have one piece of equipment that is located a significant physical distance away from your main system. They are also a good choice for eliminating a stubborn ground loop problem that nothing else seems to fix. Balanced input receivers and output drivers should always be incorporated into *professional audio equipment* that will be receiving or sending *line-level* or *signal-level* information to other pieces of equipment. Baluns are often a good option for musical instrument amplifiers as well.

Baluns can be divided into two major categories—transformer and solid-state versions. Regarding transformer-type baluns, I have to honestly admit that I don't have much personal experience with them, other than my knowledge that they function well in professional audio equipment. Based on the market specifications and the feedback I have received from satisfied users, I believe their performance in domestic applications is probably excellent. In contrast to solid-state baluns, transformer baluns do not require any type of operational power, so they are an excellent choice for anyone wanting to construct a *passive preamplifier system* (passive preamplifiers will be discussed a little later). The main disadvantage with transformer baluns is their cost—they are rather pricey.

Solid-state baluns provide excellent performance and are very inexpensive. They are easy for the home-brew hobbyist to construct, and they are very versatile and flexible for a wide variety of applications.

Baluns are sort of "trendy" right now. There have been a number of articles written about how the noise levels in domestic hi-fi systems have been lowered by −6 dB, or maybe −10 dB. But how good were the noise levels to start with? Was the improvement audible? I'm a passive spectator in this controversy—I neither recommend nor reject the installation of baluns. I'm a practical person. I recommend that if you have a problem a balun can fix, then use it. Regarding the various manufacturers of audio equipment, the only thing they're interested in is that baluns are in vogue at present, so you'll see a lot of modern hi-fi equipment with balanced input/output capabilities. You be the judge, and have fun with your experimentation!

Solid-State Balun Variations

If you happen to have a preamplifier with balanced input capabilities, the first thing the audio signal will see after it leaves the program source is the balanced receiver (providing you are using the balanced inputs rather than the conventional RCA inputs). Figure 2-1 illustrates two conventional methods of constructing balanced receiver circuits. These designs are the types that will typically be found in some professional audio equipment and budget hi-fi equipment. The effectiveness of the balun at its primary function, which is to remove common mode signals, will be determined by the matching of the external resistors and the quality of the op-amp chosen. The circuit of Fig. 2-1*A* is commonly chosen to match low-impedance microphones to an appropriate amplification system, while Fig. 2-1*B* is more general-purpose in nature. It isn't practical to go with expensive precision resistors and high-performance op-amps for these two designs, because you could spend less money and get better performance from several varieties of special-purpose baluns that will be discussed shortly.

The theory behind the Fig. 2-1 circuits is straightforward. The differential audio signals are applied through matching resistors to the inverting and noninverting inputs of the op-amp. The op-amp will

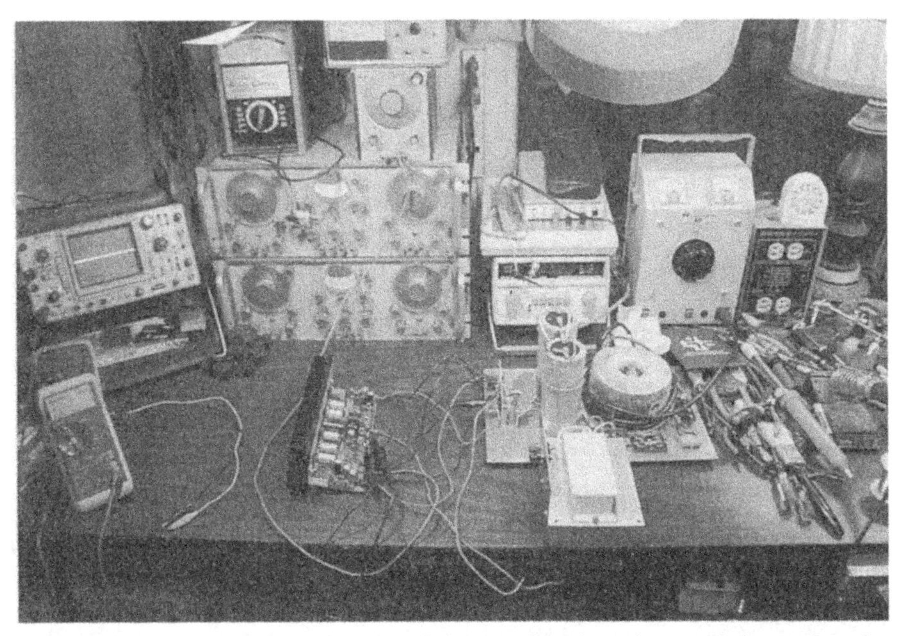

Welcome to my lab! This is my favorite of three workstations, because it has a nice window view.

cancel any significant levels of common-mode signals and the *clean* single-ended audio signal will appear at the output. The ground pin at the XLR plug can be effectively isolated if so desired.

Figure 2-2 illustrates several conventional balanced drivers that, again, can be found in some professional audio equipment and budget hi-fi systems. As in the Fig. 2-1 circuits, the external resistors should be matched as close a possible for the optimum balance of the differential output signals.

The theory behind the Fig. 2-2 circuits is straightforward also. The original single-ended signal is allowed to pass to one differential output unaltered through an appropriate impedance matching resistor. The original single-ended signal is also applied to the inverting input of a unity gain op-amp, and the inverted version of the original signal is passed on to the other differential output, through a duplicate impedance-matching resistor. Thus, an inverted and noninverted version of the original single-ended signal appears at the balanced output.

The applications for the Fig. 2-1 and Fig. 2-2 circuits can be for musical instrument amplifiers, inexpensive PA (public address) systems, and a variety of hobby and communications circuits.

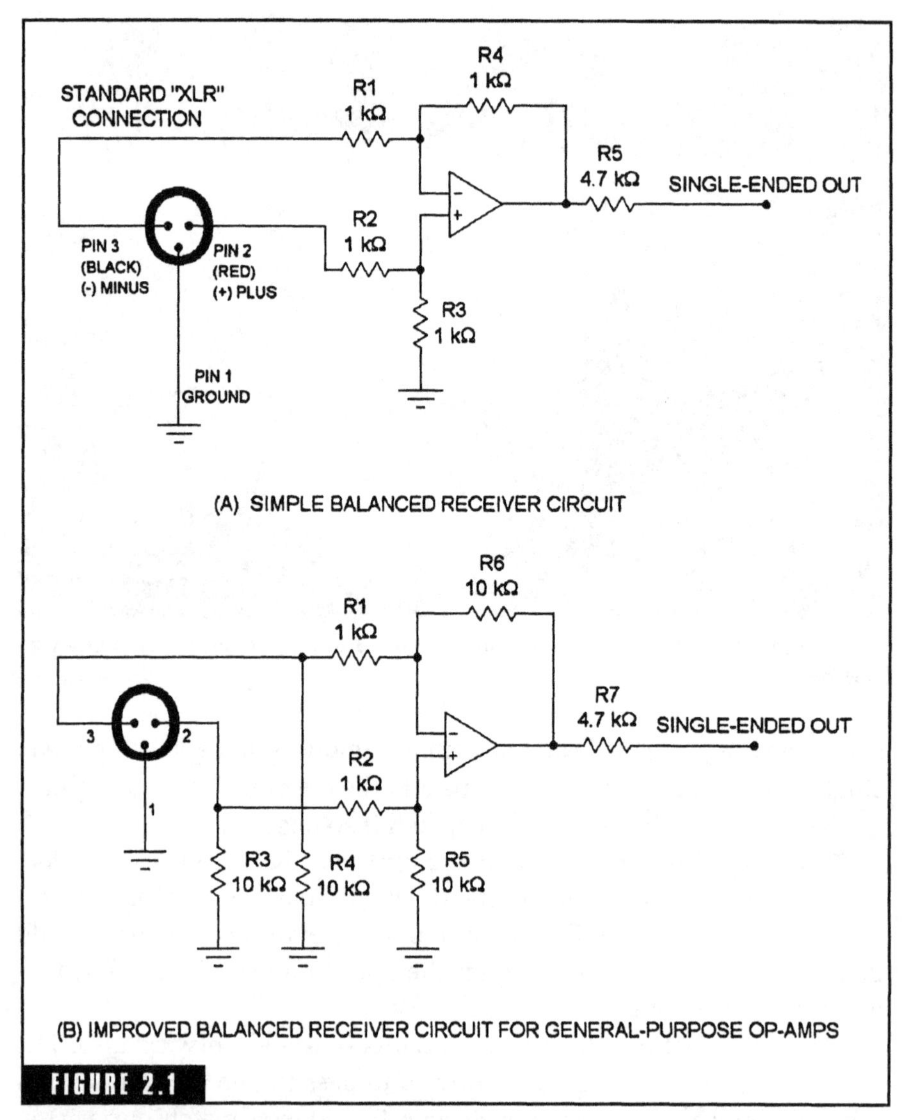

FIGURE 2.1

Conventional balanced receiver circuits.

If you happen to be the type of person who loves to build things from scratch (and you happen to be in the market for a high-performance balanced receiver circuit), you might want to consider the circuit illustrated in Fig. 2-3. In theory, this circuit doesn't differ a great deal from the previous balanced receiver circuits, but the op-amp chosen is a high-performance AD797 op-amp, with appropriate capacitors

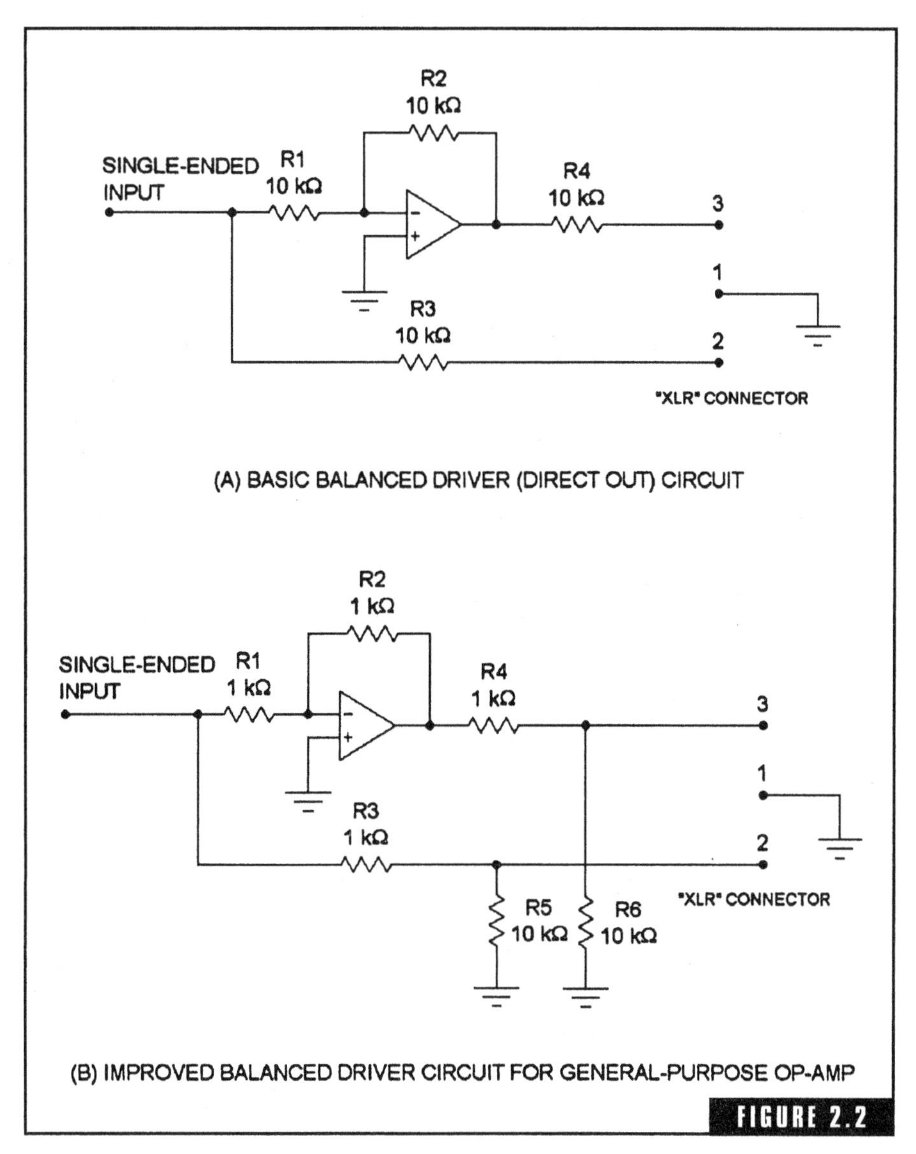

FIGURE 2.2

Conventional balanced driver circuits.

to tailor its operation for audio applications. C4, C5, C6, and C7 are power supply decoupling capacitors. As a general rule, high-performance (i.e., wide-bandwidth, low-distortion, low-noise, high slew-rate) op-amps such as the AD797, OP176, or OP184 will require special care in assuring the power supplies are adequately decoupled between stages. Otherwise, they are very likely to degenerate into high-frequency oscillations. C1 and C2 are chosen to shunt any RF frequencies that could have been superimposed on the signal lines. C3 is the compensation capacitor.

With carefully matched resistors, the Fig. 2-3 balanced receiver is capable of up to 60-dB common-mode rejection, and the noise/distortion characteristics make it applicable for a high-quality domestic or professional hi-fi application.

The Fig. 2-4 differential line driver is similar in quality and performance to the Fig. 2-3 circuit, with equivalent applications. For the sake of clarity, I have not shown the decoupling capacitors required for the OP-176 op-amps, but they should be implemented the same as for the Fig. 2-3 circuit. (*Note:* Analog Devices is currently discontinuing

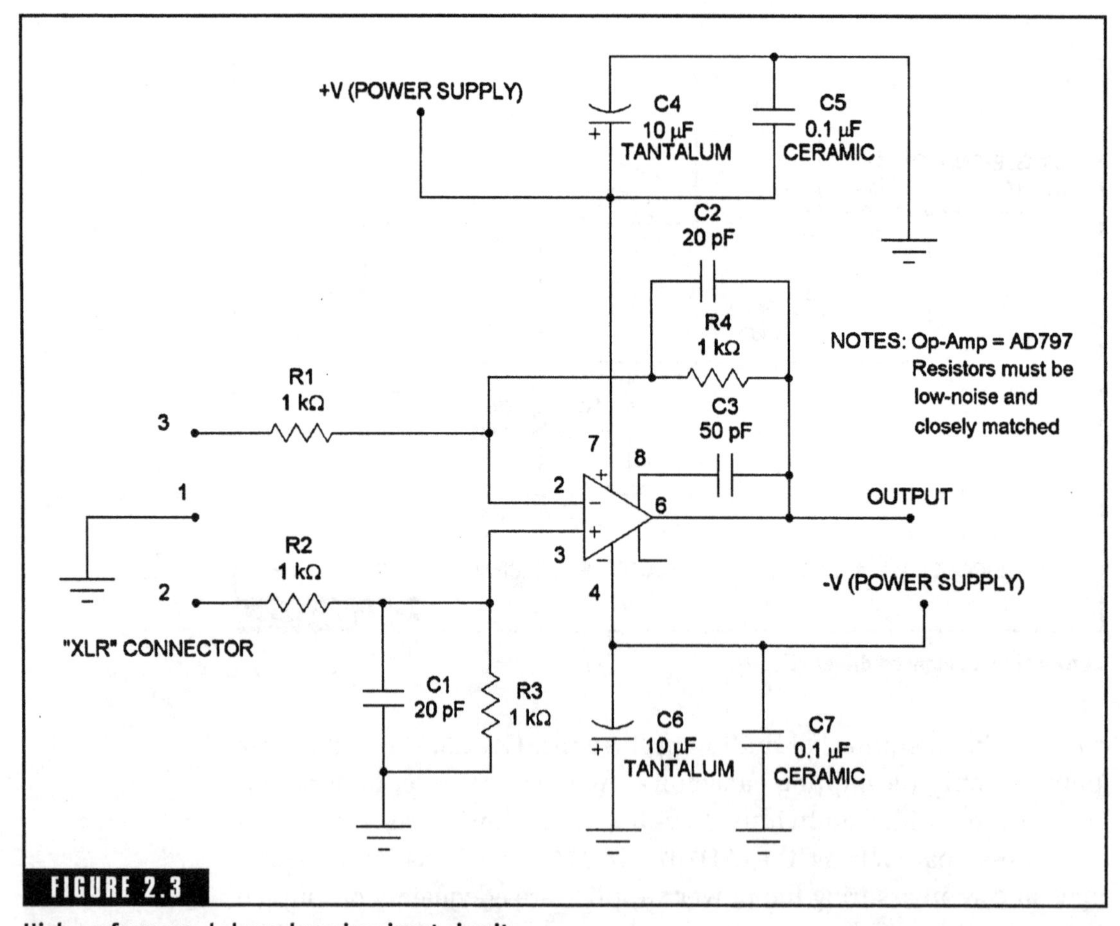

FIGURE 2.3

High-performance balanced receiver input circuit.

the production of the OP-176 op-amps. If they are not available, the newer OP-184 devices can be substituted.) Note that Fig. 2-4 utilizes two op-amps working in a complementary fashion to create the differential output, in contrast to a single op-amp only providing an inversion function in the conventional designs of Fig. 2-2. This technique provides a more evenly balanced and driven differential signal, and can be used in other differential line driver applications. P1 is a balance pot used to precisely adjust (i.e., match) the differential levels for the maximum common mode rejection on the receiving end.

The balun circuits illustrated thus far represent the conventional methods of designing mediocre and high-performance units. In recent years, several of the larger integrated circuit (IC) manufacturers have developed differential line drivers/receivers that are significantly superior in performance and specifically designed for high-quality audio applications. The reasons for their superiority are severalfold. First, the internal matched resistors are laser trimmed for ultimate matching

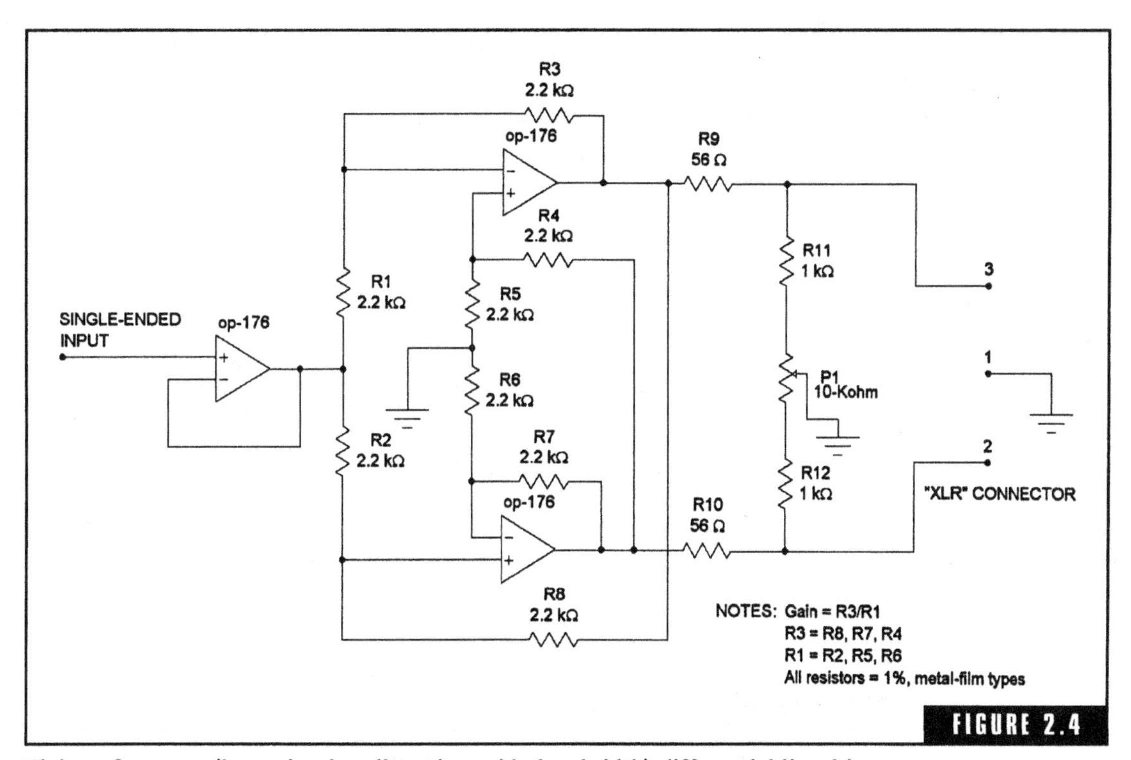

FIGURE 2.4

High-performance (low noise, low distortion, wide-bandwidth) differential line driver.

accuracy. This enables common-mode rejection factors of 80 dB or higher. Secondly, the manufacturers have taken into account that these chips will be processing signals from the "outside world," so they are designed with special protection circuitry that can better withstand the anomalies that could occur with external cables (i.e., voltage spikes, broken cable wires, operator error, etc.). The distortion, noise, and bandwidth performance will be superior to any circuit design that can be fabricated from general-purpose op-amps. For example, the Burr-Brown IN134 differential receiver chip boasts a worst-case THD+N performance of 0.0008% across the entire audio band, while producing only about 3.5-μV RMS of "broadband" noise. The bandwidth is from DC to well over 1 MHz. The other ICs detailed in Figs. 2-5 and 2-6 provide similar performance. Also, the good news is that the price of these specialized units is about the same as the high-performance op-amps of the previous circuits. Consequently, if you simply want the best performance and the least complexity in a balun circuit, these specialized versions are the ones to use.

Figure 2-5 illustrates a line driver circuit incorporating Analog Devices' SSM2142 or Burr-Brown's DRV134 chips (both of these chips

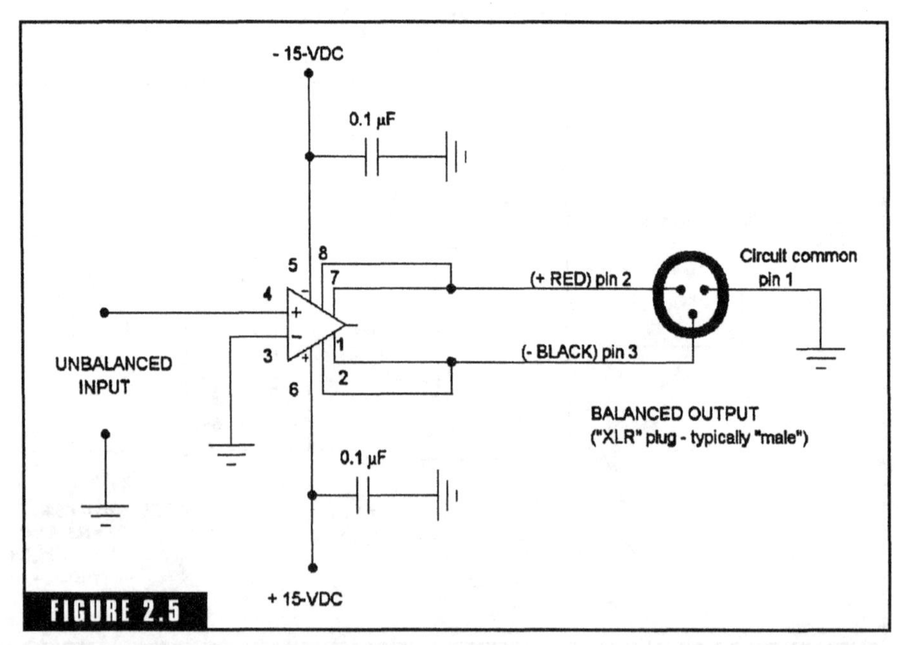

FIGURE 2.5

High-performance "balanced line driver" circuit based on Analog Devices' "SSM2142" chip or Burr-Brown's "DRV134" chip.

are direct replacements for each other). The required decoupling capacitors are illustrated.

Figure 2-6 illustrates a line receiver incorporating Analog Devices' SSM2141 or Burr-Brown's INA134 chips. (Again, both of these chips are interchangeable.) Note that the same decoupling capacitors are installed as used in the Fig. 2-5 circuit.

Figure 2-7 illustrates a simple and highly useful circuit for a person who will be doing a lot of experimenting and testing of audio equipment. It is called a "direct box." Many audiophiles like to test the quality of a preamplifier system, a new power amplifier, or a new set of speaker systems by making a direct connection from a program source to the power amplifier driving the speaker systems. The problem is that most program sources don't contain a volume adjustment for their line-level signal outputs, and most high-quality audio power amplifiers don't contain an input level adjust. Therefore, to facilitate making such direct connections, some type of signal attenuator is required. Also, it would be nice if this attenuator had the capability of accepting balanced line signals as well as single-ended (i.e., unbalanced) line signals. The Fig. 2-7 circuit will perform both of these functions. The right and left outputs can be attenuated to any desired level with P1 and P2.

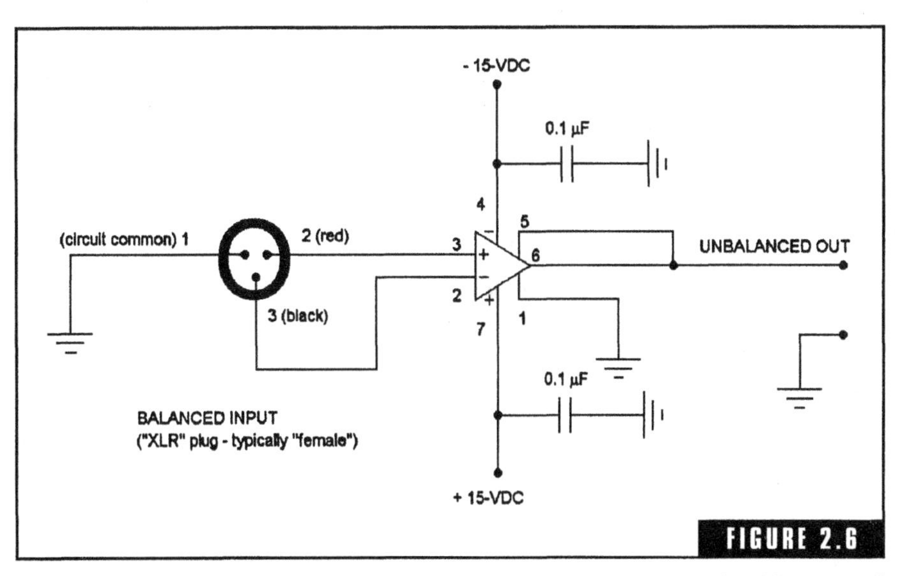

FIGURE 2.6

High-performance "balanced line receiver" circuit based on Analog Devices' "SSM2141" chip or Burr-Brown's "INA134" chip.

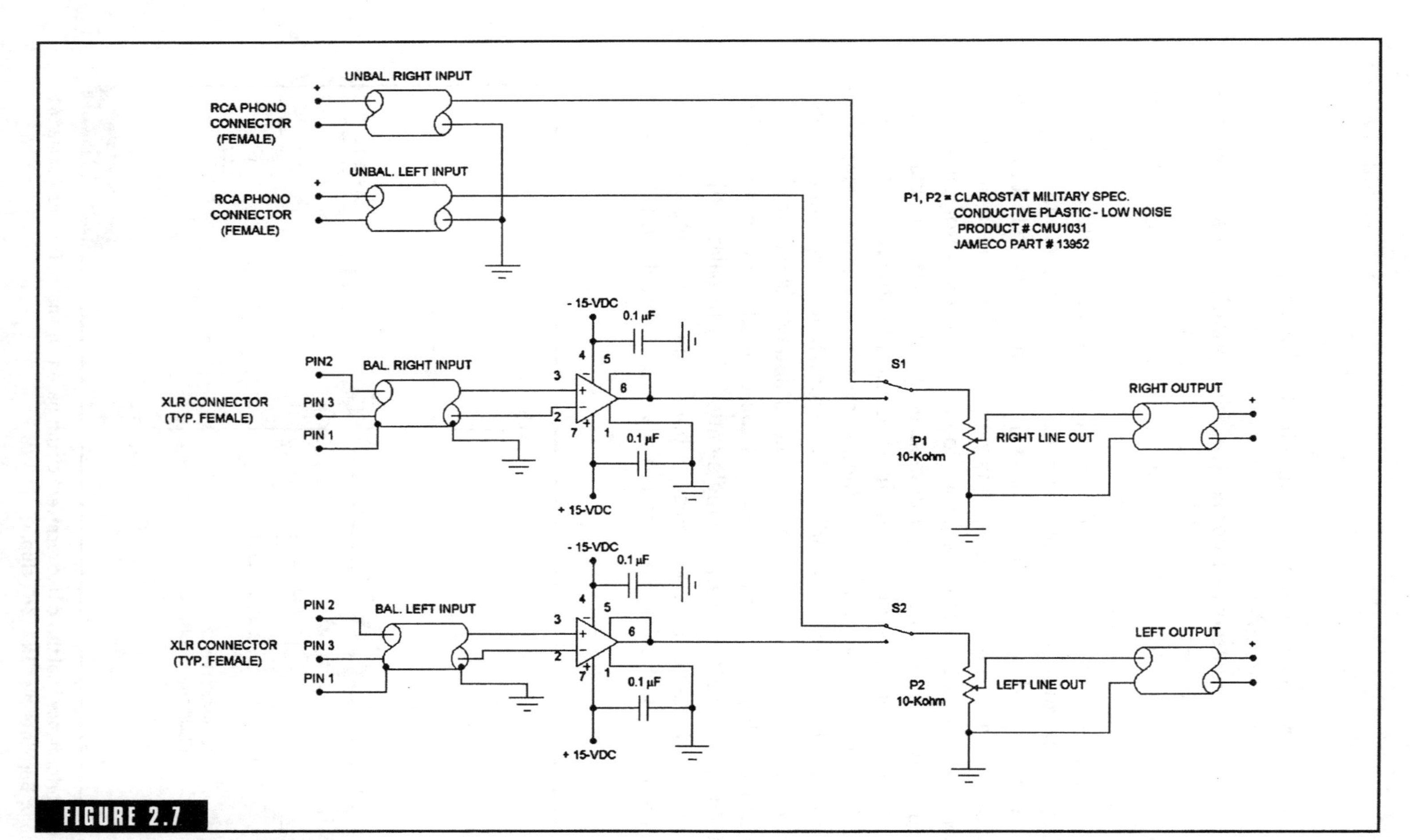

FIGURE 2.7

A simple version of a hi-fi direct box.

S1 and S2 select either balanced or unbalanced inputs. The two line receiver chips are the aforementioned SSM2141 or IN134 devices.

The entire circuit of Fig. 2-7 can be constructed in a small project box, utilizing any type of commonly available IC prototyping PC board. Conventional RCA type female phono connectors can be installed for the unbalanced inputs, and XLR type female connectors can be installed for the balanced inputs (using the pin numbers as indicated). You would probably want to use RCA-type female phono connectors for the outputs as well. I haven't illustrated any type of power supply, but you would obviously have to include a small regulated power supply to provide operational power for the balanced receiver circuits (power supply circuits will be covered in a later chapter). I also recommend that you use high-quality potentiometers for the level controls (i.e., P1 and P2), and a good grade of shielded audio cable for all of the signal wire connections.

Jumping from the Pot onto the Step Ladder

There are two schools of thought amongst audiophiles and audio electronics hobbyists today. One group searches out the newest, most innovative state-of-the-art audio processing equipment and incorporates it all together, sometimes ending up with a domestic hi-fi system that looks more like the inside of a recording studio than a stereo system. There seems to be no end to the noise reduction systems, equalizers, outboard processors, power amplifiers, speaker systems, and associated knobs to adjust. There are so many level indicators popping up and down, and lights flashing on and off, one begins to wonder if the overall effect is intended to be visual rather than aural. One hobbyist proudly named his system of tribute to modern audio technology "Lost in Spatial."

The contrasting group of modern audiophiles is continuously migrating in the opposite direction. They desire the simplicity of a bare-bones system, associating the purity of the sound with the pure simplicity of the system. Such audiophiles will spare no expense in purchasing the very best low-noise, low-distortion components, and they believe the least number of components (and the greatest percentage of gold plating) in the signal path will ultimately provide the highest quality of undiluted sonics. They also appreciate the fact that they can understand the technicalities of such systems, which results in their personal confidence and trust in the overall integrity of performance. In some cases, there is the additional advantage of the savings incurred by being capable of constructing such systems themselves.

In recent years, many audio purists have targeted the common *potentiometer* as a harbinger of all manners of sonic ills, ranging from being a source of unacceptable distortion levels to a prolific generator of high levels of noise and hum. There is a measure of truth to these claims, but in regard to high-quality potentiometers, the measure of truth is very, very small. It is no secret that ordinary consumer-grade potentiometers have their faults, including susceptibility to hum and noise pickup, noise generation from the carbon film wiper area, and all manners of pops and crackles originating from normal wear, dead spots, and contamination. The audiophile has the tendency to think that if these obvious and clearly audible faults exist, potentiometers must also be responsible for less obvious and more subtle distortions, causing significant coloration to the program material. In all probability, such suspicions are entirely unwarranted, providing the potentiometer in question is of a good quality and is relatively "quiet" and smooth as it is being adjusted. I feel compelled to remind the reader that audio programs originating from the recording studio have already gone through dozens of potentiometer stages in their route to the final mixdown, so potentiometers could not possibly be too corruptive to the sonics. Nevertheless, the practice of eliminating potentiometers in preference to *stepped attenuators* has become somewhat popular in the audiophile community.

A stepped attenuator is nothing more than a resistor-divider circuit. A switching arrangement of some sort (i.e., multiposition rotary switches, reed relays, analog switches, audio multiplexers, etc.) is used to switch various resistors in and out of the attenuator circuit, which adjusts the signal attenuation level. Simply stated, a stepped attenuator is a multiposition voltage divider. Figure 2-8 illustrates a conventional potentiometer volume control. As the wiper of P1 is moved from one adjustment extreme to the other, any desired percentage of the signal input can be tapped off and applied to the amplifier circuit. Fig. 2-9 illustrates a simple, six-position stepped attenuator. The five resistors are arranged in a series voltage divider, providing a maximum of six tap positions (i.e., from 0 dB to infinity in 10-dB per step decrements), allowing the signal input to be adjusted to a maximum of five practical levels (infinity could not be considered a level of the input signal). A choice of only five adjustment levels does not provide a sufficient *attenuation resolution* for most practical hi-fi applications, so the more common technique is to simply incorporate more resistors into the series network for improved resolution.

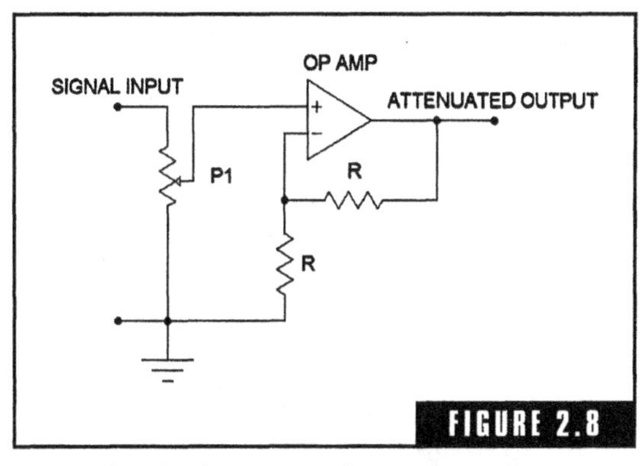

Conventional volume control.

Figure 2-10 illustrates how it is possible to construct a stepped attenuator using 15 resistors to provide an attenuation resolution of 3-dB per step. Such an attenuator is more practical and the 16-step positions coincide with many available rotary switches and audio multiplexer ICs.

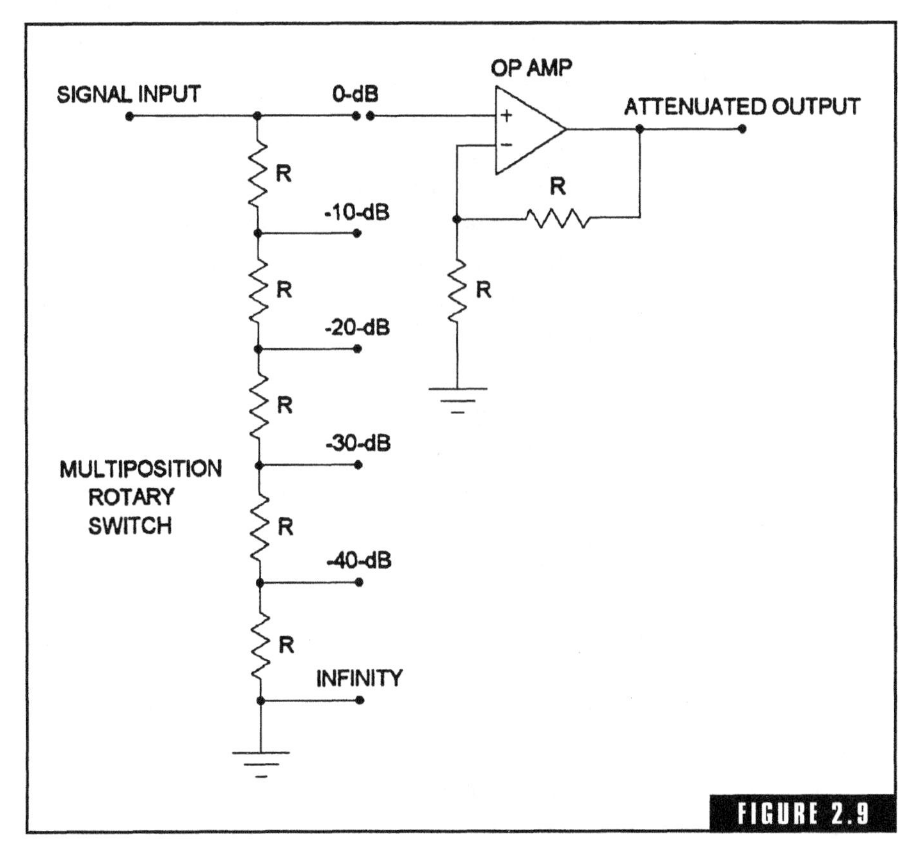

A volume control adjusted in "steps" (i.e., a stepped attenuator).

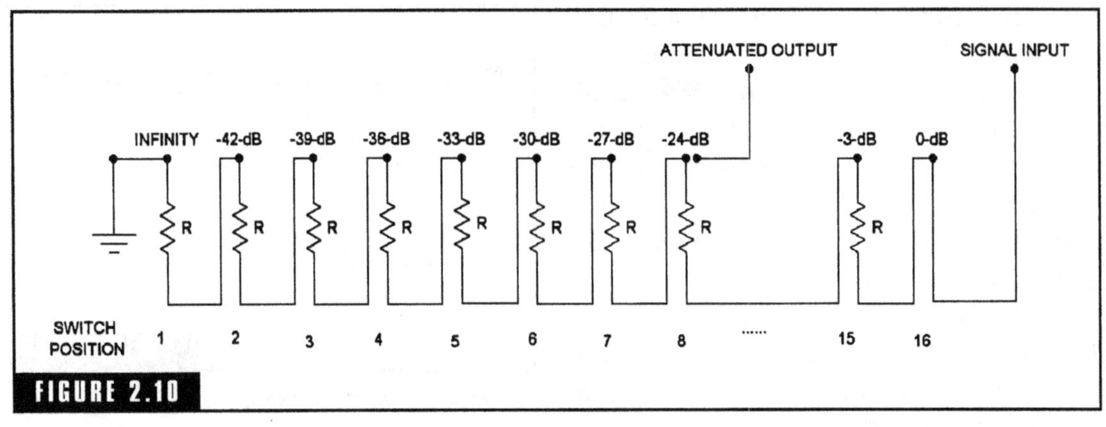

FIGURE 2.10

Diagram of a series-type stepped attenuator.

Again referring to Fig. 2-10, note that all of the divider resistors are arranged in a simple series circuit, beginning with the signal-input end and running to the circuit common point. Such an attenuator is commonly referred to as a *series stepped attenuator.* A disadvantage with this type of circuit is that the input impedance as well as the output impedance (i.e., at the attenuated output connection point) will vary radically as the attenuated output level is adjusted from one extreme to the other. (A simple potentiometer will produce the same type of impedance variation when adjusted.) Another disadvantage with the Fig. 2-10 circuit is that the noise artifacts from all of the resistors in the series string will have the tendency to be summed into the output signal.

Figure 2-11 illustrates a stepped attenuator design that greatly improves the aforementioned shortcomings of the Fig. 2-10 attenuator. An attenuator of this type is commonly called a *ladder-stepped attenuator.* In such designs, the signal input connection and the attenuated output connection are both switched coincidentally with each other, so that at any one time, there are only two attenuating resistors in the signal path. For example, in switch position #1 (as illustrated), the signal input is applied to the upper resistor, and the attenuated output is equal to the ratio of the upper resistor to the lower resistor. Thus, the noise performance is improved over series-type attenuators, and the two associative resistor values can be chosen to keep the input impedance relatively constant. Ladder-stepped attenuators are also more precise in

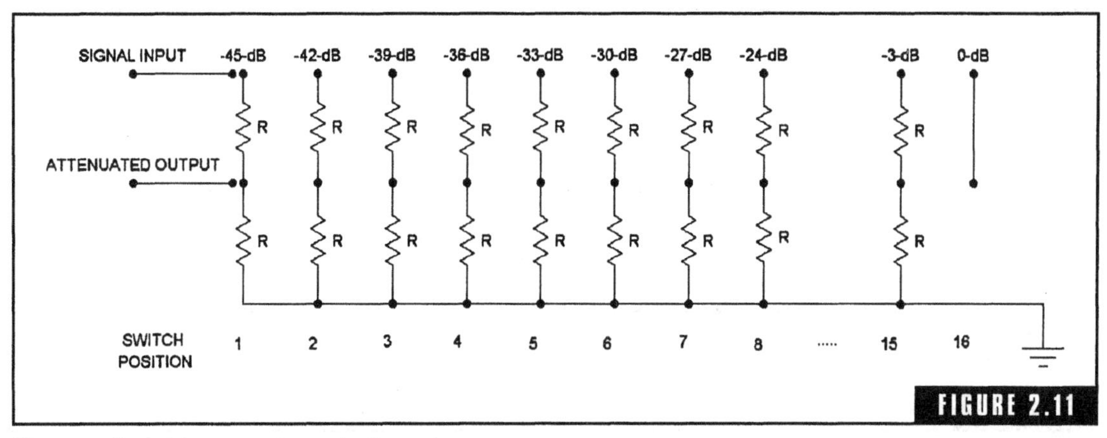

Diagram of a ladder-type stepped attenuator.

setting attenuation levels, because the tolerance errors are not added into the entire string of resistors (this characteristic also makes them easier to design and/or modify for special applications). The disadvantages of ladder attentuators are rather obvious—twice the number of switching contacts is required, along with twice the number of resistors.

Figure 2-12 illustrates a third type of stepped attenuator, commonly called a *shunt stepped attenuator.* Shunt attenuators maintain a constant-value resistor in series with the signal input and provide variable attenuation by switching a variety of dropping resistors into a series circuit with the constant-value resistance. In this way, the noise performance is about equal with ladder-type attenuators, and the same advantages of attenuation accuracy and ease of design are maintained. In comparison to series-type attenuators, only one additional resistor is needed, and the number of switching contacts required is identical. Shunt attenuators, like series attenuators, have the disadvantage of varying input/output impedances as attenuation levels are changed.

There are multiple considerations in comparing stepped attenuators with more conventional potentiometer controls. The advantages of stepped attenuators are improved sonic performance, improved reliability, and the ease of incorporation into digital volume control applications (especially important in any type of remote-control volume application). The disadvantages are increased cost, increased size, resolution limitations, and the inability to easily control channel balance functions in multichannel systems (i.e., precise balance con-

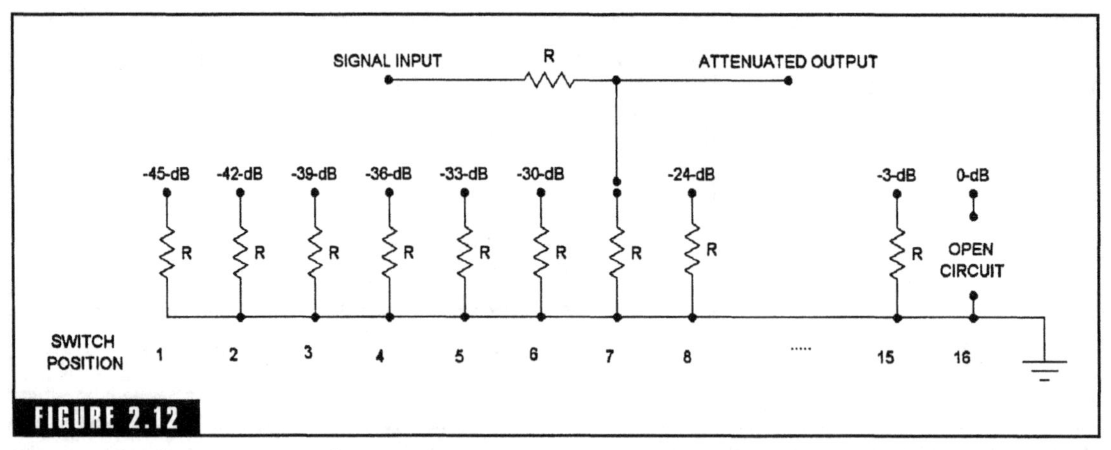

FIGURE 2.12

Diagram of a shunt-type stepped attenuator.

trol will be limited by the attenuation resolution). It should also be remembered that if any type of tone controls are desired in the preamplifier stage, they are most certainly going to be adjusted by means of conventional potentiometers, so it would seem rather pointless to go to the trouble and expense of incorporating stepped attenuators for input level control, and then expose the audio signal to the influence of potentiometers in the next step of the audio chain.

In my opinion, the decision on whether to use a stepped attenuator or a conventional potentiometer is not a difficult one. If you were constructing a preamplifier that will provide the convenience of remote control volume adjustment, a stepped attenuator system would be the logical choice. (The most common technique is to use an analog multiplexer, such as the CD4067, to control the switching of the stepped attenuator.) For most types of manual adjustment volume controls, I believe a high-quality audio potentiometer would be a more practical choice.

For those audiophiles who construct high-quality stepped attenuators, the series types are, by far, the most popular. Figure 2-13 illustrates a good design for a 24-position (−2 dB per step) stepped attenuator. The attenuated output signal is taken from the illustrated tap positions, and the design can be used with the common varieties of switching techniques (i.e., reed relays, rotary switches, etc.). The resistors are all 0.1% tolerance, low-noise types, available from several resistor manufacturers and audio supply houses. The Fig. 2-13

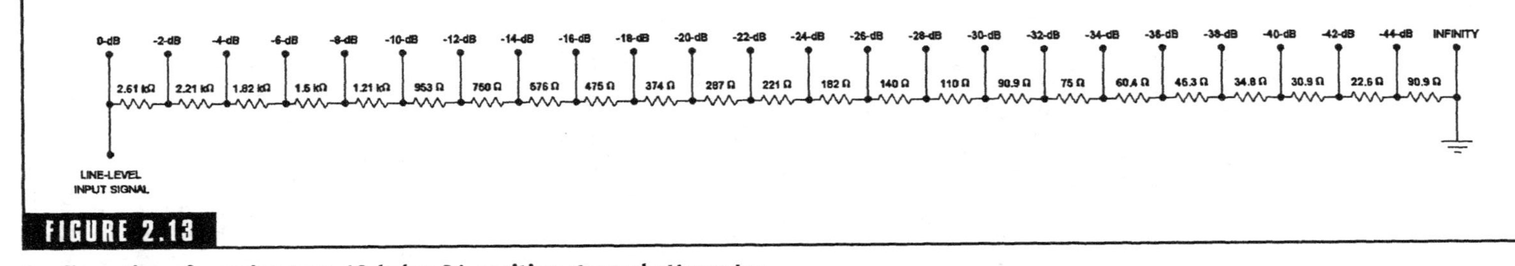

FIGURE 2.13

Configuration of a series-type 10-kohm 24-position stepped attenuator.

attenuator design is called a 10-kohm attenuator because the approximate impedance from the input signal connection to circuit common is 10 kohm. Obviously, if any significant loading is presented by the attenuated output connection, the input impedance will vary.

Figure 2-14 illustrates a technique for constructing a lower-impedance type (i.e., approximately 2 kohm) series stepped attenuator. Lower input impedance for a stepped attenuator may be desirable, since the noise performance will be a factor of the total resistance of the string. It would be difficult to accomplish such a task by simply incorporating lower resistance values in the Fig. 2-13 design, because the low-resistance-value resistors close to the circuit common connection would turn out to be very low, nonstandardized types. Figure 2-14 illustrates how it is possible to divide the resistor string into two (or more) sections, use standardized resistor types, and accomplish almost any input impedance desirable. Of course, low-input impedances are easily obtained in either the ladder type or shunt type attenuators.

I included the attenuator designs of Figs. 2-15 and 2-16 for those who want to construct stepped attenuators of less complexity and cost. The resistor values have been calculated with a computer program and are nonstandard values, so you will need to choose resistor values as close as possible to the ones illustrated. If you would like to make your own evaluations of any sonic differences between stepped attenuators and potentiometers, I recommend constructing one of the simpler attenuator designs (using 1% low-noise resistors and commonly available rotary switches) and compare it directly with a good quality potentiometer.

As a final entry into this chapter, Fig. 2-17 illustrates a block diagram of a typical remote control stepped attenuator system. The remote control receiver would probably be the more typical infrared type, and the information processor could be one of the specialized ICs designed for such applications. There are a variety of choices for the step-control system, including dot/bar display drivers, up/down counters, or a variety of digital decoders. SW1 through SW16 could be reed relays or high-performance analog switch ICs (such as Analog Devices' SSM2402 or SSM2412 models). An analog multiplexer, such as a CD4067B IC, could perform the functions of both the step-control system and the 16 switches. The new generation of versatile

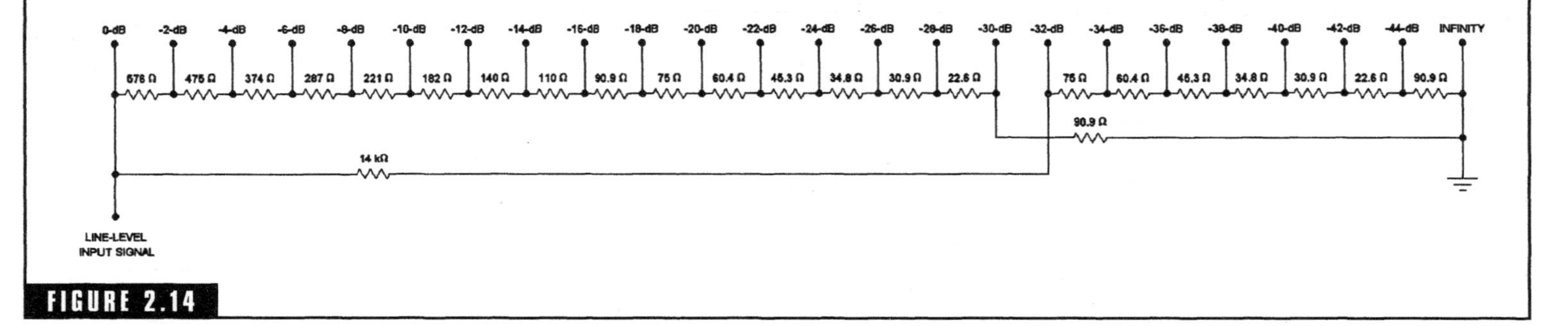

FIGURE 2.14

Configuration of a series-type 2-kohm 24-position stepped attenuator.

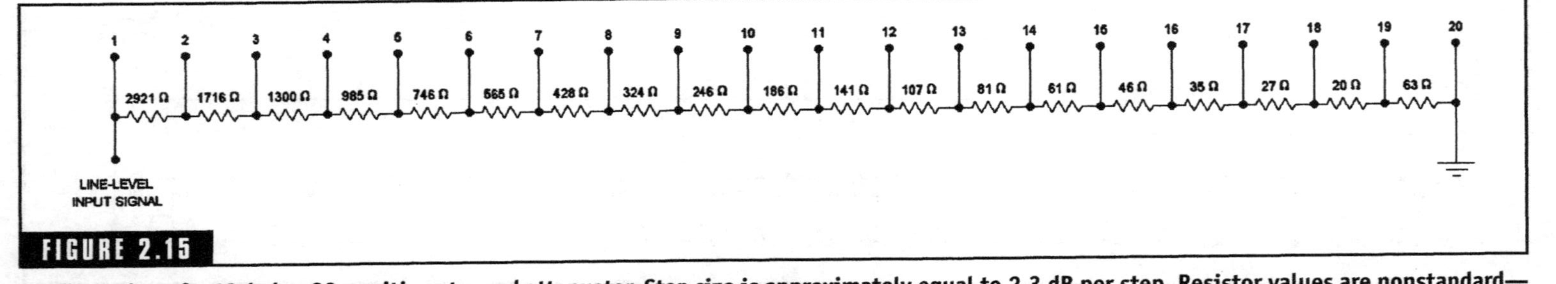

FIGURE 2.15

Configuration of a 10-kohm 20-position stepped attenuator. Step size is approximately equal to 2.3 dB per step. Resistor values are nonstandard—choose the closest available values.

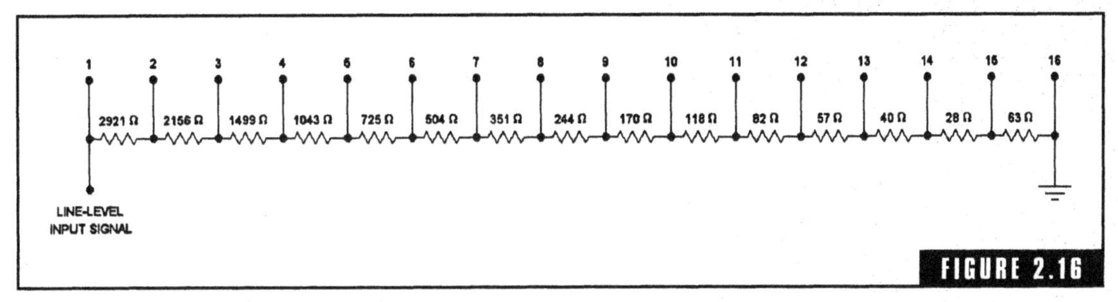

FIGURE 2.16

Configuration of a 10-kohm 16-position stepped attenuator. Step size is approximately equal to 3 dB per step. Resistor values are nonstandard—choose the closest available values.

microprocessors, such as the "PIC" or "BASIC STAMP," could be programmed to combine the functions of the information processor and the step-control system.

As you can easily perceive, there are many methods and options in the design of a remote controlled/digitally controlled stepped attenuator system. If you consider using analog switches instead of relays or other switching devices, you should be careful to insure that the *dynamic range* parameter of the analog switch can meet the performance demands of high-performance audio.

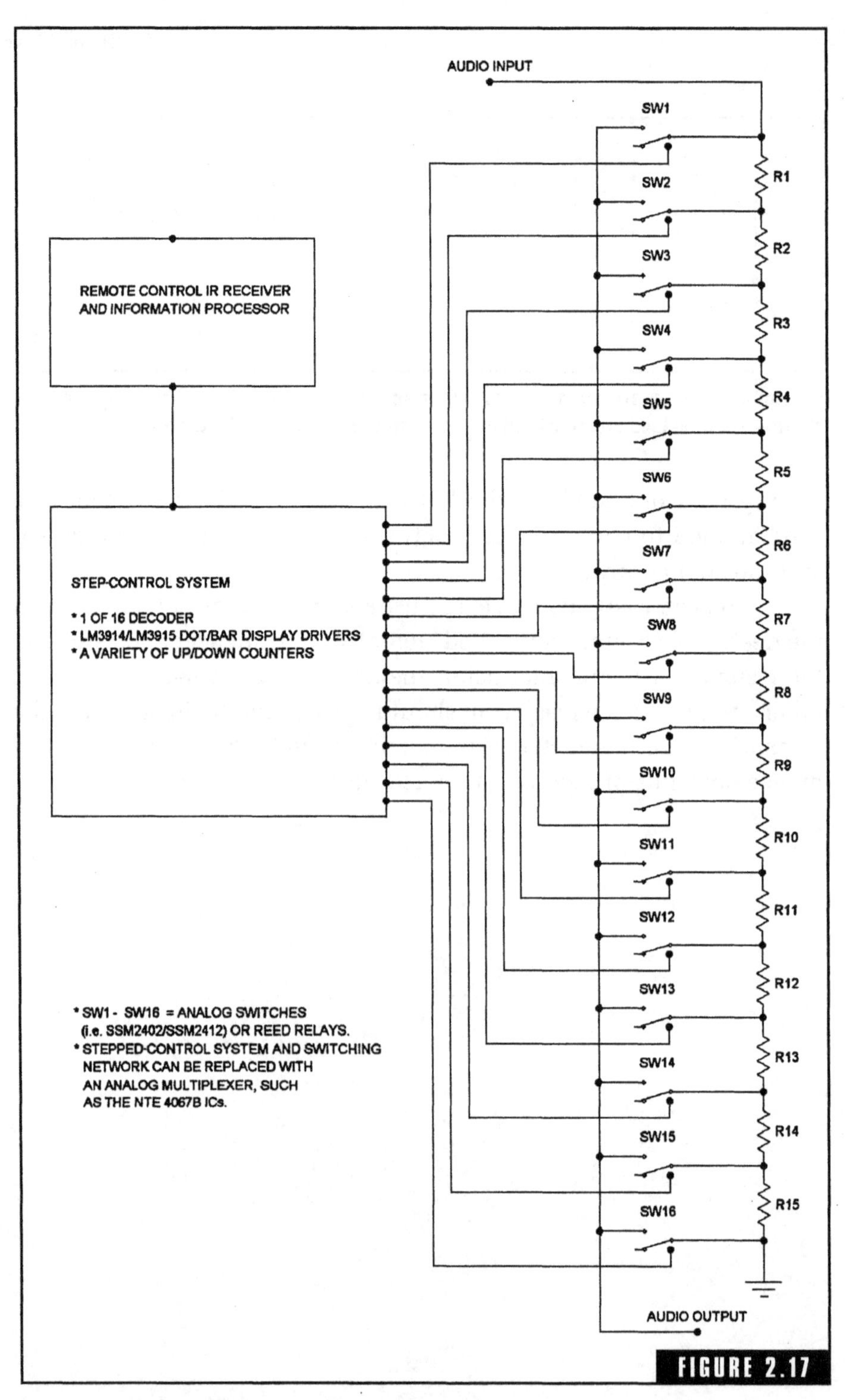

FIGURE 2.17

Typical 16-position remote control stepped-attenuator system.

CHAPTER

3

AUDIO PREAMPLIFIER SYSTEMS

Have you ever noticed that there seems to be a sense of urgency to change every form of standardized terminology within our modern society? Many within the audiophile community are now referring to preamplifiers as *control amplifiers.* I'm not sure of the rationale behind the change, but it seems that most control amplifiers do not have many of the options or capabilities of traditional preamplifier units.

As I write this text, I am looking at a catalog advertisement of two new high-end *control amplifiers* in the $400 price range. One unit contains an RIAA equalization network (the other doesn't), and neither unit incorporates any type of tone control system. They provide a reed-relay switching system for input program selection, attractive panel lights to indicate the current program selection, and a high-end attenuation potentiometer (not a stepped attenuator) for volume control. They advertise great performance specifications (which should be an easy task, since they don't do anything but switch the program inputs and route the signal through an attenuation potentiometer), and I must admit that the enclosures are very attractive, with their brushed aluminum finish and oversized platform feet. Except for the RIAA equalization network in the one unit, both units could be considered *passive preamplifiers* (or *passive control amplifiers,* if you're the type of person who likes to use the most up-to-date buzz words). Quite honestly, neither unit should cost anywhere near $400, and if such a unit is all you need (or want) for your audio reproduction system, you can easily construct a "performance-comparable" unit for around $50.

Passive Preamplifiers in an Active World

Passive preamplifiers are really attenuator switch boxes. They are called *passive* because they do not contain any *active components* (i.e., components capable of gain, such as transistors or op-amps). A typical passive preamplifier will provide the normal variety of stereo program input connectors (CD, tape, tuner, AV, and AUX) on the rear panel. The stereo input signals will be routed to a two-pole, multiposition rotary switch (for program selection), and the selected program signals will be applied to a high-quality level potentiometer or stepped attenuator. The attenuated program signals are then connected to (usually) two output connectors; "line" and "monitor." An optional feature on some passive preamplifiers is a transformer-type balun for the accommodation of any balanced input lines. Some passive preamplifiers may utilize reed relays or analog switches for input program selection. And that's all there is to it!

If you would like to construct a passive preamplifier for your audio system, you already have all of the information you need from the previous chapter. If you want to go first class all the way, you could use two stepped attenuators for stereo volume control, and a couple of transformer baluns for balanced inputs. If you use high-quality shielded cable for all of your wiring connections, and choose a really nice enclosure to house your project, the end result will be very impressive to any esoteric audiophile. But before jumping headlong into such a project, take a few minutes to ask yourself the question, "Is a passive preamplifier what I really want?"

A passive preamplifier will lock the user into a *straight-through* listening experience; the output of the program source will run directly to the power amplifier system (except for the level adjustment). If your speakers sound too "bassy," you're out of luck. If the room acoustics emphasize an irritable midrange frequency, you're out of luck. If you want to listen to low-volume music and still be capable of hearing a good, rich bass, you're out of luck (there isn't any loudness circuitry in typical passive preamplifier systems). If you want to listen to some of your old phonograph records, you're out of luck (there won't be any signal amplification stages and no RIAA equalization network). (I should mention in this context that vinyl records are making a comeback, so it might be wise to ensure that you have a phono-input capability on any preamplifier design that you intend on keeping for a while.) If you'd like to get a good idea of what your system would sound like with a passive preamplifier, you could use the "direct box" I detailed in the previous chapter to run a program source directly to your power amplifier system. Try the direct box for a few days, and if you're really satisfied with it, build a nice passive preamplifier for yourself!

Some *purist* audiophiles truly prefer a passive preamplification system, and I don't offer any criticism toward them. I honestly believe such individuals are more infatuated with the mental assurance of unaltered sonics rather than any audible performance improvements, but music appreciation is a *mental* thing, so who am I to try to change anyone's perspectives? I would simply like everyone to have the type of equipment that provides them with the audio experience they desire. 'Nuff said.

Conventional Preamplifier Systems

I believe most audiophiles desire a good, clean preamplifier that provides them with smooth volume control, a reasonably versatile tone control system (to compensate for speaker deficiencies, room acoustics, and program variations), loudness correction, precise balance control, RIAA equalization (just in case they want to blow the dust off the turntable), and a versatile quantity of selectable inputs and outputs to meet their system needs. With the availability of modern high-performance components, it is relatively easy to construct such a preamplifier system in the home-based lab, and the overall performance can be significantly superior in comparison to the bulk of the available commercial units. In addition, there is the advantage of being able to customize your system according to your personal desires, such as adding a headphone amplifier or installing optional balanced input/output circuitry. Some individuals like to incorporate a microphone input into their preamplifiers, so they can "karaoke" along with their music! (Guitar players can use the same microphone input to play their electric guitars through the audio system as well.)

Before getting into the first preamplifier cookbook project, I'd like to discuss a few details of preamplifier construction in general. Back in the early days of op-amp development, when the only commonly available types were the '741 and '709 families, there were performance advantages in designing preamplifiers with discrete components. However, in these modern times, discrete construction is pointless. It is certainly possible to construct high-performance preamplifiers with discrete components, but the overall expense and complexity is greater, and there aren't any significant performance advantages to be gained. Consequently, the majority of preamplifier designs in this chapter will be based on high-performance operational amplifiers. For those who want to construct a discrete-version preamplifier system "for the heck of it," I have included one such design in this chapter.

Again, generally speaking, the performance parameters of distortion, bandwidth, and slew rate are easily met in most preamplifier designs by using high-performance op-amps, incorporating a good circuit design, and maintaining good construction procedures. If you construct a preamplifier project and experience any *difficult* problems, they will probably be related to *noise*. This is true of most audio projects involving mostly low-level, low-power signal amplification. In contrast, when constructing audio projects involving relatively high power gains, such as audio power amplifiers, it is usually more difficult to obtain excellence in terms of *distortion*. Therefore, when constructing preamplifiers and other types of signal processing equipment, it is wise to focus on using components and construction techniques designed to provide the best low-noise performance. Of paramount importance is the assurance of excellent grounding integrity, the appropriate use of decoupling capacitors (on power supply lines and the power supply inputs of most operational amplifiers), the utilization of shielded cable for interconnection purposes, and the physical placement of any power transformers well away from sensitive circuitry.

Figure 3-1 illustrates my favorite preamplifier design for high-performance domestic audio applications. The two-stage RIAA equalization network, consisting of AR1, AR2, and their associated components, is accurate to within 0.2 dB of the RIAA curve. I especially like the fact that the loudness correction circuit (C10, R14, R15, C11, R16, and the "loudness" switch) does not require the use of a

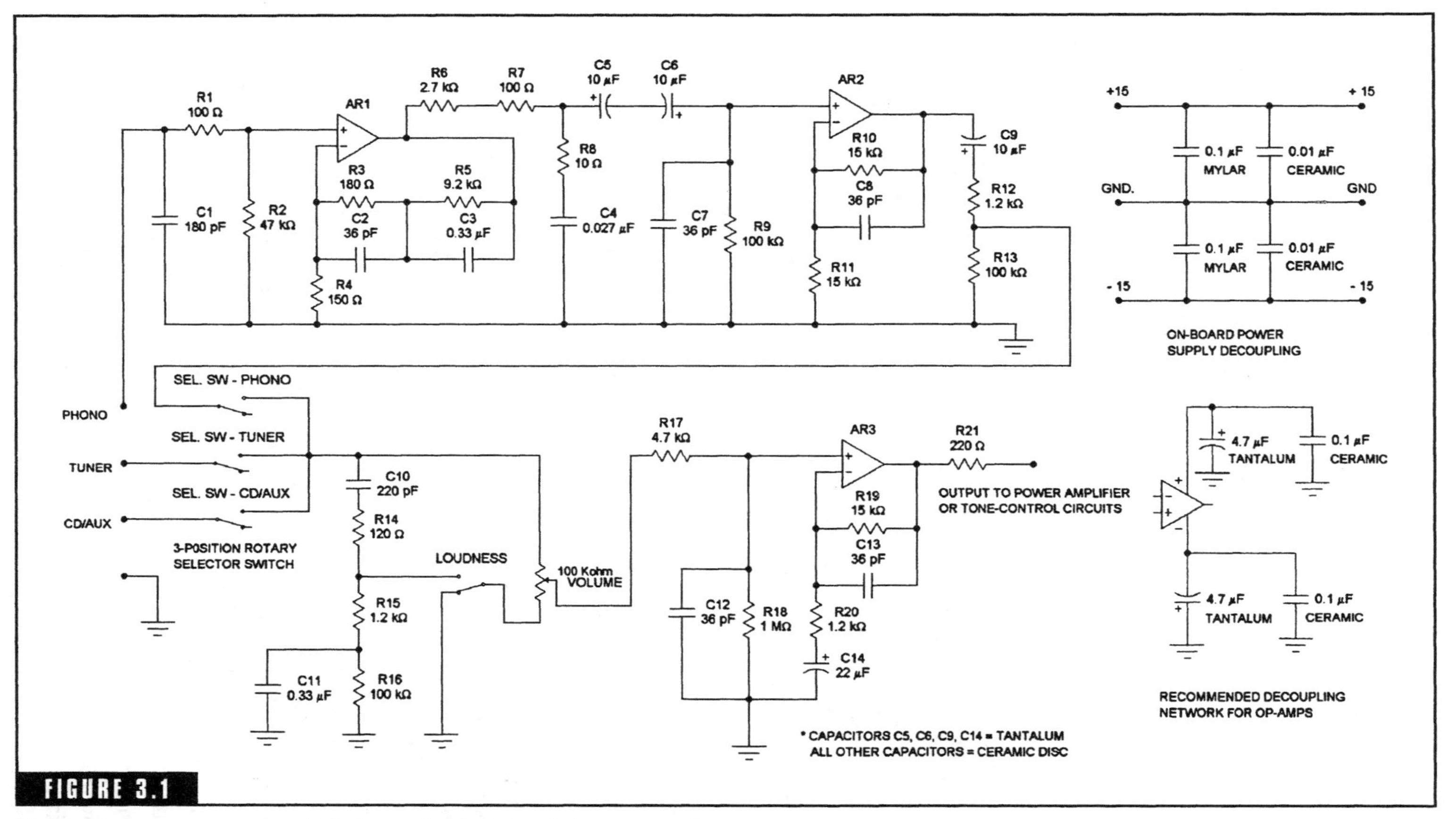

FIGURE 3.1

High-performance preamp and RIAA equalization circuit (one channel).

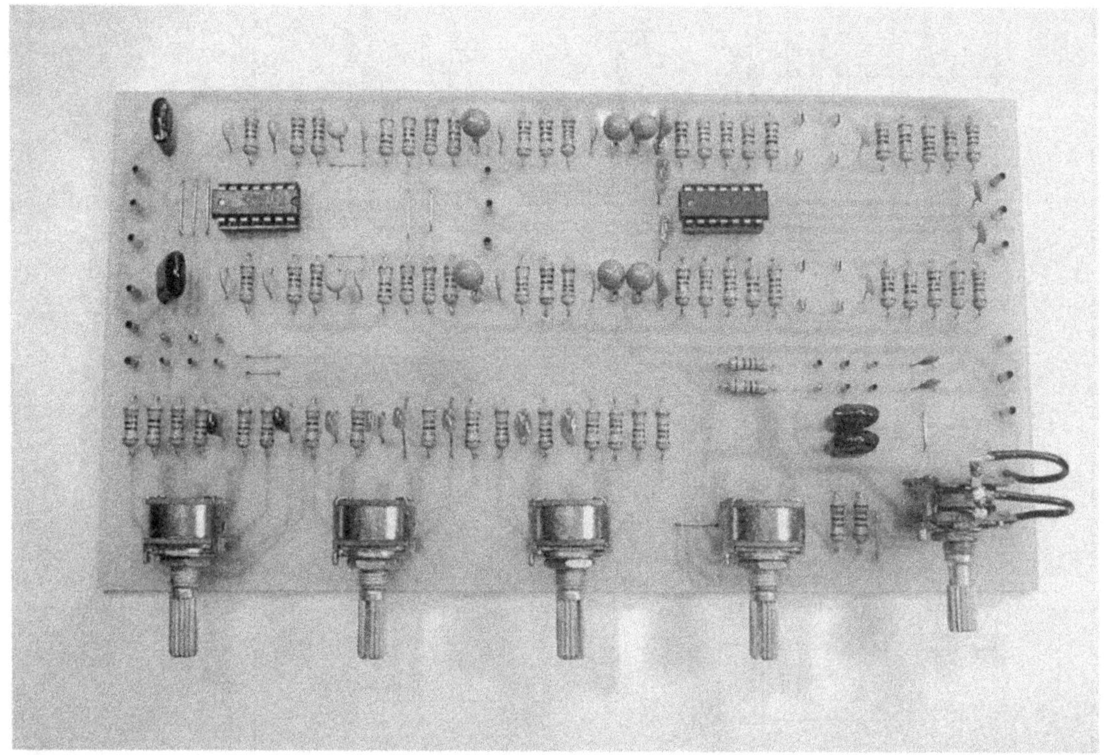

A stereo version of the Fig. 3-1 preamplifier circuit combined with the Fig. 4-1 tone control circuit.

"center-tapped potentiometer," which can be difficult to find in the general marketplace. I also like the convenience of having an approximate 10-dB voltage gain capability for the line-level inputs, which is provided by the ratio of R19/R17.

If you decide to use the common TL-074/TL-084 variety of J-FET operational amplifiers for AR1, AR2, and AR3 in Fig. 3-1, the broadband THD performance should be about 0.005%, or better, and the SNR performance should be at about −90 dB for the line-level inputs and about −75 dB for the phono input. Frequency response is 10 Hz to 100 kHz, at ±0.5 dB, and the sensitivity is approximately 350 mV for the line-level inputs and 2.5 mV for the phono input. By incorporating high-performance op-amps and low-noise metal-film resistors in the Fig. 3-1 design, the SNR performance can be improved to about −110 dB for the line-level inputs and −85 dB for the phono input. Broadband THD performance will improve to about 0.003%.

The construction of a preamplifier system based on the Fig. 3-1 design is relatively straightforward. Figure 3-1 represents a single channel, so you would need to duplicate this design for as many channels as required. The volume potentiometer can be incorporated individually for each channel, thus providing a balance adjustment by adjusting the individual volume controls for each channel (i.e., for multichannel home theater applications). Another option is to incorporate a two-gang potentiometer for simultaneous "stereo" volume adjustment, and install a separate "balance" potentiometer between the volume potentiometer "tap" and the left-side connection point of R17. Figure 3-2 illustrates how a balance potentiometer could be incorporated into the Fig. 3-1 design. The balance potentiometer would have to be a two-gang, 100-kohm variety (the same as the volume potentiometer), and you would have to wire one stage of the balance potentiometer "backward" relative to the other stage. Thus, as you rotate the balance potentiometer, you would be increasing the volume of one channel while simultaneously decreasing the volume of the other.

As illustrated in Fig. 3-1, the power supply inputs to the completed preamplifier board should be bypassed (i.e., decoupled) with 0.1-μF Mylar and 0.01-μF ceramic capacitors. In addition, the power supply

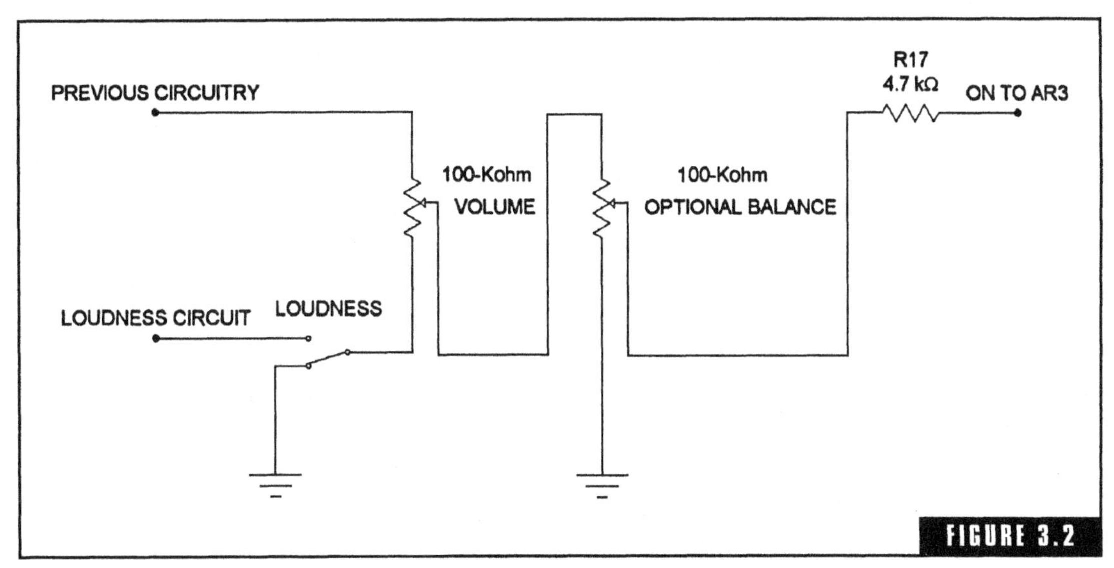

FIGURE 3.2

Optional balance control for a stereo version of the Fig. 3-1 preamplifier design.

pins of each operational amplifier should be bypassed with 4.7-μF tantalum and 0.1-μF ceramic capacitors. The op-amp bypass capacitors should be physically located as close to their associated op-amps as possible. (The three-band tone control system for this preamplifier design will be detailed in the next chapter.)

Figure 3-3 is a very good preamplifier design for most professional applications. (*Note:* The term *professional* does not denote *superior performance.* It defines the application as being well suited for professional, rather than domestic, usage. Many professional audio systems do not perform nearly as well as their domestic counterparts.) Professional audio equipment is subjected to a great deal of *field abuse,* and the risk of operator error (i.e., erroneous connections) is greatly increased over domestic systems. Note how the input section of Fig. 3-4 is protected. R1 is often specified with a 1- or 2-W rating, so the input can withstand the accidental application of speaker-output cables or even power line voltages for a short duration. C2 is typically rated for 250 V. The diode overvoltage protection provided by D1 and D2 is a very good protection scheme for a wide variety of op-amp applications. Most op-amps can be damaged if the input signal voltage exceeds the op-amp's power supply voltages. By connecting D1 and D2 in a reverse-biased configuration as illustrated, any peak input voltages applied to the input of U1 that exceed the power supply levels will cause one of the two protection diodes to become forward-biased, shorting the peak signal through the power supply, and protecting the input of U1.

The action of C4 and C5 isolates the AC gain factor from the DC gain factor, improving the DC temperature stability and offset factors. Volume is adjusted by the 10-kohm "LEVEL" potentiometer. An optional "treble boost" is incorporated, which is a feature that is highly desirable by musicians and keyboard players. The remainder of the design is a pretty conventional voltage amplifier with a three-stage active tone control circuit.

The Fig. 3-3 design is an excellent choice for musical instrument amplifiers or other professional applications. The use of the TL074 op-amps, as illustrated, will provide a SNR of about −80 dB. The slightly higher noise characteristic as compared to the Fig. 3-1 design is due to the higher gain factors. The broadband THD performance is better than 0.01%, and both the SNR and THD performance can be significantly improved by using one of the higher-performance op-amp families.

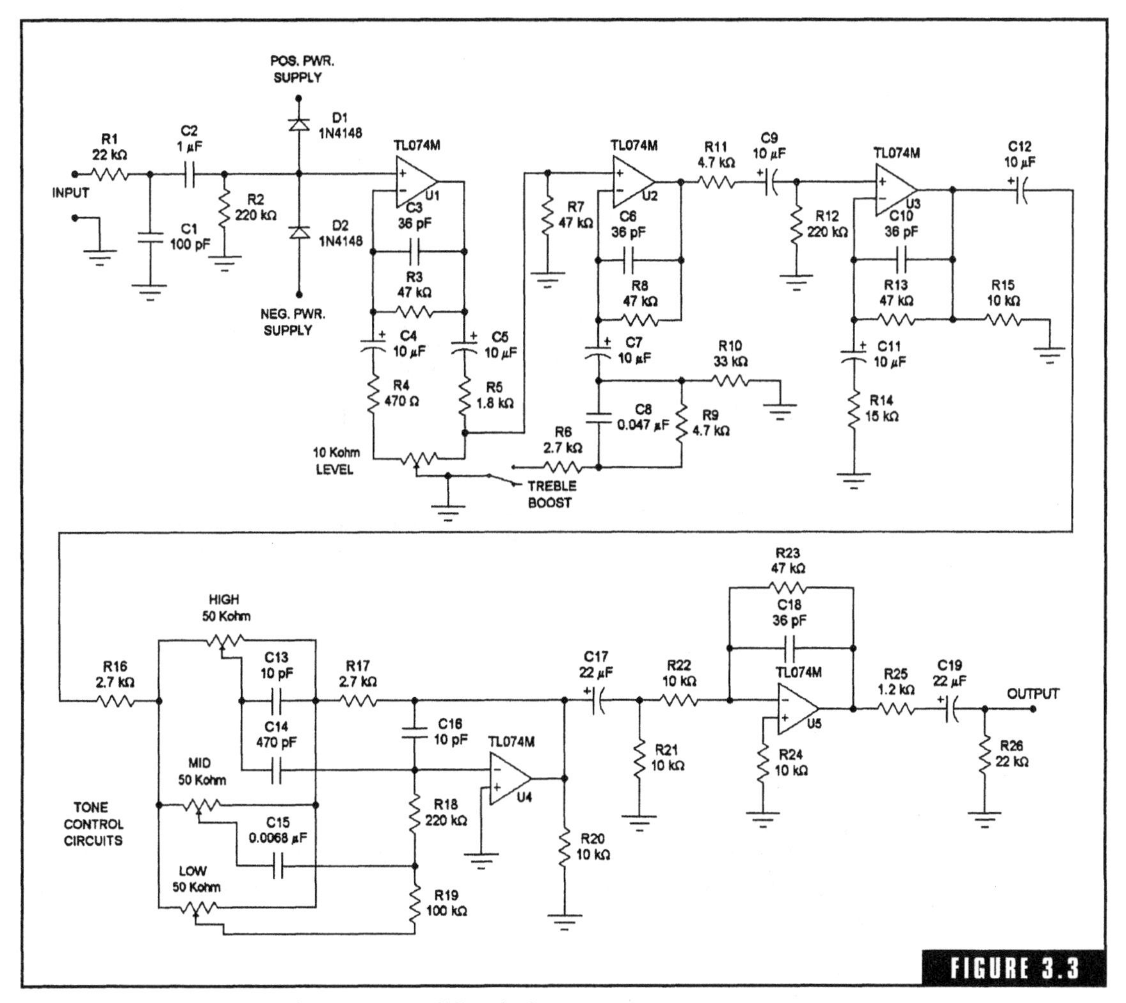

A general-purpose "professional" preamplifier design.

Remember to incorporate appropriate decoupling capacitors (not shown) for the operational amplifiers, as previously detailed for the Fig. 3-1 design.

Some modern preamplifier units do not include RIAA equalization networks for phonograph reproduction. The circuit of Fig. 3-4 often comes in handy if you need an external method of providing such RIAA equalization and amplification. Its RIAA curve deviation is only slightly inferior to the two-stage design of Fig. 3-1, and I seriously doubt that there could be any audible differences. The use of one of the high-per-

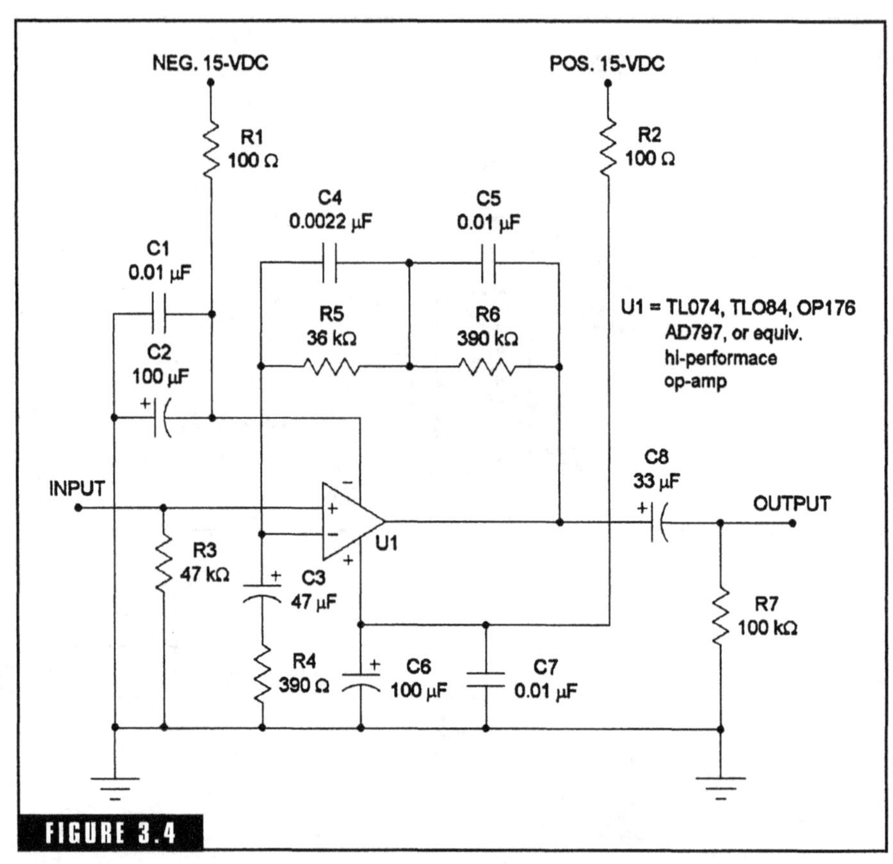

FIGURE 3.4

A simple, high-performance single op-amp RIAA equalization amp for phonograph inputs.

formance op-amps listed together with low-noise metal-film resistors will result in audiophile-quality results. Note that the decoupling capacitors and power supply connections are included in the Fig. 3-4 illustration. Resistors R1 and R2 may be eliminated if you use a dual-polarity power supply of relatively high quality and performance.

It is often desirable, depending on a variety of applications, to include low-level inputs for dynamic microphones into a preamplifier system (the output signal level of dynamic microphones is only about 3 to 6 mV, which is much too low to be processed through normal line-level preamplifier circuits). The Fig. 3-3 preamplifier design will perform very well with dynamic microphones, but dynamic microphones typically need a "wee" bit of help in the lower frequency realm. The

Fig. 3-5 dynamic microphone preamplifier circuit provides a slight bass boost that flattens the response of most high-impedance-type dynamic microphones. (Low-impedance microphones will also perform well with the Fig. 3-5 circuit, but they should be processed through an appropriate differential line receiver stage beforehand.) The output signal appearing at the negative side of C5 can be applied directly to the line-level inputs of most typical preamplifier systems. Because the frequency response characteristics of dynamic microphones will vary quite a bit from one model type to another, the illustrated tone control circuit will aid in adjusting the overall circuit for a wide variety of microphone models. The Fig. 3-5 design will also perform quite well with most musical instruments using dynamic pickups (i.e., colloquially called "electric pickups") for amplification purposes. If you construct this design, don't forget to include the appropriate decoupling capacitors (not illustrated) on the op-amp power supply lines.

As promised earlier, Fig. 3-6 illustrates a discrete-version audio preamplifier design. This particular type of audio preamplifier is commonly called a *line amplifier* within professional recording circles. Q1, Q2, and their associative circuitry, make up a high-gain inverting amplifier stage. R5 and C4 provide power supply decoupling. R4, working in conjunction with the voltage divider of R6 and R7, applies a percentage of the output signal back to base input of Q1, providing the stabilizing negative feedback. C3 is a compensation capacitor, and D1 and D2 provide input overvoltage protection in the same manner as previously described for the design of Fig. 3-3. C1 is incorporated to shunt any undesirable ultrasonic or RF signals that happen to be picked up by the input signal. R3 is the collector load for Q1.

Q3, Q4, and their associative circuitry make up an inverting amplifier stage of the same basic design as the Q1/Q2 amplifier stage, so the functional description given previously for the Q1/Q2 stage applies to this one as well. C18 is a coupling capacitor, and the AC output voltage from C18 is applied to the voltage divider network of R22 and R23. A certain percentage of the AC output voltage obtained from the R22/R23 voltage divider is applied back to the tone control circuitry, and the outputs of the tone control network are summed and applied back to the base input of Q3 through coupling capacitor C16. Thus, the *active* tone control network is actually the AC feedback loop for the Q3/Q4 amplifier stage.

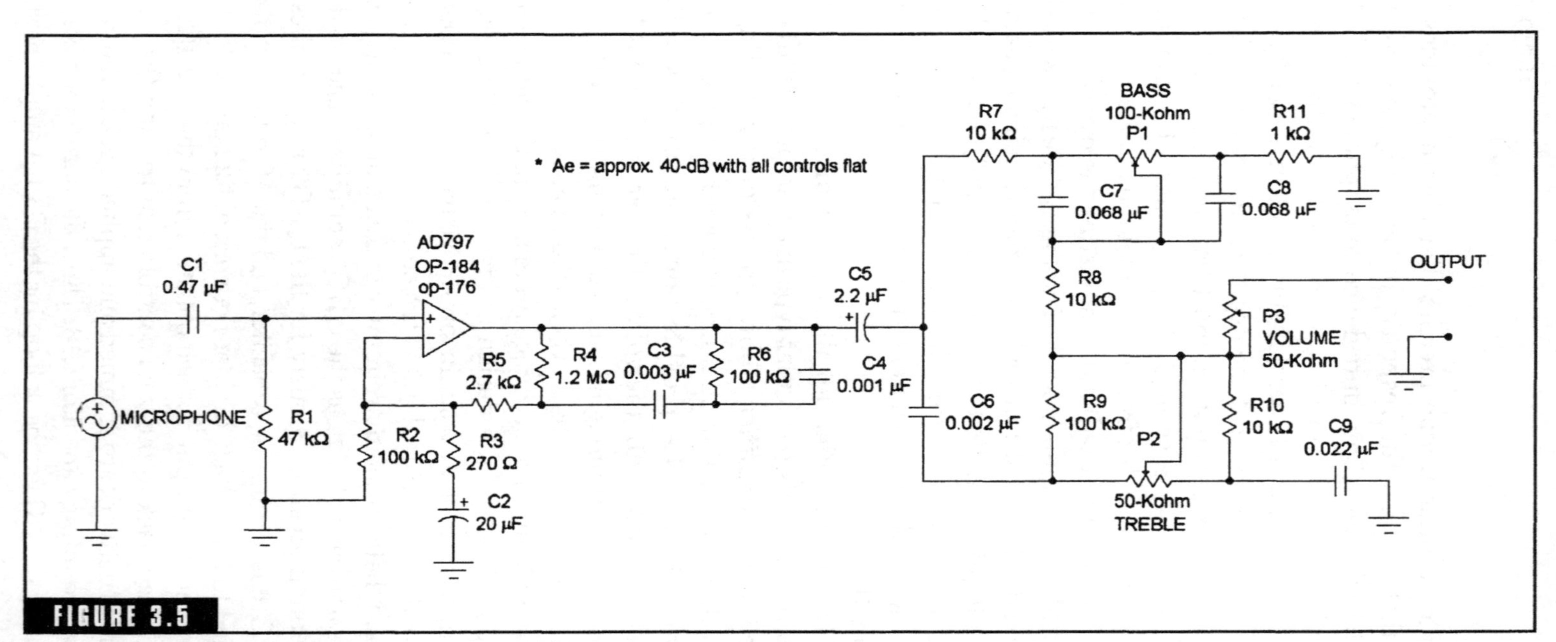

FIGURE 3.5

A high-performance microphone preamplifier with tone control.

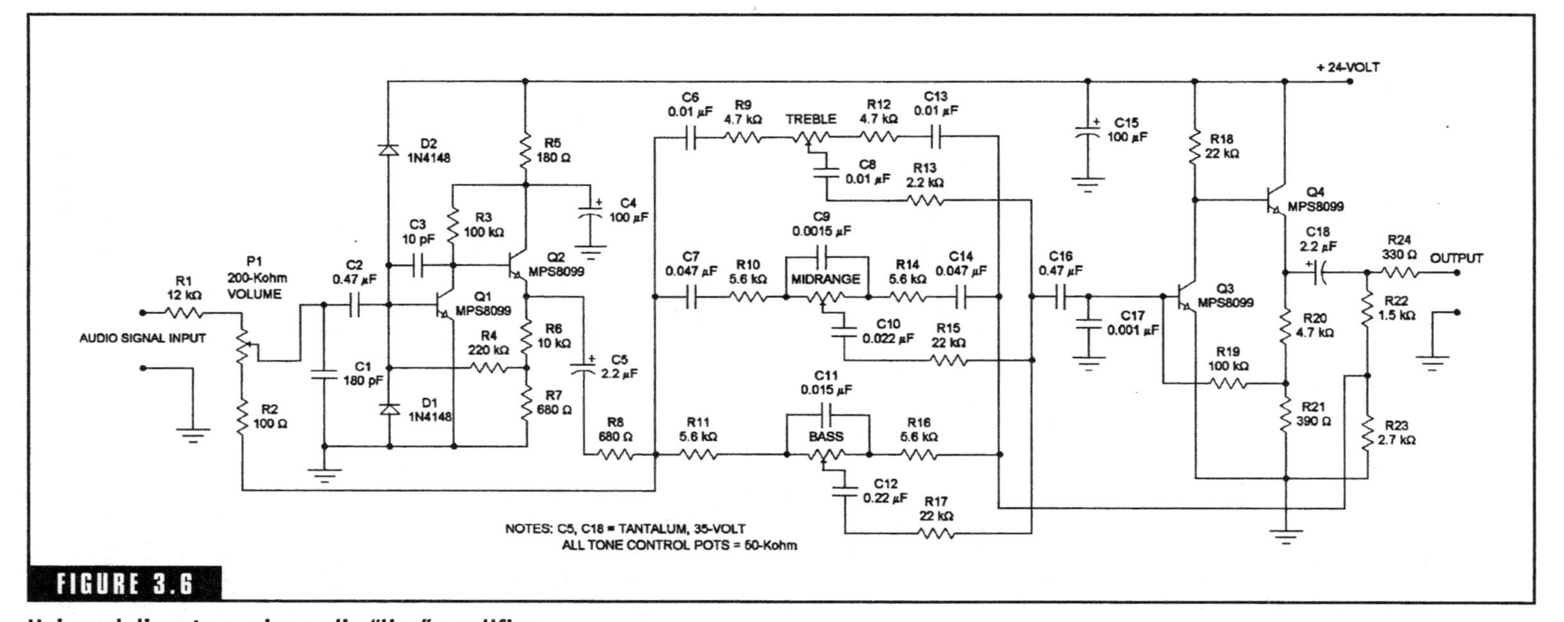

FIGURE 3.6

Universal discrete-version audio "line" amplifier.

Note the negative feedback loop connecting from the right side of R8 and running back to the bottom side of P1 (through R2). Since the Q1/Q2 amplification stage provides an *inverted* output, this feedback applied to the bottom side of P1 will be 180 degrees out of phase with the input signal. Consequently, as the applied audio signal input level rises in amplitude, the negative feedback applied to the bottom of the P1 volume control will also rise proportionally, greatly *reducing* the signal level as seen by the base of Q1. Such a technique results in an enormously wide range of audio signal levels that can be processed through this line amplifier. For example, the overall high voltage gain factors are more than sufficient to process dynamic microphone inputs, but the unique negative feedback loop connected to the volume potentiometer will provide for input levels as high as speaker-level inputs. Simply stated, the Fig. 3-6 line amplifier has a usable input signal range of 82 dB! Consider the versatility and multiple uses for such an amplifier circuit.

The Fig. 3-6 line amplifier is a high-performance unit, and if you decide to construct one of these, it will probably surprise you with its overall performance quality. It can process almost any type of audio input signal for application to a typical audio power amplifier, and the three-band active tone control circuit is very effective at tailoring the frequency characteristics of the input source. However, it is absent of any differential stages, and the operational power supply is a single polarity (such types of amplifiers are commonly referred to as *single-ended*). Therefore, it is extremely important to insure that the power supply is a very high-quality one, producing very low levels of ripple, and extraordinarily stable.

Figure 3-7 illustrates very versatile and general-purpose preamplifier/tone-control circuit combination. This design illustrates how it is possible to operate a preamplifier from the power supply rails of a typical audio power amplifier. Note that the power supplies are dual-polarity 38-V inputs, which are regulated down to dual-polarity 15-V levels by the simple zener supplies consisting of ZD1, ZD2, and their associated components. For use with higher power amplifier rail voltages, the resistance values of the two series 1.5-kohm resistors may have to be increased (and you will probably have to use 1-W models).

Referring to Fig. 3-7, IC1 and its associated components make up a simple inverting amplifier with a voltage gain of 10. For decreased

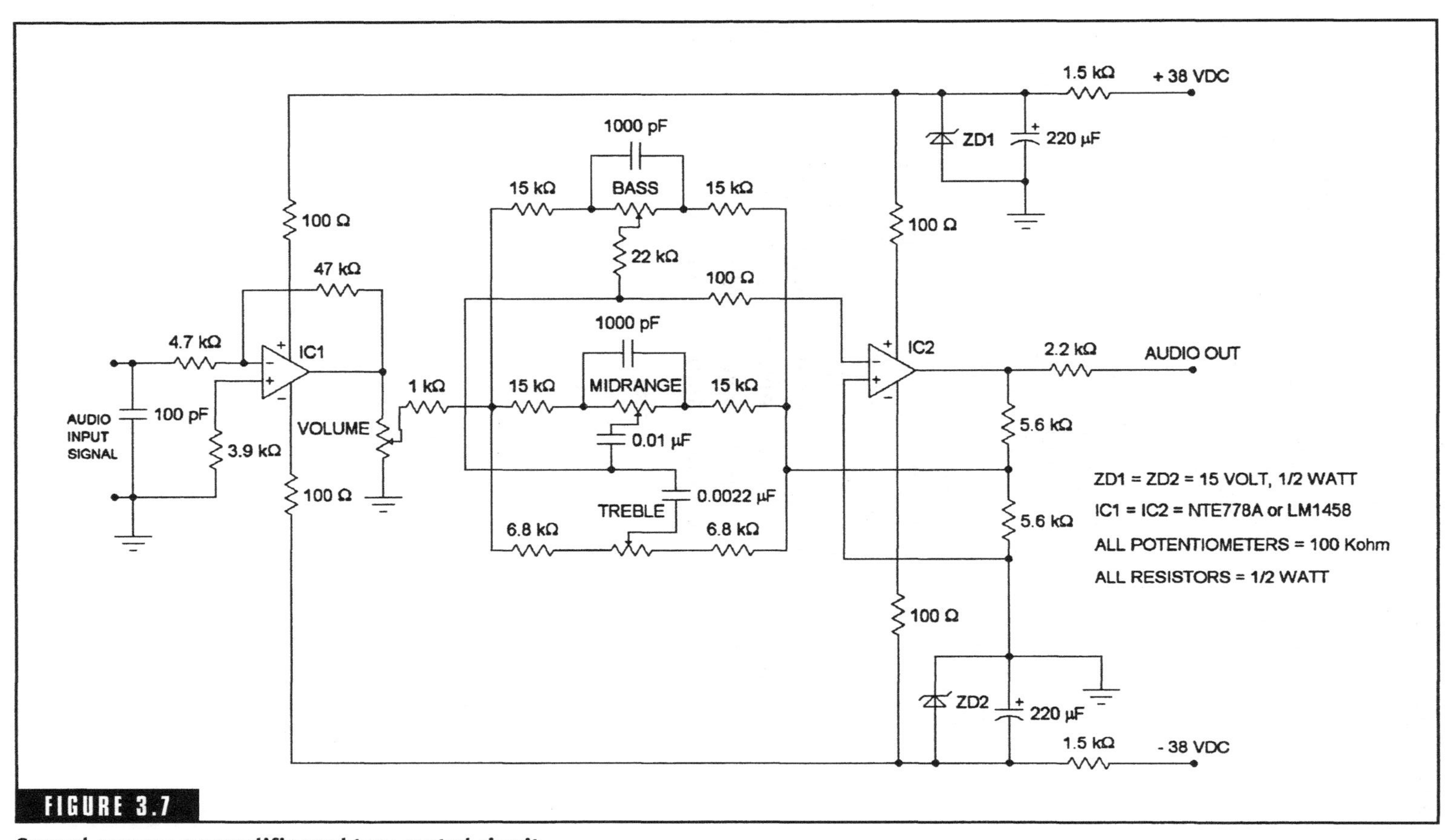

FIGURE 3.7

General-purpose preamplifier and tone control circuit.

voltage gains, the 4.7-kohm input resistor can be increased in resistance value. Also, for stereo applications, a dual-gang 100-kohm "balance" potentiometer can be incorporated after the "volume" potentiometer using the same technique as illustrated in Fig. 3-2.

The tone control circuits in Fig. 3-7 are conventional in design, and incorporated into the negative feedback loop of IC2, thus constituting a three-band active tone control network (tone control circuits will be detailed further in the upcoming chapter). The specified operational amplifiers are general-purpose types, requiring no further power supply decoupling. If you would like to use higher-performance op-amp types, you should properly decouple the power supply pins, as explained previously in this chapter.

Although the Fig. 3-7 preamplifier/tone control circuit could probably not be considered "high-performance" (primarily due to the inadequacies of the power supplies), it certainly provides good performance, and is well-suited for musical instrument amplifiers and other general-purpose audio amplifier applications. It can be incorporated on the same PC board as a medium-sized audio power amplifier, thus providing the preamplifier and amplifier system in one compact package.

CHAPTER 4

AUDIO FILTER CIRCUITS

Audio filter circuits are designed to be frequency selective throughout the audio bandwidth. As the name implies, filter circuits tend to *filter,* or *screen,* certain frequencies (or frequency bands) in preference of other frequencies (or frequency bands). The concept of selectively filtering, screening, or attenuating a variety of frequency components from a program signal seems to contradict the primary goal of high-quality audio, which is *sonic accuracy.* In addition, it is quite impossible to pass a program signal through any type of conventional audio filter without affecting the phase relationships of some components within the program signal. Therefore, it is only reasonable to pose the obvious question, "Why should we strive for sonic accuracy if we intend on coloring the audio program with filter circuits?"

Domestic Hi-Fi Considerations

Suppose you were to connect your CD player directly to your audio power amplifier through an appropriate attenuation network. You then put on some of your favorite music, and listen to it for about a half-hour, just long enough to get a good *feel* for the sonic qualities. Then suppose you were to take this same music to a professional recording studio, put your music into the studio's CD player, connect this CD player directly to the studio's power amplifier (through appropriate attenuation), and listen to your music through the studio's reference speakers in an acoustically *dead* room. Without a doubt, the studio system is going to sound profoundly different than your system at home. Providing you have a good-quality CD player and audio power amplifier in your home system, the program output signal from your domestic audio power amplifier should have been virtually identical to the program output signal from the studio's audio power amplifier. So what would cause such radical differences in the sonic qualities of the two systems? The answer is twofold—the *speaker systems* and the *room acoustics*.

Most domestic speaker systems are very far from optimum in regard to their performance capabilities. Several contributing factors to speaker system shortcomings are *cost* and *practicality.* From the cost perspective, a high-performance pair of studio monitors can cost up to $2000, or more. And even such high-priced speaker systems will vary significantly in sonic qualities from one manufacturer to another. Such differences will be due to the number of driving elements, the physical sizes of the driving elements, the design of the crossover networks, and the overall size and acoustical properties of the speaker cabinets. In terms of practicality, most consumers don't want to fill up their living rooms with speaker cabinets the size of refrigerators, so the modern trend is to manufacture the smallest, least conspicuous speaker systems that can still provide acceptable performance. The issues of cost and practicality are issues of compromise, meaning that even relatively expensive speaker systems often need a little "electronic" help for the best performance.

While the enthusiastic audiophile has some control over the problems of speaker performance (you always have the option of picking up a set of JBL exponential horn drivers for $8000, and then rearranging all of your living room furniture to one side of the room—such options usually work out better if you're not married), there is little that can be done about the room acoustics of your listening area. I have heard of audiophiles installing acoustic wall panels and gluing Styrofoam egg crates on the ceilings, but most people aren't willing to invest in a major remodeling job for the sake of improving the room acoustics. Consequently, the highly individualistic acoustics of a listening area are always going to represent a sonic anomaly that will require some type of compensation.

Music is an art form, and as such, it will always be subject to the individual tastes of the listener. It is not uncommon to pick up a CD of your favorite artist and later discover that you don't particularly like the mixdown tastes of the studio engineer. Maybe the bass is too low, or the highs are too brassy. Such variances in musical programs are especially prominent in "live concert" CDs or tapes.

All of the shortcomings and acoustic anomalies I have discussed thus far illustrate the need for some degree of *corrective filtering* to obtain the best performance from almost any audio reproduction system. (To recap, most speaker systems will require some help, room acoustics are totally unpredictable, and the original program material is not always optimum according to your personal tastes.) However, the concepts that the use of tone controls or other types of corrective filtering corrupts the program material, or somehow diminishes the need for sonic accuracy, are entirely untrue. Unpleasant types of distortion relate to problems of linearity and faithfulness of reproduction, whereas the coloration produced by tone controls is typically an enhancement of desirable musical attributes and benign in nature. Allow me to provide a practical example.

It is common for some types of music to begin with an introduction of brushes or light tapping on a cymbal to set the beat. Such introductions are typically low in volume level (compared to the remainder of the musical selection) and the harmonics produced by the click of the drumsticks or the swishing of the brushes are very high frequency. Such low-level, high-frequency passages are often "worst-case" conditions for an audio reproduction system, and significant levels of harmonic distortion will make such a passage sound harsh and "gritty." From a practical consideration, tone controls can neither help nor hurt such an undesirable situation. If you turn the treble control down to where the grittyness is no longer audible, you will lose the rest of your high-frequency response, causing the music to sound muffled and unnatural. Turning the treble up will only accentuate the problem. This hypothetical situation illustrates why it is necessary to *begin* with the highest degree of sonic accuracy possible. Tone controls are essentially impotent at correcting problems resulting from harmonic distortion or excessive noise (i.e., most types of noise problems are broadband, except for hum).

I know of audiophiles who have gone to great trouble and expense to eliminate any tone controls in the signal path by customizing their speaker systems to match their room acoustics and personal tastes. From my personal perspective, such a methodology is very impractical and inefficient. What happens if the audiophile moves to a different house or apartment? What happens if the audiophile decides to sell the audio system to someone who doesn't share the same musical tastes? How does the audiophile compensate for the normal variations in program material? Many modern preamplifier systems contain a *tone bypass* (often referred to as a *tone defeat*) switch. You always have the option of switching the tone control system out of the signal path. However, if you're like most audiophiles, you'll discover that they come in *pretty handy.*

Professional Considerations

Audio filter circuits in professional audio applications are an absolute necessity, and their usefulness has never been disputed by anyone that I am aware of. Professional recording studios are filled with multichannel tone control circuits, graphic equalizers, parametric equalizers, and a host of all types of outboard processors containing a multitude of specialized filtering circuits. Musicians *demand* high-quality audio filtering circuits to achieve personalized sounds from their instruments. All professional public address (PA) systems require multichannel tone controls (to compensate for microphone and musical instrument variations), as well as high-quality graphic equalizers (to compensate for acoustic properties and reduce regenerative feedback problems). In short, all areas of professional audio make prolific use of audio filter circuits.

Audio Filter Circuit Projects

The audio filter projects detailed in this section provide a good selection of filter circuits for a variety of applications. There are a few generalities that apply to all of the designs. First, always remember to properly decouple any of the high-performance op-amps that you decide to utilize for these designs. Secondly, regarding capacitor types, aluminum electrolytics and ceramic types are good choices for any kind of power supply decoupling purposes. For op-amp decoupling, tantalum and ceramic capacitors are recommended. For all capacitors placed in the signal path, either polystyrene, aluminized polypropylene, or tantalum types are usually the best choice. And finally, to achieve good noise performance, it is best to use metal film resistors in all of the forthcoming projects (metal film resistors typically have a 1% tolerance rating).

Figure 4-1 illustrates an excellent complement to the preamplifier design of Fig. 3-1 (of the previous chapter). It represents a typical three-band (i.e., bass, midrange, and treble) active tone control circuit. You would probably want to use the same op-amp type for this circuit as you incorporate into the Fig. 3-1 circuit. If a tone defeat switch is desired, it would be installed so that the preamplifier output signal is either applied to the tone control circuit (i.e., "tone in" position) or diverted directly to the preamplifier output jack (i.e., "tone defeat" position).

For those of you who occasionally get the urge to construct discrete component circuits, Fig. 4-2 illustrates a two-band (i.e., bass and treble) discrete-version active tone control design. If you happen to be

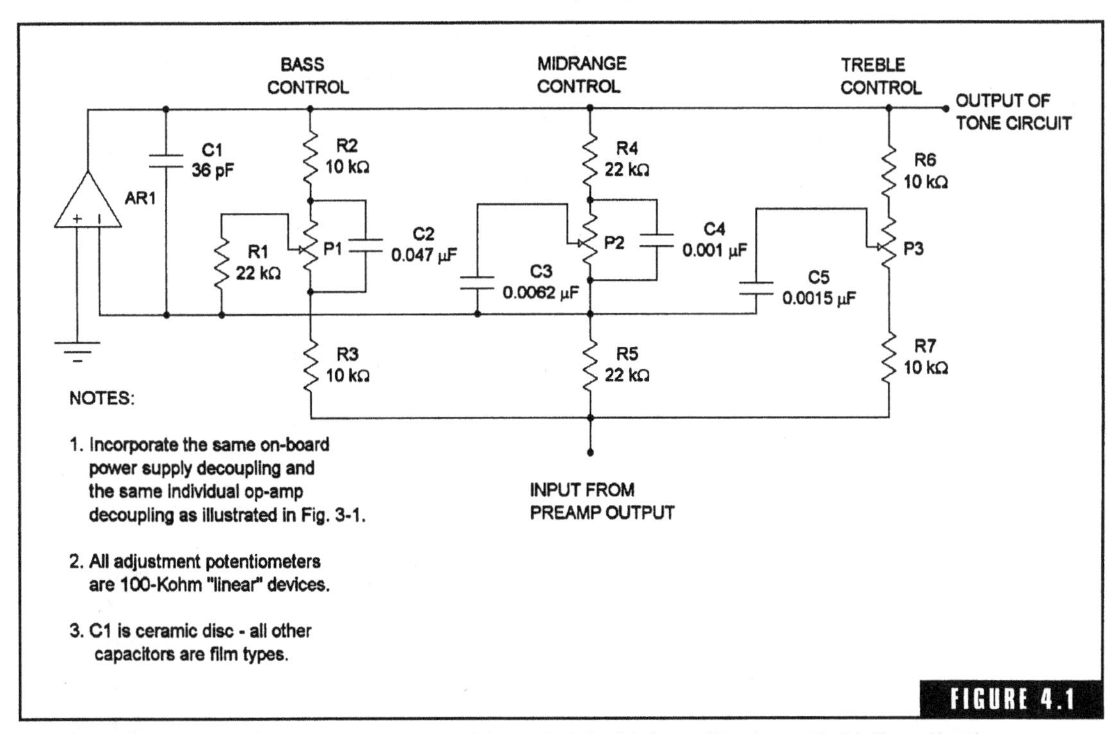

FIGURE 4.1

A high-performance active 3-way tone control intended for high-quality domestic hi-fi applications.

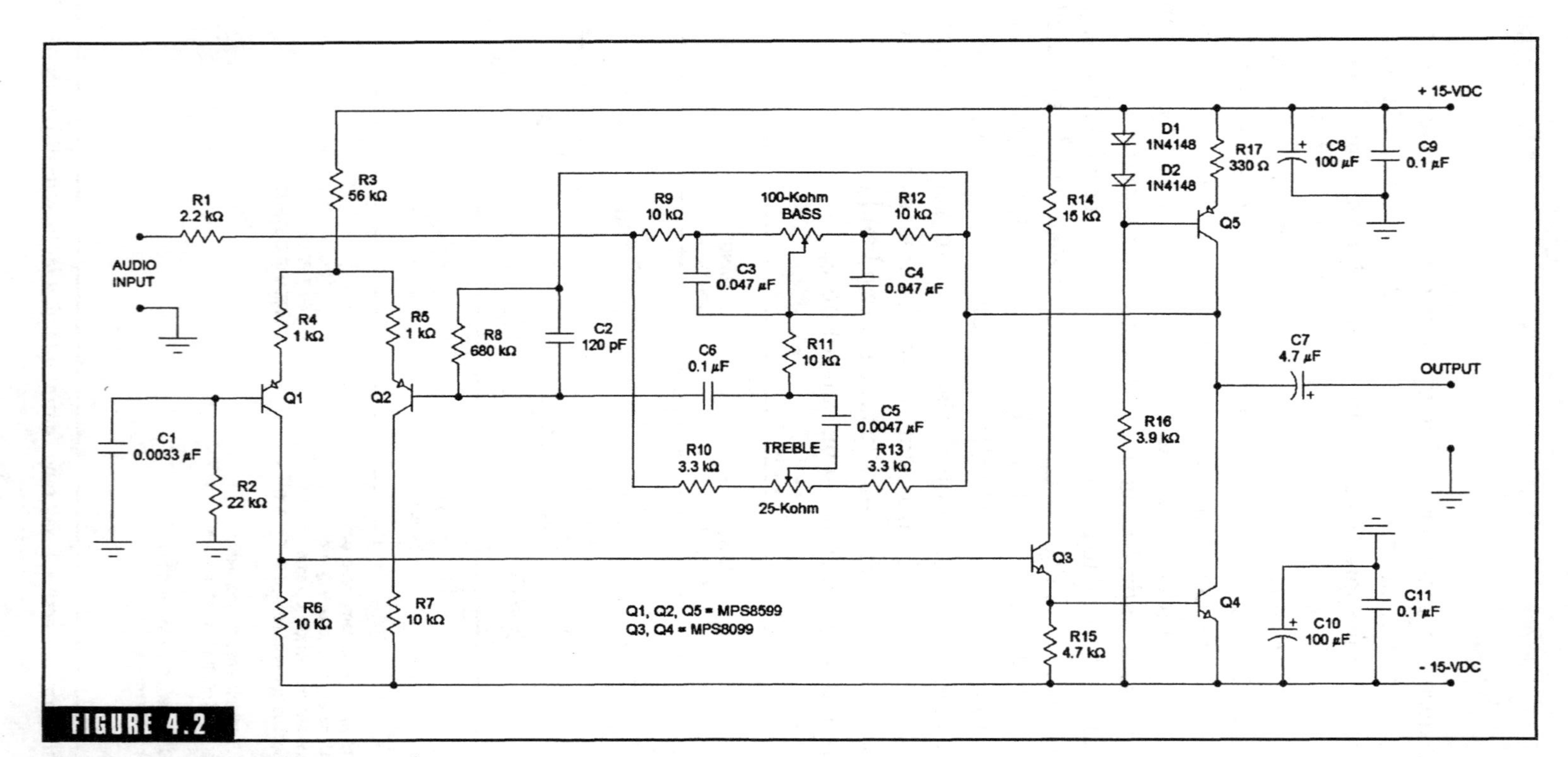

FIGURE 4.2

High-quality, high-level tone control.

familiar with audio power amplifier design, you may notice that the Fig. 4-2 circuit is essentially the input and voltage amplifier stages of a conventional power amplifier, with the tone control circuitry implemented as a function of the negative feedback loop. This design is preferable to most discrete-version tone control designs, because the differential input stage and actively loaded output stage provide a good deal of power supply immunity (commonly defined by the *power supply rejection ratio* parameter; abbreviated PSRR).

Figure 4-3 illustrates a versatile subwoofer low-pass filter or a two-way active crossover network. For subwoofer applications, the right and left channel line-level signals are summed by U1, and the low-pass output level is adjusted by P1. Of course, the high-pass output would not be needed for subwoofer applications. If you decide the use this design for an active crossover application, the summing network of U1 can be deleted, with the line-level signal applied directly to the R5/R7 junction point. Figure 4-3 details a variety of CF capacitors that can be incorporated to provide different low-pass and crossover responses.

Figure 4-4 represents a better approach to designing a low-pass filter specifically for subwoofer applications. It consists of an input summing amplifier and attenuation amplifier (U1 and associative circuitry), an inversion circuit (U2, SW1, and associative circuitry), and a third-order low-pass Butterworth filter section (U3 and associative circuitry). (A continuously variable phase control circuit that can be used with this design is detailed in Chap. 9, Miscellaneous Audio Circuits.) With the filter capacitor values illustrated (i.e., CF1, CF2, and CF3), the cutoff frequency response will be according to the RF resistor values detailed in the illustration. If you desire a different cutoff frequency from those detailed, the illustration also includes the equations for calculating the required filter capacitor values (based on a chosen RF value).

If you happen to be really ambitious toward excellence in a subwoofer system, you might consider constructing the "subwoofer equalizer" design of Fig. 4-5. Subwoofer speakers often have the tendency to produce a frequency response curve that wobbles up and down due to speaker resonance and cabinet construction. In addition, the low-frequency response characteristics often need a little electronic "boost" to push the performance into the most optimum realm.

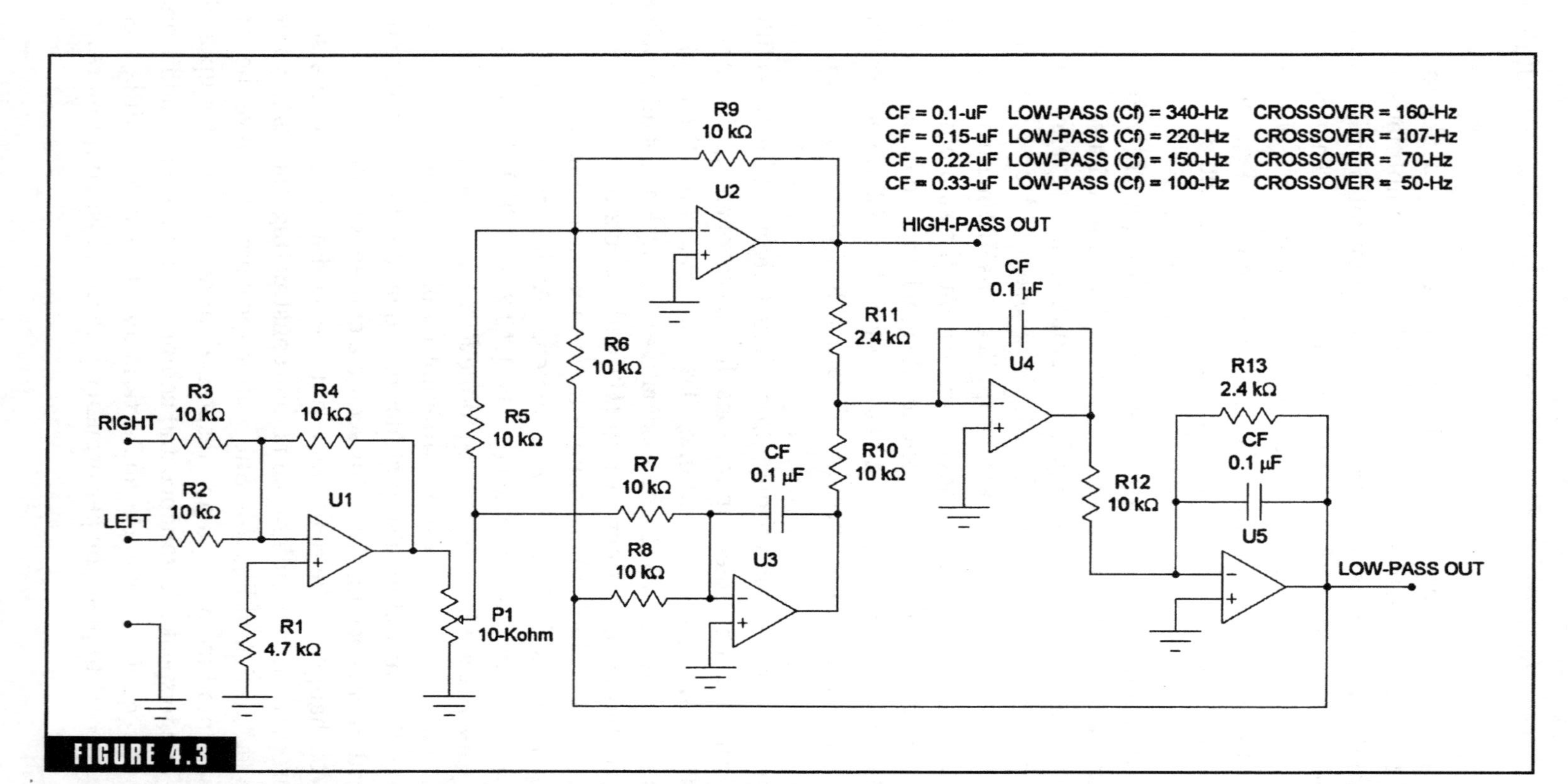

FIGURE 4.3

A versatile subwoofer low-pass filter or 2-way active crossover network (first order).

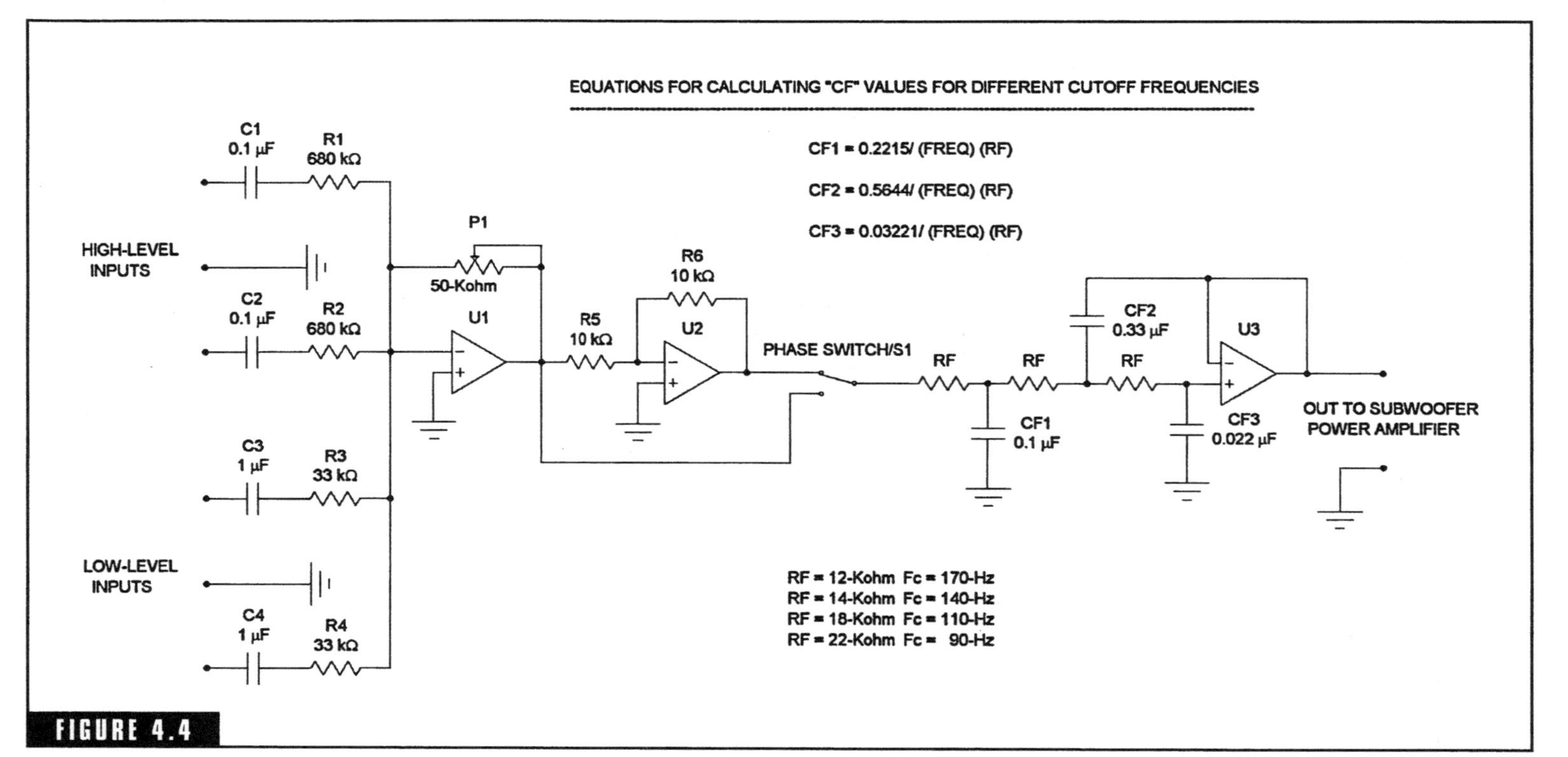

FIGURE 4.4

Complete schematic of a high-performance subwoofer filter circuit.

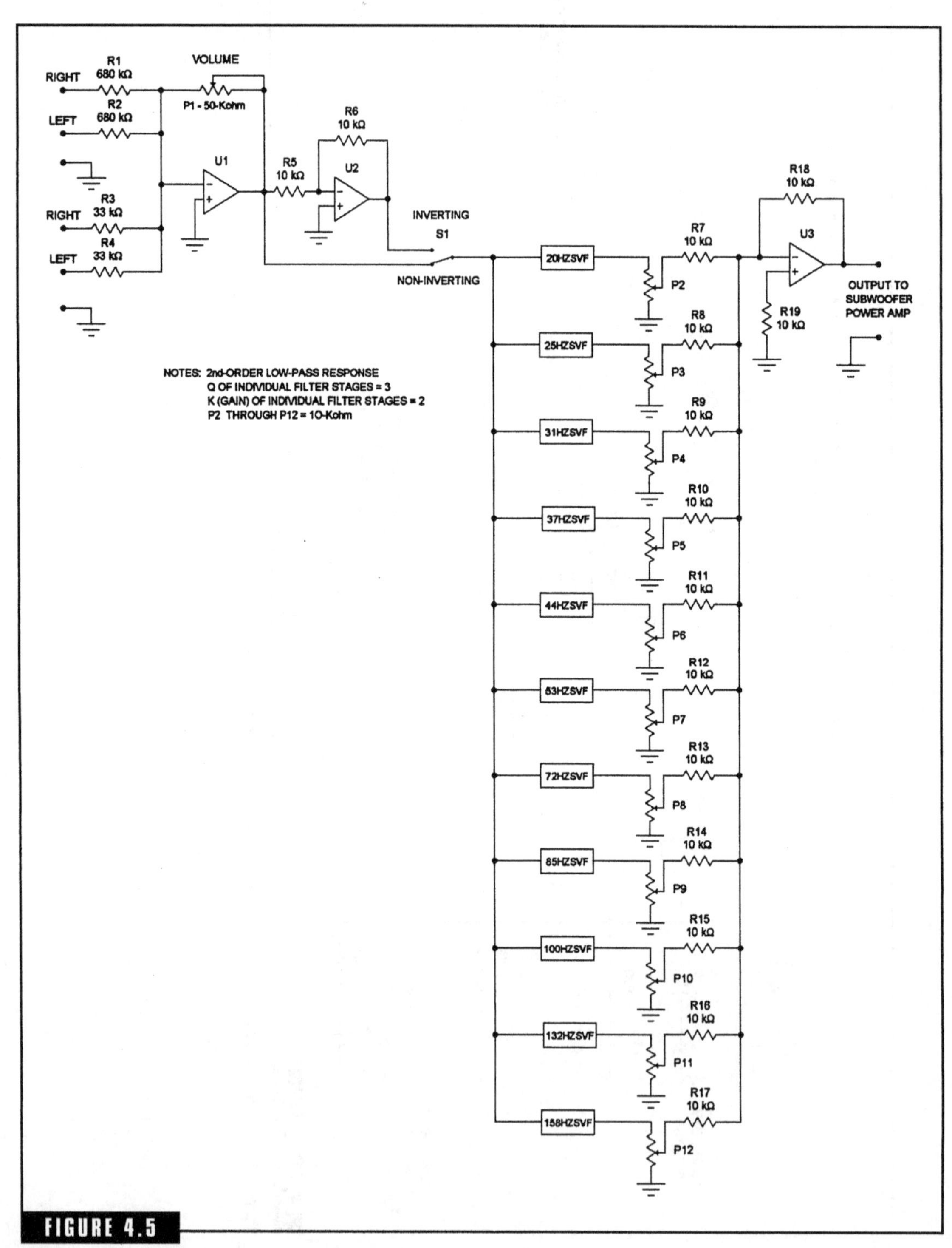

FIGURE 4.5

A subwoofer equalizer design.

The Fig. 4-5 subwoofer equalizer will provide the user with precise control over the amplitude characteristics of subwoofer systems.

The front end of the Fig. 4-5 design is similar to the previous subwoofer filter project. However, the signal output from the phase inversion circuit is fed to a parallel network of eleven second-order state variable filters. The filters are tuned to eleven specific frequency bands, ranging from 20 up to 158 Hz, with the outputs applied to the level control potentiometers P2 through P12. If all of the level control potentiometers are set to their "mid" position, any of the selected frequency bands can be cut or boosted by approximately 12 dB. The outputs from the level control potentiometers are summed by U3, and the final output is applied to the appropriate subwoofer audio power amplifier.

Figure 4-6 illustrates one of the eleven state variable filter subcircuits used in the subwoofer equalizer, together will the applicable equations for modifying the filter characteristics if so desired. With the component values provided in Table 4-1, the equalizer performance is pretty well optimized for almost any subwoofer application. All of the resistor values specified are standard values of 1% metal film resistors. (The state variable filter of Fig. 4-6 is extremely versatile. You may want to use the illustrated equations to modify the basic state variable filter circuit for a wide variety of audio applications.)

A general-purpose five-band audio equalizer is illustrated in Fig. 4-7. The actual filter stages incorporated (i.e., U2, U4, U6, U8, U10, and their associated components) are a type of *infinite gain multiple feedback bandpass filters,* set to a unity gain and a Q of approximately 0.8. The overlapping bandwidth coverage of the five frequency groups is approximately 20 Hz to 29 kHz. The output of each filter stage is applied to a simple inverting amplifier circuit (i.e., U3, U5, U7, U9, U11, and their associated components) with a variable gain, provided by potentiometers P1 through P5. The variable gain amplifier circuits provide a "cut" or "boost" to each of the five frequency bands, and eliminates the need for a "master" gain control. The outputs of the five variable gain amplifiers are summed in the summing amplifier of U12 and its associated components.

The disadvantage with the Fig. 4-7 equalizer is that the center frequencies of each bandpass filter will create a "peak" in the wideband frequency response. If the passband of each filter stage is widened to

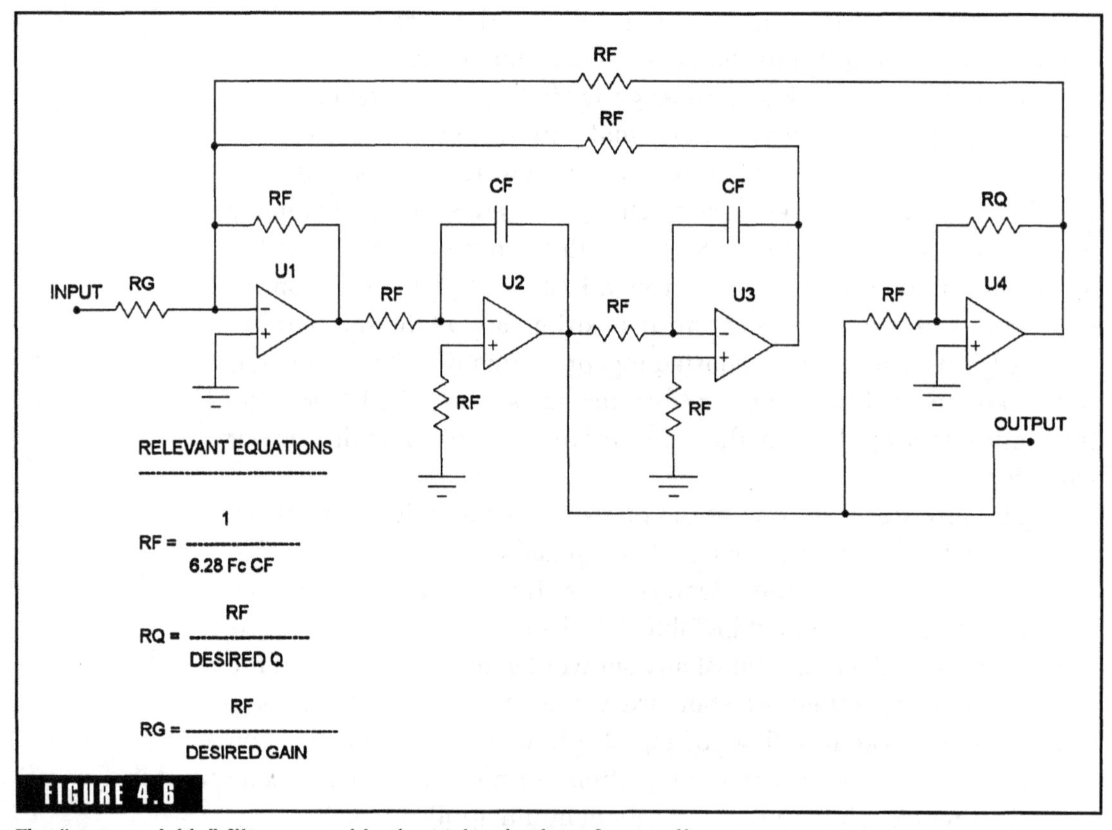

FIGURE 4.6

The "state-variable" filter as used in the 11-band subwoofer equalizer.

reduce the peaking, the effective control is sacrificed. Nevertheless, the overall performance of the Fig. 4-7 equalizer is rather good, and should function well in most audio applications.

Figure 4-8 illustrates a professional-quality 24-band equalizer design. The filter response of this design is more desirable for audio applications than many other types of filter circuits that are often used in graphic equalizers. When PS is adjusted to its center position, the entire passive filter section (consisting of R3, R4, R5, R6, C2, C3, and PS) reaches a state of reactive balance, setting the gain of U2 at unity, and causing its response to be perfectly flat. Thus, with all 24 PS potentiometers adjusted to the center position, the "overall" response of the graphic equalizer is flat, without any significant phase shifts or "peaking." However, by adjusting the P2 potentiometers up or down, a full 12 dB of boost/cut is available for each selective frequency band.

TABLE 4-1 Component Values for the State Variable Filters in the 11-Band Subwoofer Equalizer*

Frequency band, Hz	Schematic reference	RF value, kohms	RQ value, kohms	RG value, kohms
20	20HZSVF	82	27	39
25	25HZSVF	62	20	30
31	31HZSVF	51	17	27
37	37HZSVF	43	15	24
44	44HZSVF	36	12	18
53	53HZSVF	30	10	12
72	72HZSVF	22	7.5	10
85	85HZSVF	18	6.2	9.1
100	100HZSVF	16	5.1	6.8
132	132HZSVF	12	3.9	6.2
158	158HZSVF	10	3.3	5.1

*All state variable capacitors (CF) = 0.1 μF (polystyrene or metal-film types).

If properly constructed with premium quality components, this design is suitable for high-end audiophile or professional applications. (It will probably perform better than commercial units, since they seldom utilize high-performance op-amps due to cost compromises.)

U1 and its associated components serve as an input buffer to the 24 parallel filter circuits. R7, R8, and U3 serve to isolate each filter section and provide necessary buffering. C4, R10, and U4 make up an "AC" summing amplifier for the 24 filter outputs. Switch S1 provides a bypass function of the equalizer circuits. Note that if you don't want the line-level input signal inverted when switched to the "flat" position, you can simply connect the "flat" signal line to the left side of R1 (i.e., the line level signal input line). The required C2 and C3 values, along with the associated center frequencies, are listed in Table 4-2.

The Fig. 4-8 equalizer is not overly difficult to construct, but it is of relatively formidable *size*. If you construct the typical stereo version, it

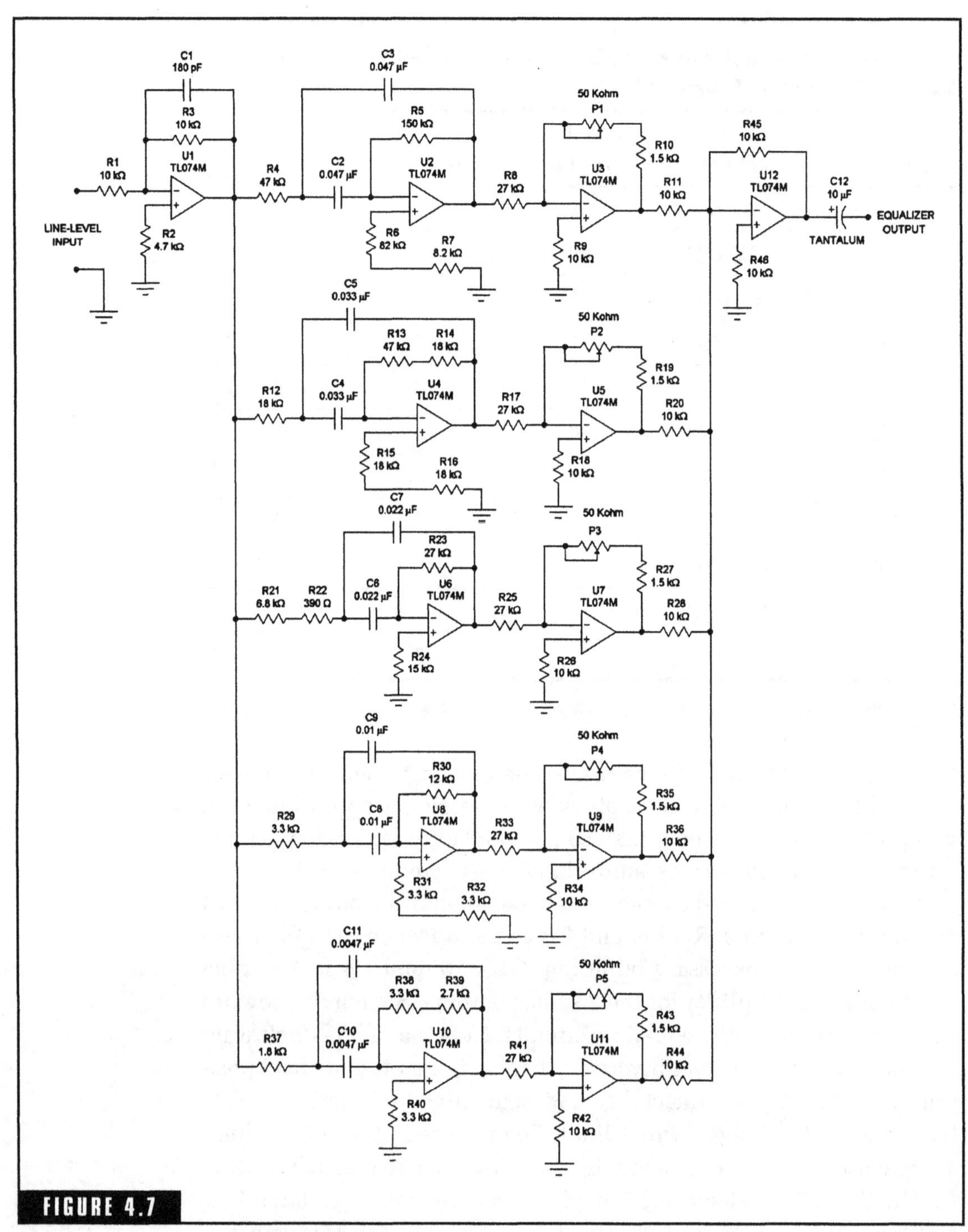

FIGURE 4.7

A general-purpose 5-band audio equalizer.

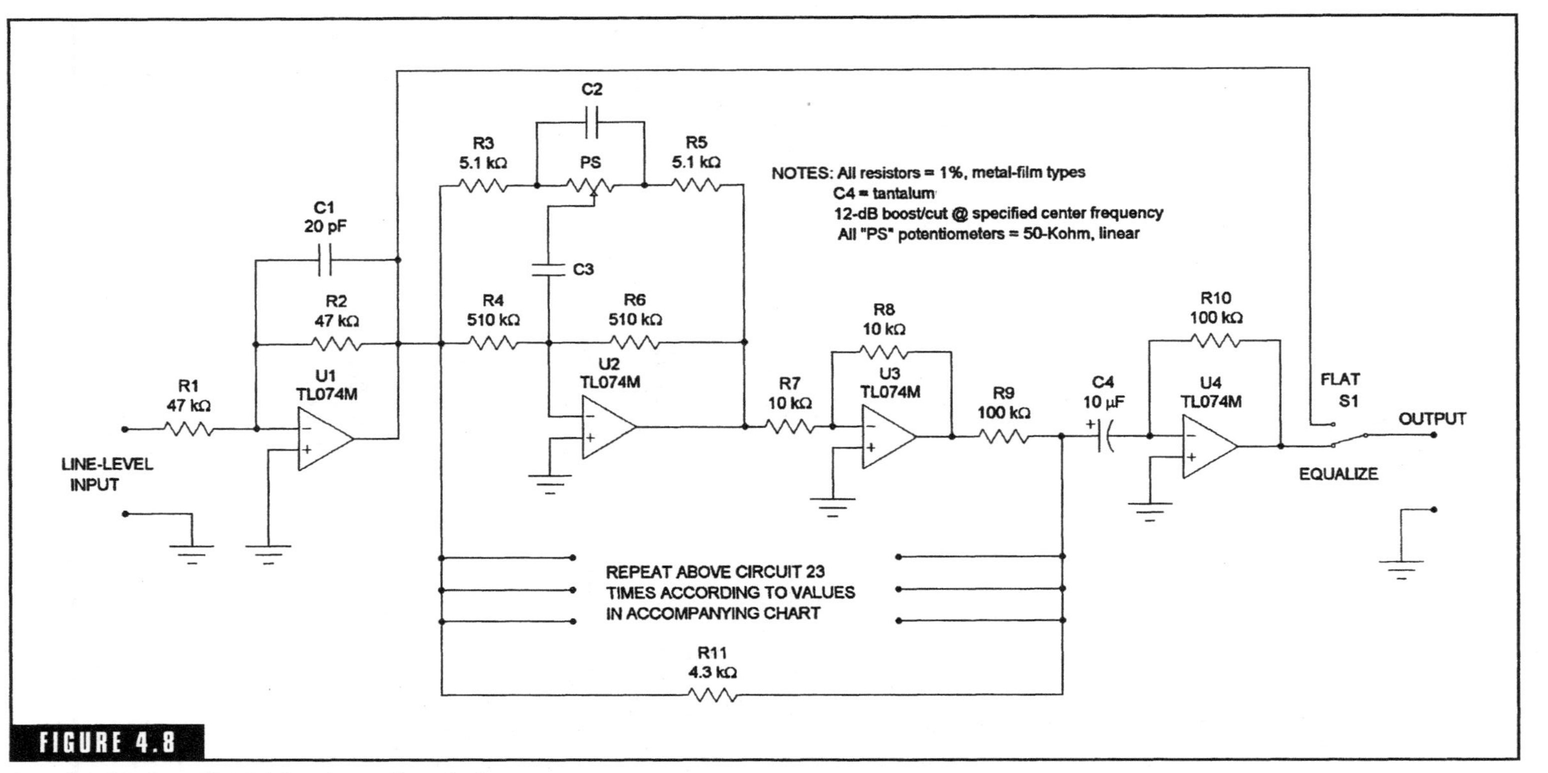

FIGURE 4.8

A professional quality 24-band equalizer design

TABLE 4-2 Component Chart for 24-Band Audio Equalizer

Frequency out, Hz	C2	C3
1. 22	0.47 μF	0.047 μF
2. 33	0.33 μF	0.033 μF
3. 50	0.22 μF	0.022 μF
4. 75	0.15 μF	0.015 μF
5. 110	0.1 μF	0.01 μF
6. 135	0.082 μF	0.0082 μF
7. 160	0.068 μF	0.0068 μF
8. 200	0.056 μF	0.0056 μF
9. 230	0.047 μF	0.0047 μF
10. 333	0.033 μF	0.0033 μF
11. 500	0.022 μF	0.0022 μF
12. 740	0.015 μF	0.0015 μF
13. 1100	0.01 μF	0.001 μF
14. 1350	0.0082 μF	820 pF
15. 1650	0.0068 μF	680 pF
16. 2000	0.0056 μF	560 pF
17. 2300	0.0047 μF	470 pF
18. 3160	0.0033 μF	330 pF
19. 4800	0.0022 μF	220 pF
20. 7300	0.0015 μF	150 pF
21. 11000	0.001 μF	100 pF
22. 12780	820 pF	82 pF
23. 15200	680 pF	68 pF
24. 19770	560 pF	56 pF

will require a total of 48 filter circuits, 48 potentiometers, and 98 operational amplifiers. In some of the commercial versions of this type of equalizer, 48 small PC board assemblies are manufactured, with each PC board containing one complete filter circuit, the associated buffer circuit, the required decoupling capacitors, and the control potentiometer (PS). The potentiometer is usually a "slide" type, and mounted on a side of the small PC board, so that when the potentiometer is mounted to the enclosure face, the small PC board extends out behind it (due to the small size of the PC boards, they can be held in place by the mounting of their associated potentiometer). The small PC boards are designed so that all external wiring connections can be made on the end opposite to the potentiometers. Thus, when all 48 PC board assemblies are mounted to the enclosure face (by means of the potentiometer mounting), the power supply and signal lines can be "daisy-chained" to all 48 boards. A final PC board must be constructed to contain the input circuit and summing amplifier circuit.

Regarding the next audio filter project, I debated on whether to include it in this chapter, or place it in the chapter devoted to "effects" circuits. It is called a *parametric equalizer,* as illustrated in Fig. 4-9. A parametric equalizer, as its name suggests, permits user control over various parameters of the individual filter stages in the equalizer. For example, the filters utilized in the previous designs allowed user control over output amplitude (i.e., termed *gain, attenuation,* or *cut/boost* functions). In contrast, a parametric filter allows user control over Q (or filter bandwidth), center frequency, and amplitude (or gain). The applications for parametric filters are about evenly divided between professional uses in recording studios (to remove or enhance narrow frequency bands) and by musicians (to provide special tonal effects for stringed instruments and synthesizers). If you happen to be a musician, you will be amazed at the sonic effects you can achieve by playing your instrument through the Fig. 4-9 circuit. Parametric filters have few applications in domestic audio systems.

Two of the op-amps in the Fig. 4-9 circuit are a special type of op-amp referred to as *operational transconductance amplifiers* (abbreviated OTA). A few of the commonly available OTAs are detailed in the illustration. An OTA is simply an op-amp with an additional input that allows the output transconductance (i.e., gain factor) to be varied with a control current (usually deemed I_{BIAS}). Basically speaking, the

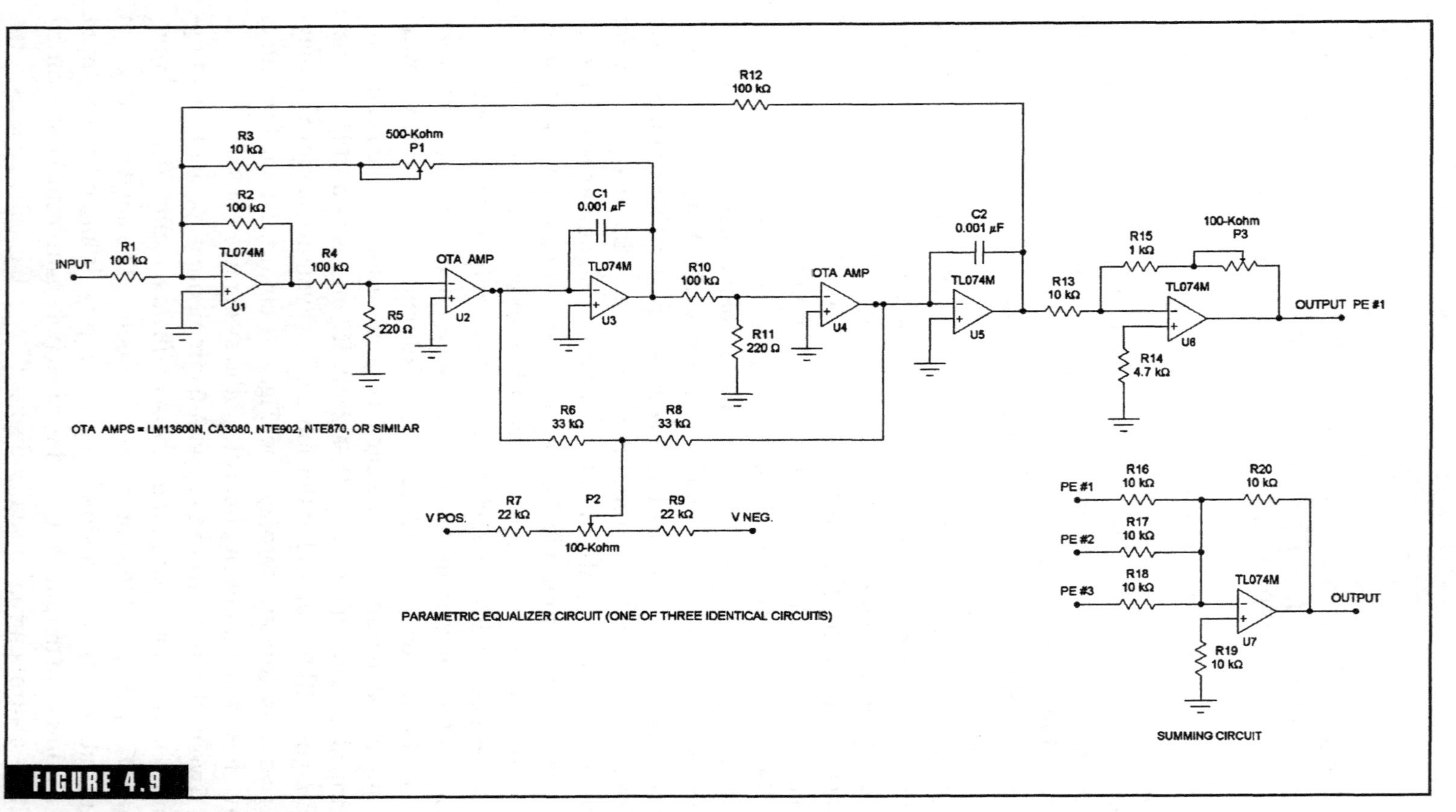

FIGURE 4.9

A 3-band parametric equalizer.

OTAs in the Fig. 4-9 circuit (i.e., U2 and U4) are serving the function of current controlled amplifiers.

In the Fig. 4-9 design, potentiometer P1 controls the Q (filter bandwidth), potentiometer P2 controls the center frequency, and P3 controls the gain factor. Note that P2 is part of a simple voltage divider operating from the positive and negative power supplies used to provide operational power to the op-amp circuits. As illustrated, the complete parametric equalizer project requires the construction of three parametric filters, with their outputs summed by the simple summing circuit consisting of U7 and its associated circuitry. The input signal is applied in parallel to all three parametric filters, and for best results, it should be at an amplitude level of 100 mV, or higher.

In recent years, the technique of *bi-amping* and *tri-amping* has become more popular within the audiophile community, but it has been utilized for much longer by the professional audio fields. Bi-amping and tri-amping are the practices of using a crossover network to separate the individual frequency bands *prior* to the power amplifier stage, applying the frequency bands to multiple audio power amplifiers, and driving the individual speaker elements directly from the multiple audio power amplifier system. In other words, one two-way (i.e., two-element) speaker system would require two power amplifiers in a bi-amped configuration—one power amp would drive the tweeter and the other would power the bass driver. A stereo bi-amped system would require four power amplifiers; two power amplifiers for each two-way speaker system. Tri-amping is the same basic technique, but it is applied to speaker systems incorporating three elements (i.e., tweeter, midrange, and bass elements).

Why bi-amp or tri-amp an audio reproduction system? Originally, the practice of bi-amping was used in high-power public address systems. Passive crossover networks internal to most multielement speaker systems are somewhat inefficient. Such passive inefficiency is considered negligible in most domestic hi-fi applications, but when you begin driving professional speaker systems at multikilowatt levels, the power waste can become substantial. Also, the power distribution within a multielement speaker system is very unbalanced. A typical exponential-horn speaker system will require about 60 to 80% of the applied power to drive the bass driver, while the midrange horn will operate much more efficiently with the remaining power. Audio professionals discovered that they could remove the internal crossover network from the speaker system, drive the individual speaker elements directly from multiple power amplifiers, and improve the overall efficiency. In addition, they discovered that they had more "control" of the system (i.e., independent control over midrange and bass driver levels), and they suffered from fewer regenerative feedback problems (which were due to the excessive phase shifts introduced by the original passive crossover networks).

Audiophiles initially began experimenting with bi-amping and tri-amping techniques as a means to gain better sonic control over a multi-element speaker system. However, in later years, it has come to light that the *real* advantage lies in maintaining a proportional phase relationship between the individual speaker elements. All passive crossover networks will introduce excessive phase shifts around the crossover points. These phase shifts create a literal canceling of sonics between the various speaker elements at certain frequency points, and results in the radiation pattern of the speaker system becoming frequency dependent. Implementing a bi-amped system utilizing conventional active filter crossover networks improved this situation significantly, because the active filter networks did not introduce such extreme phase shifts. However, there was still great room for improvement.

The goal of controlling phase shifts in an active crossover network is to maintain a consistent radiation pattern from the speaker system, and to keep the radiation pattern from being frequency dependent. To accomplish this goal, it isn't necessary to keep phase shifts from occurring (which is a good thing, since such a goal would be quite impossible to accomplish with conventional filter circuits). Rather, it is to keep all of the phase shifts seen by the individual speaker elements equal and proportional to each other. For example, if the bass driver sees a 100-degree phase lag at 300 Hz (relevant to the phase of the input signal), we want the midrange speaker and the tweeter to see a 100-degree phase lag as well. Of course, the tweeter isn't going to be reproducing 300-Hz frequencies, but it is necessary to cause the phase relationships of the individual speaker elements to "track" each other to avoid any cancellation or frequency-dependent effects at the crossover points.

An active crossover filter that *keeps all of its output frequency bands in phase* with each other is commonly called a *phase-linear fil-*

ter. If you happen to be familiar with conventional filter design, you might jump to the conclusion that such a goal is impossible, but to understand the physics behind phase-linear crossover action, it is necessary to dispel conventional thoughts for a few moments.

Begin by visualizing a conventional low-pass filter (of any fundamental design). If a wideband (i.e., 20-Hz to 20-kHz) audio signal is applied to such a filter, the output phase lag will vary with frequency, becoming more extreme around the cutoff frequency, and continuing to increase as the input frequency increases. For example, the phase lag of the output signal (compared to the input signal) at 20 Hz might be close to unity, but at 20 kHz, the output signal may be lagging by as much as 250 to 300 degrees. Hypothetically speaking, suppose it were possible to construct another filter circuit receiving the same input signal as the aforementioned low-pass filter, and suppose the input/output phase relationship happened to be *exactly the same.* Now, suppose this second filter circuit passed *all frequencies equally well.* Under these conditions, it would be possible to create a high-pass response from the second filter circuit by *subtracting the low-pass response from its output.* Obviously, the process of subtracting a portion of an *all-pass* response to create a filter response will only be possible if the phase relationship between the conventional filter response and the all-pass response is exactly the same. In basic principle, this is how a phase-linear crossover network functions.

A high-performance two-band phase-linear filter is illustrated in Fig. 4-10. U1, C1, and P1 provide the function of a simple AC-coupled buffer/attenuation input circuit. R1, R2, C2, C3, and U2 make up a conventional second-order low-pass filter. Note that R3, R4, C4, C5, and U3 make up another identical second-order low-pass filter. Therefore, the output of U3 is actually a fourth-order low-pass response. This low-pass output is AC-coupled through C8 and C9, and applied to P2, which serves as the low-pass output amplitude adjustment.

Referring to the lower portion of the Fig. 4-10 phase-linear filter, C10, C11, R14–R19, and U6 make up a special type of filter usually referred to as an *all-pass filter* (sometimes called a *phase shift filter*). All-pass filters exhibit the unusual characteristic of producing a phase shift of exactly twice that of the same order in more conventional filters. Therefore, since the all-pass filter illustrated is a second-order design, it will produce the phase characteristics of a fourth-order conventional filter. If

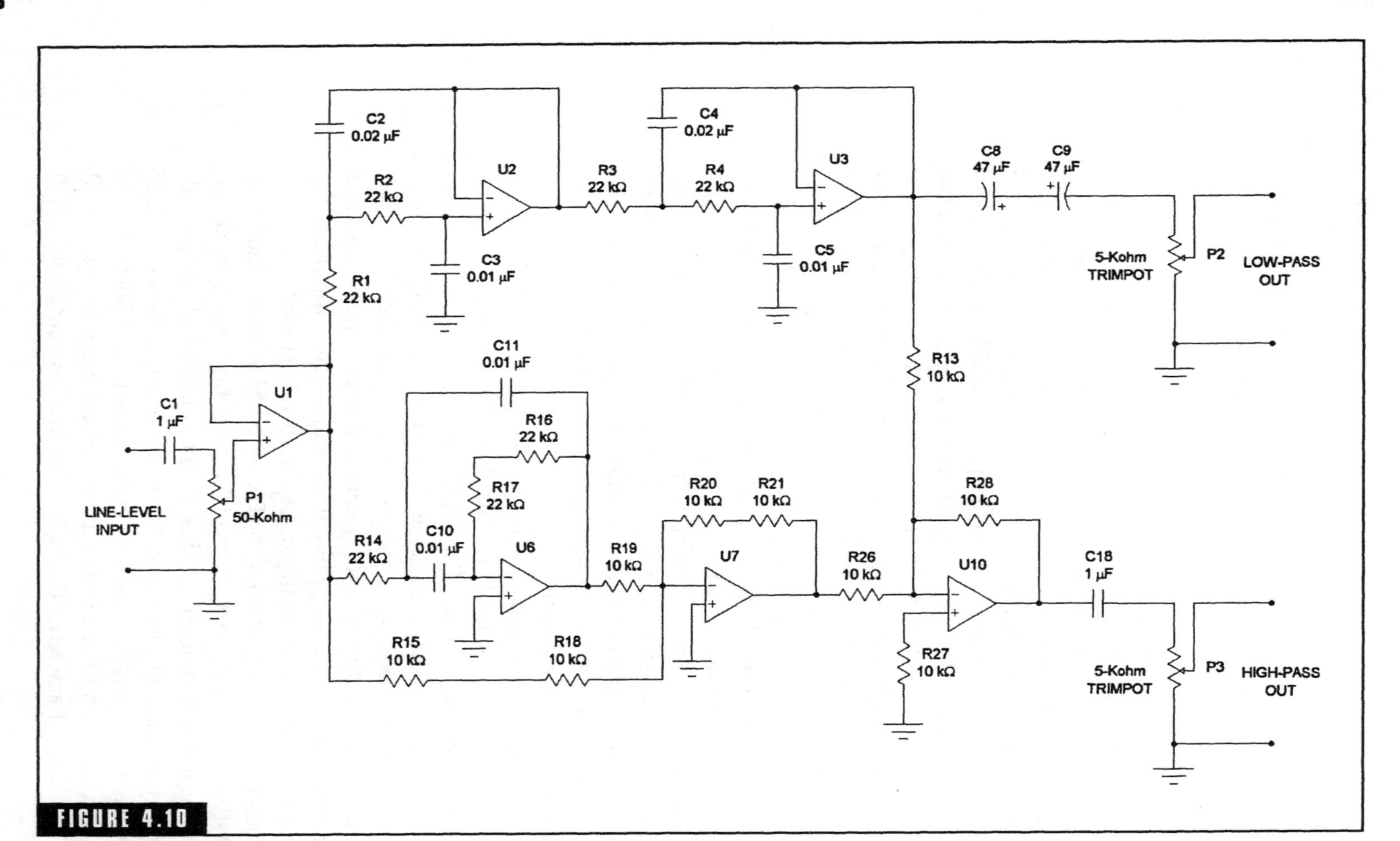

FIGURE 4.10

A high-performance 2-band phase-linear filter network.

you recall, the low-pass filters of U2 and U3 are both second-order, so by combining them, the output of U3 is a fourth-order response. Therefore, the phase relationships at the output of U3 and the output of U6 are identical. Note that the output of U6 is applied to the simple inverting circuit of U7, causing the output of U7 to be 180 degrees out of phase with the output of U3. U3 and U7 are summed at U10 (through R13 and R26), and since they are 180 degrees out of phase, they will cancel (or subtract) from each other.

Let us now "walk" through the operation of the Fig. 4-10 phase-linear filter. Beginning with an input frequency of 20 Hz and increasing the frequency up to about 200 Hz, the output of U3 (the low-pass filter) is high in amplitude. This high-amplitude signal is mixed with the inverted output of U7 at the summing amplifier (U10), causing a complete cancellation of the two signals. Thus, the output of U10 is near zero. However as the amplitude of the low-pass filter begins to roll off above 200 Hz, it is no longer capable of fully canceling the maximum amplitude output of U7, so a portion of U7's output signal will begin to appear at the output of U10. Note that U10 is an *inverting amplifier,* so whatever portion of U7's signal is applied to the input of U10 will be inverted at the output of U10, which will return it to an *in-phase* condition with the output of U3. As the input frequency continues to rise above 200 Hz, the amplitude rolloff of the low-pass filter will cause a proportional increase of the output amplitude at U10. When an input signal frequency of about 2 kHz is attained, the output of the low-pass filter drops to negligible levels, causing the output of U10 to be at maximum level. The output of U10 is the *high-pass* portion of this filter design. Its output is AC-coupled via C18 and applied to P3, which serves as the high-pass output amplitude adjustment.

Figure 4-11 illustrates the actual AC response of the Fig. 4-10 phase linear crossover filter. The upper graph illustrates amplitude versus frequency characteristics. Note how the low-pass response begins to roll off in amplitude at about 200 Hz and intersects the high-pass response (rising up from the bottom of the graph) very close to the 500-Hz line. The high-pass response has to be a "mirror-image" of the low-pass response, because it is created by subtracting the low-pass response from its original all-pass response. The lower graph of Fig. 4-11 illustrates the phase relationships between the high-pass and low-pass outputs. The small "wobble" in the high-pass response between 20 and

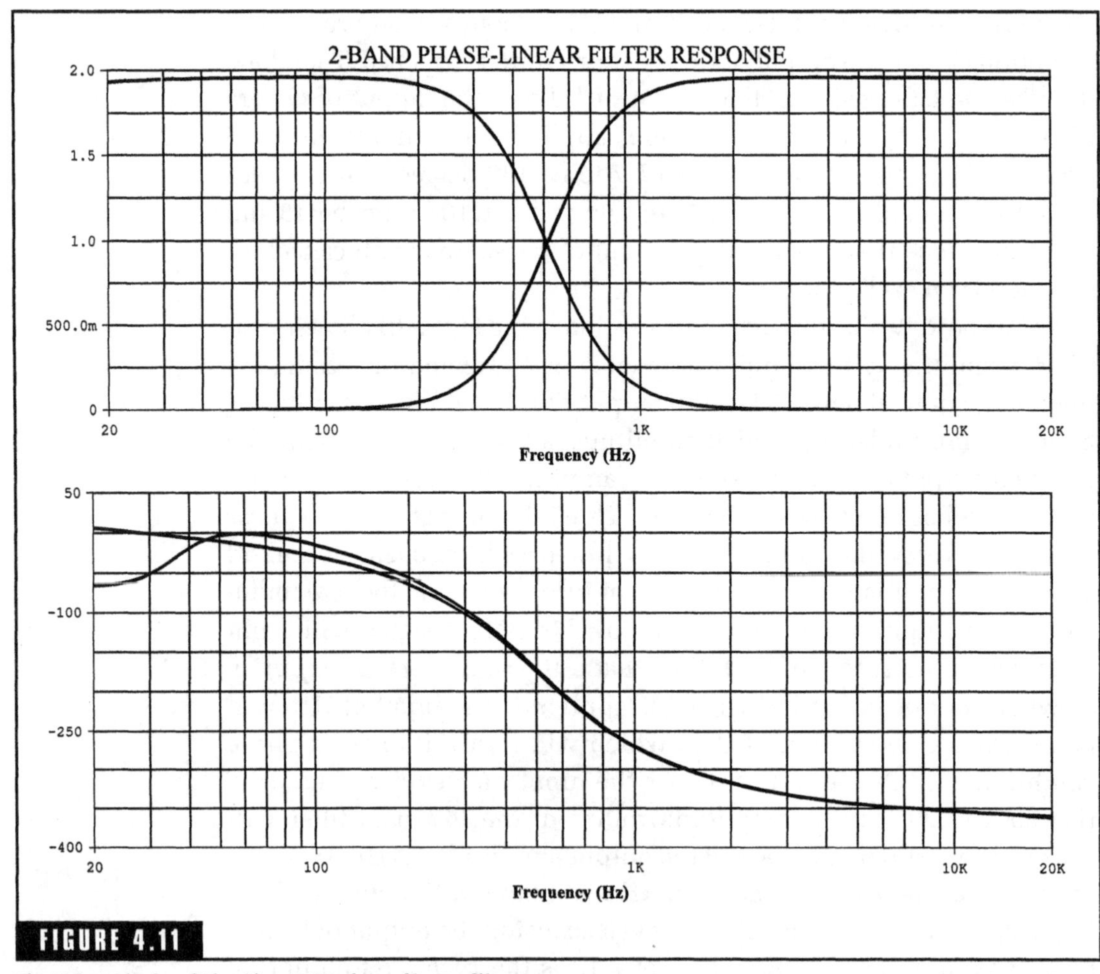

FIGURE 4.11

The AC analysis of the Fig. 4-10 phase-linear filter.

40 Hz is caused by the capacitance value of C18 (Fig. 4-10) being too low to efficiently pass such low frequencies. This is an insignificant anomaly, since the output of the high-pass section is at negligible levels until the input frequency reaches at least 200 Hz. The important point is to note how the phase relationships of the low-pass and high-pass outputs stay in almost perfect phase relationships throughout the crossover region. Consequently, a phase-linear filter of this type incorporated into a bi-amped audio reproduction system will maintain a totally stable and "full-bodied" speaker system radiation pattern. The individually adjustable output levels of the two frequency bands also

allow for greater control of the individual driver responses of the speaker system.

Figure 4-12 illustrates a high-performance three-band phase-linear crossover network. As you can readily see, it is just a "wee" bit more complicated that the two-band version, but the actual construction of this design is not overly difficult. If you compare the design of Fig. 4-10 with the Fig. 4-12 design, you will notice the same individual circuits; there are just more of them in the Fig. 4-12 design. The circuitry of U2, U3, U8, and U9 make up four second-order low-pass filters. The circuitry of U4, U6, and U11 comprise all-pass filters. The remaining op-amp circuits are inverters, buffers, or summing networks.

U1, P1, and C1 make up the same input buffer/attenuator circuit as detailed in Fig. 4-10. The second-order low-pass filters of the U2 and U3 circuits combine to form a fourth-order low-pass response. This is applied to the all-pass filter of the U4 circuitry, causing the *phase response* to look like an eighth-order filter. The output of U4 is applied to the inverter stage of U5, and the output of U5 is AC coupled to the low-pass level adjust potentiometer P2. The bandpass string of circuits begins with the all-pass circuit of U6, followed by the inverter circuit of U7, followed by the two low-pass filter circuits of U8 and U9. Collectively, all of these circuits add up to an eighth-order phase response, exactly the same as for the low-pass string of circuits. Thus, the phase response of both the low-pass and bandpass circuit strings should be identical, which they are. The only difference between the two strings is the higher cutoff frequency of the U8 and U9 low-pass filter circuits.

Note that the output of U5 is applied to the inverting input of U10, and the output of U9 is applied to the noninverting input of U10. As a low-frequency is applied to the input of the phase-linear filter, and gradually increased in frequency, both low-pass responses of the low-pass string and the bandpass string will cancel each other through the differential action of U10, so the output of U10 will be insignificantly low. When the cutoff frequency of the low-pass string is reached, the output of U5 will begin to roll off, but since the cutoff frequency of the bandpass string is higher (i.e., the cutoff frequency of the U8 and U9 filters is higher than the cutoff frequency of the U2 and U3 filters), the output amplitude of U9 will stay at maximum, causing the output of the U10 subtraction circuit to begin to rise. The bandpass response from the output of U10 will be capacitor coupled to the bandpass level

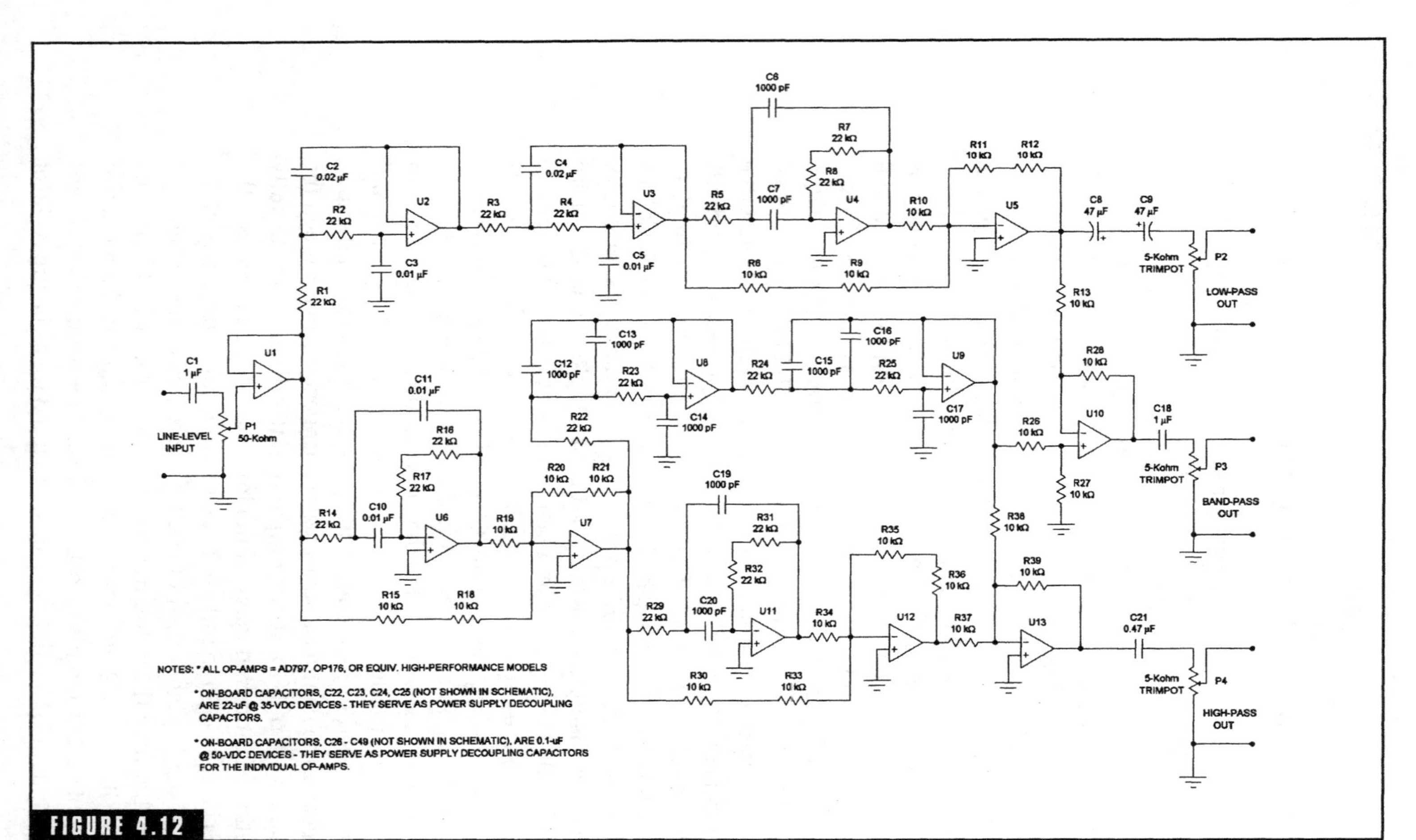

FIGURE 4.12

A high-performance 3-band phase-linear filter network.

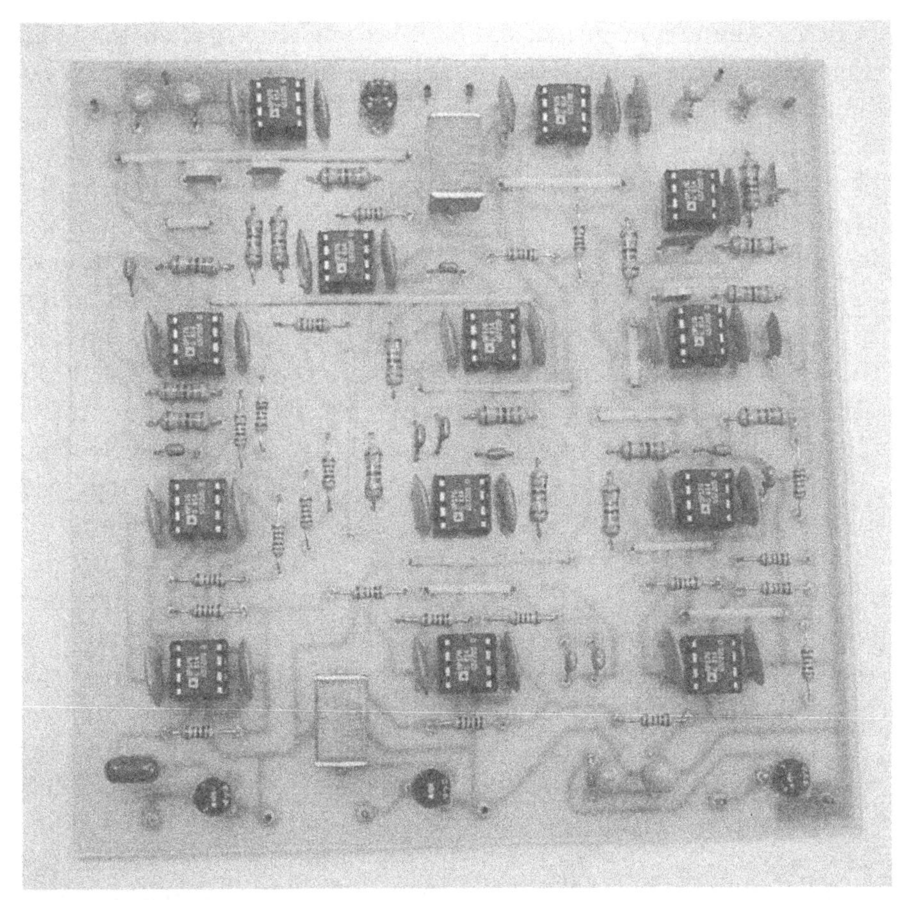

The completed 3-band phase-linear crossover filter of Fig. 4-12.

adjustment potentiometer P3. This process will continue until the cutoff frequency of the U8 and U9 filters is reached, causing the output of U9 and U10 to begin to roll off.

If you have been able to follow the abstract (and admittedly confusing) functional description of the Fig. 4-12 phase-linear filter thus far, it should become evident that the overall principle involved is actually rather simple. We are simply creating a bandpass response from a lowpass filter by subtracting a certain portion of its *lowest-frequency response.* For example, if the cutoff frequency for the U8 and U9 lowpass filter circuits happened to be 5 kHz, and we subtracted all of the *lowest frequencies,* say from 500 Hz down, we would then be left with a *bandpass response,* ranging from 500 Hz to 5 kHz.

Taking the aforementioned principles one step further, the technique for creating a high-pass response should come as no surprise—simply subtract the low-pass and bandpass responses from an all-pass filter, and you're left with the high-pass response. That is precisely what the Fig. 4-12 phase-linear filter does. The output of the inverting circuit of U7 is fed to the all-pass circuit of U11 (which mimics the phase response of the U8 and U9 low-pass filters), and the output of U11 is applied to an inverter circuit, which causes the phase relationship between the output of U9 and U12 to be 180 degrees out of phase. These two outputs are applied to the summing circuit of U13, and this results in a total cancellation of the two signals until the cutoff frequency of the bandpass string is exceeded. Thus, all higher frequencies appear at the output of U13, and are AC-coupled to the high-pass adjustment potentiometer P4.

In real life, the phase-linear circuit design of Fig. 4-12 functions incredibly well. It is much more forgiving of op-amp types and component tolerances than one would imagine at first consideration of its complex functions. Figure 4-13 illustrates the AC analysis graphs of the actual performance. The upper graph shows how ideally the low-pass, bandpass, and high-pass responses complement each other at the crossover points, which are 500 Hz and 5 kHz with the values illustrated. The lower graph illustrates the wideband phase response of the three outputs. You'll note the high-pass output (the top line on the graph) is showing some significant phase deviation from the other two outputs. This is due to the low capacitance value of the output coupling capacitor (C21 in Fig. 4-12). This situation is totally irrelevant, because the high-pass phase response matches the other two outputs by the time the bandpass/high-pass crossover frequency is reached. If you modify the Fig. 4-12 phase-linear filter for a lower high-frequency crossover point, you would probably want to increase the value of C21 up to about 1 μF.

Table 4-3 provides a listing of component values to incorporate into the Fig. 4-12 phase-linear filter circuit for a variety of the most common crossover frequencies. Also, you will note that I kept the component numbers in the Fig. 4-10 phase-linear circuit comparable to its three-band big brother. Therefore, if you wish to modify the crossover frequency for the two-band design, incorporate the component values in Table 4-3 according to the *low-frequency crossover point.*

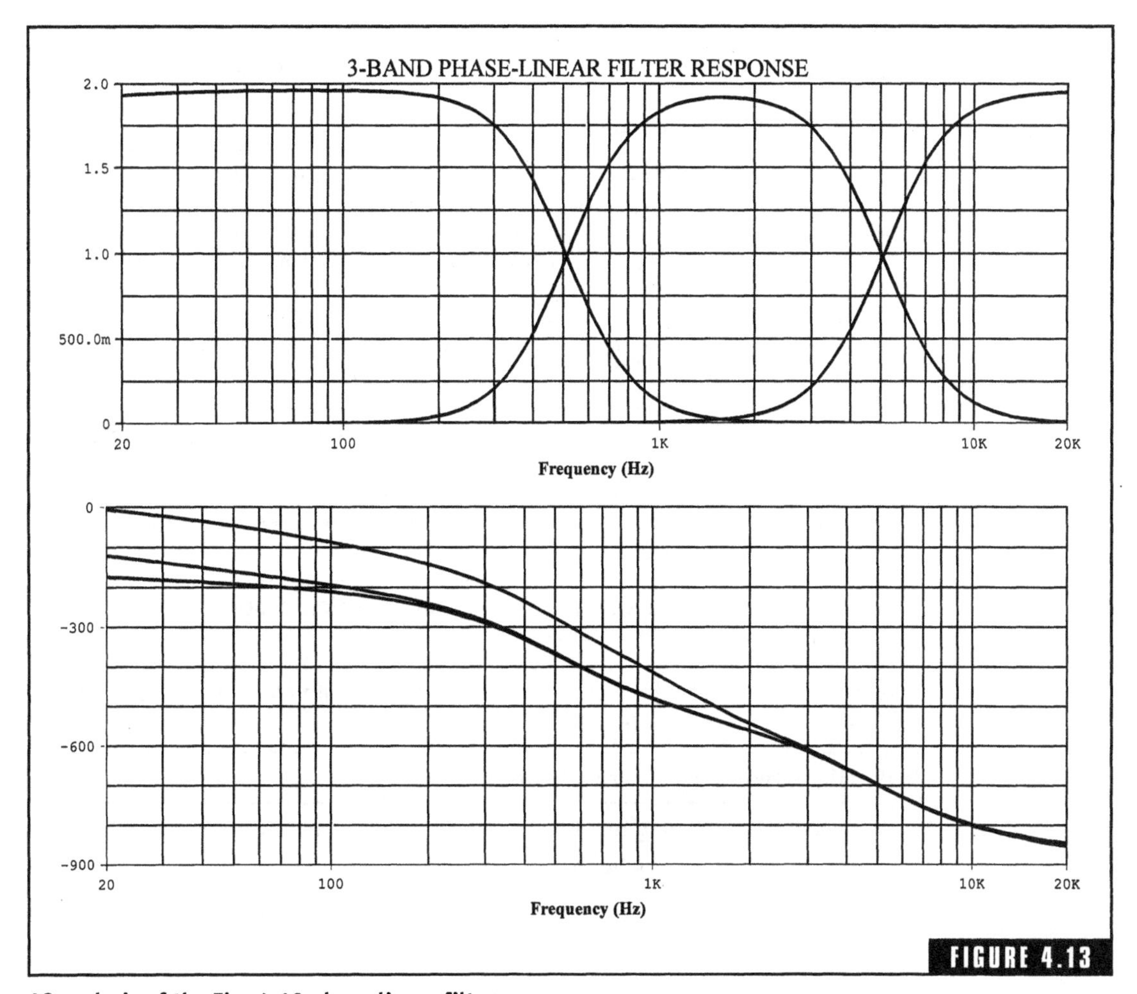

AC analysis of the Fig. 4-12 phase-linear filter.

Figures 4-14 through 4-16 illustrate the full-size layout and artwork diagrams of the Fig. 4-12 phase-linear filter. If you plan on etching a PC board from these illustrations, use the *bottom view "reflected" artwork* for your pattern. (Projects don't function too well when you get the PC board artwork upside down.) I used Analog Devices' OP-176 op-amps for this project, and I consider them to be an excellent choice. (Analog Devices' OP-184 op-amps should function equally well.) Capacitors C26 through C49 are the individual op-amp decoupling capacitors. They should be 0.1-μF @ 50-V ceramic devices, or equivalent (don't try to use mylar capacitors for this application). There are four on-

TABLE 4-3 FREQUENCY CHART FOR PHASE-LINEAR 3-BAND FILTER

A. Low-frequency crossover point

Frequency-determining resistors for low-frequency crossover point = R1, R2, R3, R4, R14, R16, and R17.

1. For a 250-Hz low-frequency crossover point, the aforementioned resistors must all be 47-kohm in value.
2. For a 300-Hz low-frequency crossover point, the aforementioned resistors must all be 36-kohm in value.
3. For a 350-Hz low-frequency crossover point, the aforementioned resistors must all be 33-kohm in value.
4. For a 500-Hz low-frequency crossover point, the aforementioned resistors must all be 22-kohm in value.
5. For a 750-Hz low-frequency crossover point, the aforementioned resistors must all be 15-kohm in value.
6. For an 800-Hz low-frequency crossover point, the aforementioned resistors must all be 14-kohm in value.

B. High-frequency crossover point

Frquency-determining resistors for high-frequency crossover point = R22, R23, R24, R25, R29, R31, and R32.

1. For an 1800-Hz high-frequency crossover point, the aforementioned resistors must all be 62-kohm in value.
2. For a 2000-Hz high-frequency crossover point, the aforementioned resistors must all be 56-kohm in value.
3. For a 3500-Hz high-frequency crossover point, the aforementioned resistors must all be 33-kohm in value.
4. For a 5000-Hz high-frequency crossover point, the aforementioned resistors must all be 22-kohm in value.
5. For a 5600-Hz high-frequency crossover point, the aforementioned resistors must all be 20-kohm in value.
6. For a 6200-Hz high-frequency crossover point, the aforementioned resistors must all be 18-kohm in value.

board power supply decoupling capacitors utilized (i.e., C22, C23, C24, and C25). They should be 22-μF @ 35-V devices, and I recommend tantalum types for this application. RP1 through RP4 are appropriately sized trimpots.

Before you run out and purchase six new audio power amplifiers, allow me to add a few final comments on the practice of bi-amping, tri-amping, and phase-linear crossover filters. If you decide to experiment with bi-amping techniques, don't expect to hear an "amazing" difference. The effects are subtle, and the typical person would probably not perceive them unless you specifically called their attention to the differences. A major portion of the overall impact will depend on the quality of the speaker systems. As a general rule, the effects of bi-amping (or tri-amping) will be more pronounced when applied to less-expensive speaker systems, and more subtle in conjunction with high-quality speaker systems. You probably already have two audio power amplifiers to start with, so if you want to experiment with the benefits of bi-amping, why not construct the Fig. 4-10, two-band phase-linear filter, and try some "before" and "after" testing with a single two-element speaker system?

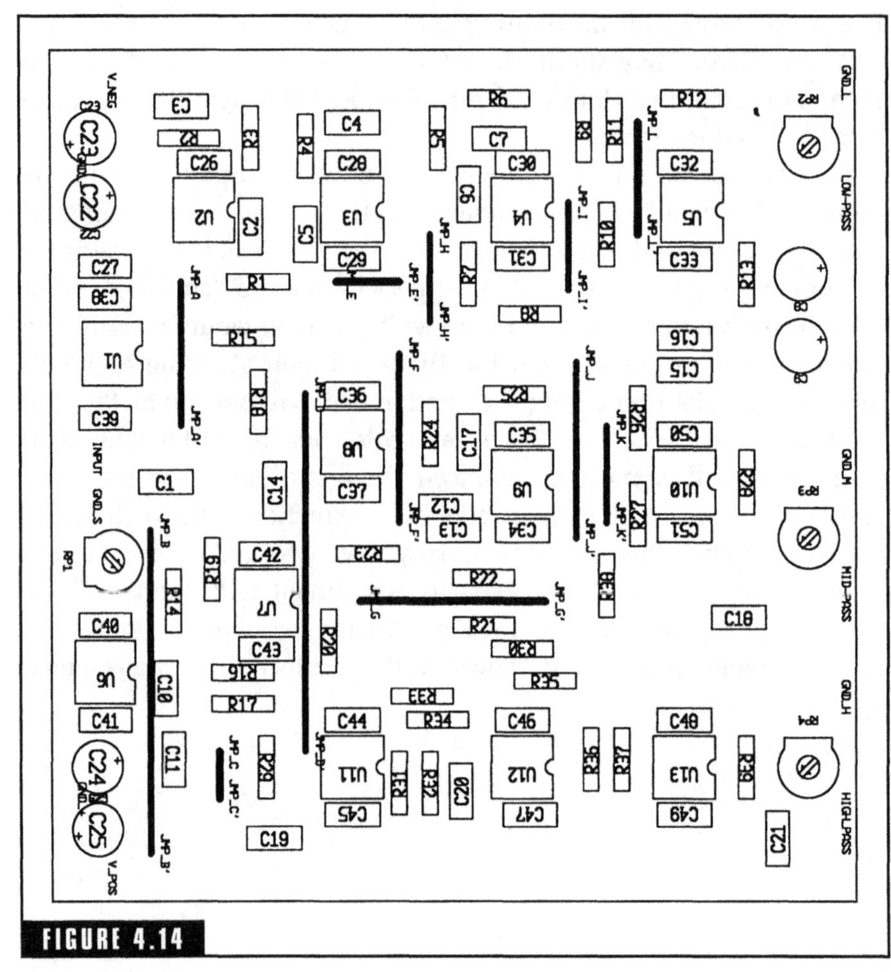

FIGURE 4.14

Component layout of the Fig. 4-12 phase-linear filter.

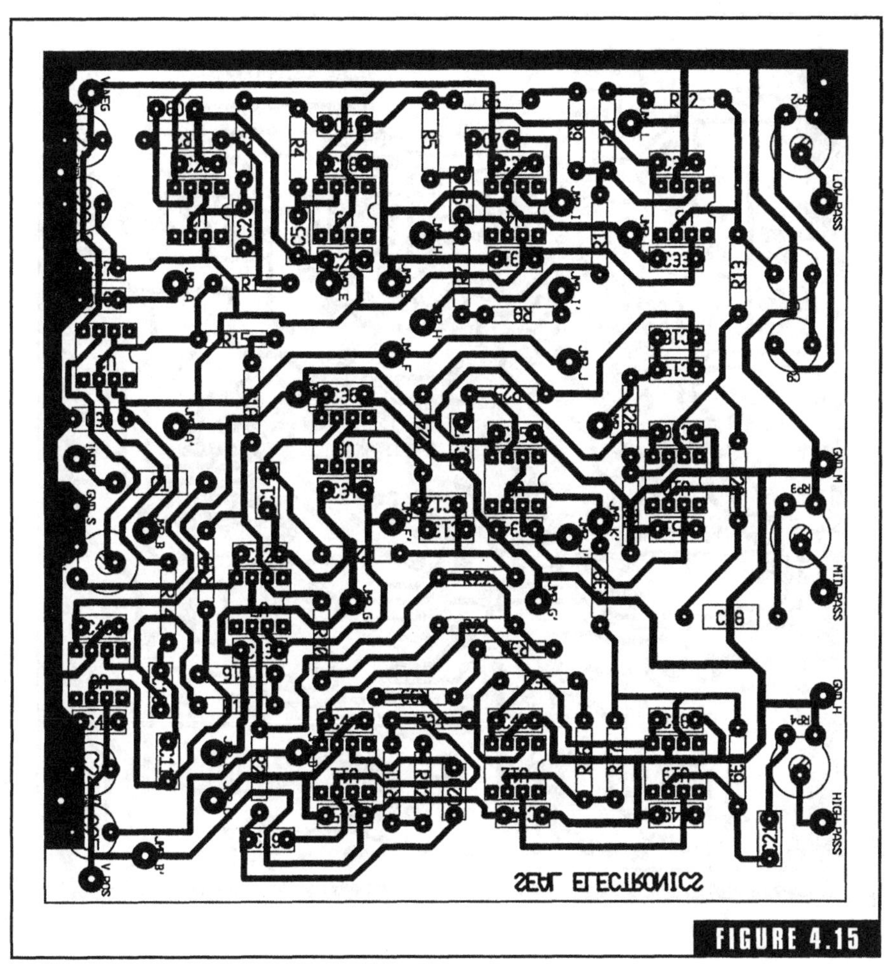

FIGURE 4.15

Top view layout of the Fig. 4-12 phase-linear filter.

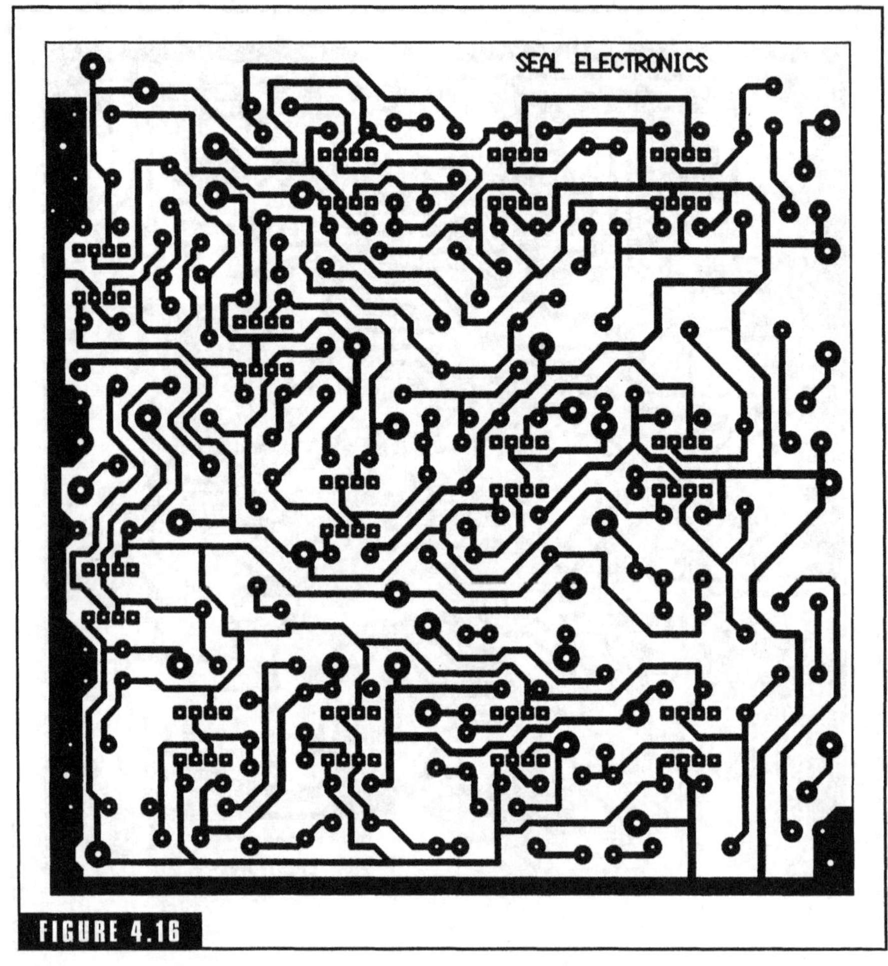

FIGURE 4.16

Bottom view "reflected" artwork for the Fig. 4-12 phase-linear filter.

CHAPTER 5

HEADPHONE AMPLIFIERS

Why Listen to Music through Headphones?

I doubt if anyone could refute the fact that we live in a high-pressure, fast-paced world. Most everyone will experience occasional periods of excessive frustration, and during these times, we long to get away somewhere peaceful. I know many professionals who have developed destructive habits in the effort to cope with stress, and I also know an equal number who suffer from the eventual physical ramifications of chronic stress, such as ulcers and heart problems. I have a suggestion for you to try the next time you feel really *uptight* and stressed. Obtain a quality set of studio-type headphones (i.e., the type that provide almost total outside noise isolation), plug them into your stereo system, put on some of your favorite music, relax in your favorite easy chair, close your eyes, and let the music take you where you want to go! Many people often forget how easy and inexpensive it is to shut out the world, unwind, and go somewhere without leaving home. Maybe you'd like to go back to the 70s for a visit, so you put on some Eagles or James Taylor favorites, or maybe you'd like to take a brief trip to the tropics with some calypso music. Headphones provide the unique experience of shutting out the world, and providing a temporary environment according to your pleasure. Try it the next time you feel the need to get away for a little while.

Generalities of Headphone Systems

In the case of most domestic preamplifiers or integrated stereo systems containing a *headphone output* jack, the stereo headphone signal is obtained from the loudspeaker outputs via some form of attenuation network (typically a simple series resistor or resistive voltage divider). Such a technique is not optimum if you desire the best possible performance from your headphones. Since the input sensitivity of most modern headphone sets is very high (typically about 100 dB for a 1-mW input), they have a tendency to accentuate the imperfections inherent to audio power amplifiers that would not normally be considered significant when driving high-power loudspeaker loads. For this reason, many audiophiles, who spend a considerable time listening to music through a headphone set, prefer to construct a separate headphone amplifier.

Generally speaking, a headphone amplifier is a type of low-gain line amplifier. Assuming the headphone amplifier is going to receive its input from standard line-level inputs, the overall voltage gain need only be a factor of three to six times to produce ear-deafening volume levels. It should be noted that the headphone amplifiers provided in this chapter are capable of producing 40 to 50 mW of power into the more common types of commercial headphones. *If you listen through typical headphones at such power levels it will permanently damage your hearing.* I designed the forthcoming amplifier circuits for maximum versatility with a very wide range of headphone types. Therefore, provide appropriate input attenuation to match the maximum output levels suitable to the headphones you will be using—*and keep the volume down.* It isn't any fun being a deaf audiophile!

I might mention that the headphone amplifier projects in this chapter are not suitable for driving any type of *electrostatic headphones.* Such headphones require much higher energy levels and are normally driven directly from the loudspeaker outputs.

Headphone Amplifier Projects

If you want a basic headphone amplifier design that is relatively easy to construct and provides superb performance, the Fig. 5-1 circuit is an excellent choice. With a 1-V RMS input signal, the output power to typical 600-ohm loads is approximately 26 mW (which will be much too loud for most headphone sets). At this power level, the THD performance is better than 0.001% @ 1 kHz, and a very respectable 0.011% @ 20 kHz. This design is equally suited to low-impedance headphone sets, with only a slight increase in THD performance with loads down to 16 ohms. Unlike conventional audio power amplifier designs, the THD levels will not increase at lower volume levels, because the output stage (i.e., Q3 and Q4) is biased in class A mode. If a high-performance op-amp is incorporated (as illustrated), the SNR performance will be very good. The maximum output power into high-impedance or low-impedance headphone designs is well over 60-mW.

The operational description of Fig. 5-1 is not overly complicated. R1 provides an appropriate matching impedance for most line-level

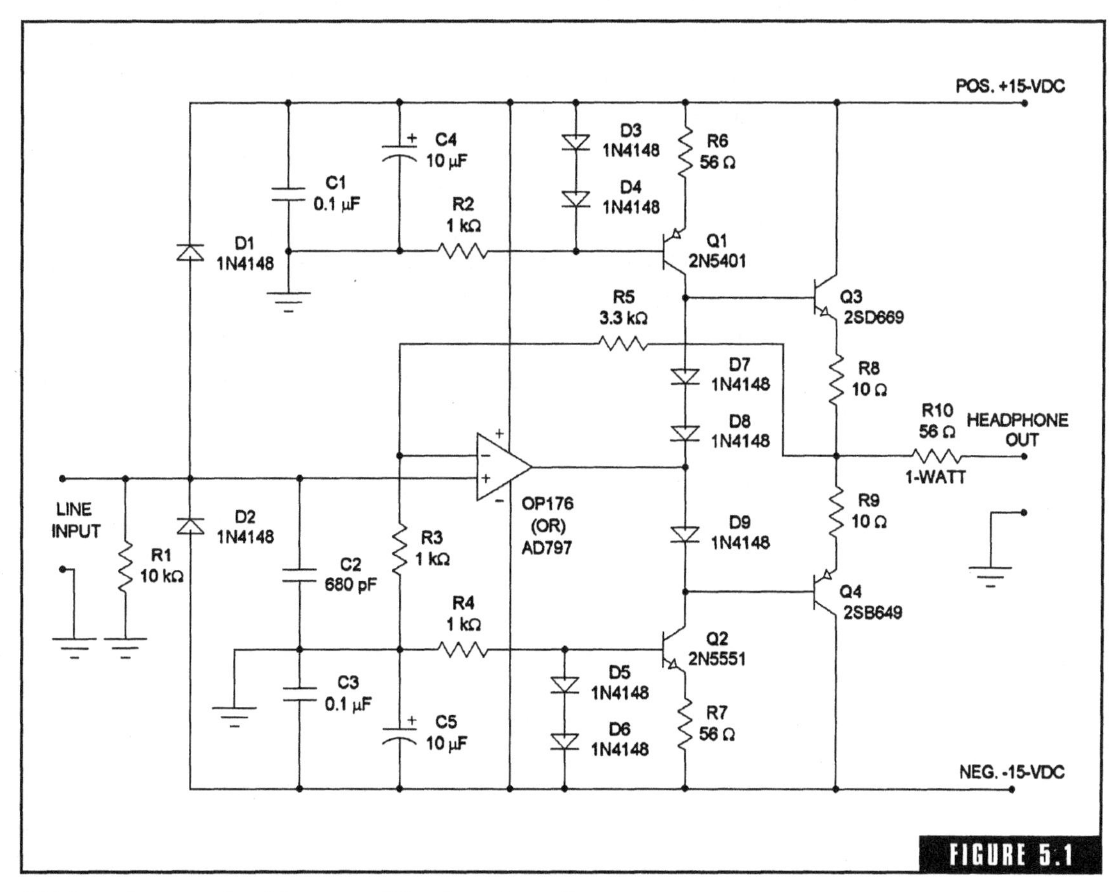

FIGURE 5.1

A high-performance headphone amplifier (one channel).

signals. D1 and D2 are reverse-biased protection diodes, which prevent the op-amp from being destroyed if the input signal amplitude exceeds the power supply voltages. Capacitors C1, C4, C3, and C5 are power supply decoupling capacitors. C2 filters RF frequencies within the signal content. R2, D3, D4, Q1, and R6 make up a constant current source operating from the positive power supply rail. A complement constant current source, operating from the negative power supply rail, consists of R4, D5, D6, Q2, and R7. The purpose behind the two constant current sources is to provide a highly-regulated and "power supply immune" bias current for the output stage bias diodes (D7, D8, and D9). These bias diodes set the quiescent output stage current to approximately 64 mA. Q3 and Q4 are configured in a typical push-pull complementary class A output stage design, with R10 providing sufficient output impedance to protect the output stage from destructive current levels in the event of a short-circuit at the output. Global negative feedback is provided by R5, which also sets the voltage gain. If desired, increasing the value of R5 will increase the input sensitivity.

Typically, two of the Fig. 5-1 headphone amplifiers would be constructed on a single PC board for stereo headphone operation. It is best to remotely locate the power supply to eliminate any possible hum resulting from EMI radiation of the AC power transformer. It would also be desirable to install some type of input attenuation circuit (i.e., a high-quality potentiometer or stepped attenuator) for a volume control. Several of the power supply designs in Chap. 7 would perform well with this design.

A discrete version high-performance headphone amplifier is illustrated in Fig. 5-2. As in the previous design, this is essentially a low-power class A amplifier circuit. The distortion performance when driving a 600-ohm load is 0.0024% @ 1 kHz, and rises to 0.028% @ 20 kHz. This design is well suited for low-impedance headphones also, with only slight increases in the THD levels. Utilizing the components shown in the illustration, the SNR was better than −110 dB.

C1 and C2 are input coupling capacitors. AC input coupling may not be necessary if you will be using this amplifier in conjunction with most line-level signal sources. R1 and C3 form a simple low-pass filter to attenuate any RF frequencies. Q2 and Q3 make up a differential amplifier, consisting of emitter degeneration resistors R4 and R5, and the constant current source of R3 and Q1 (note that this constant cur-

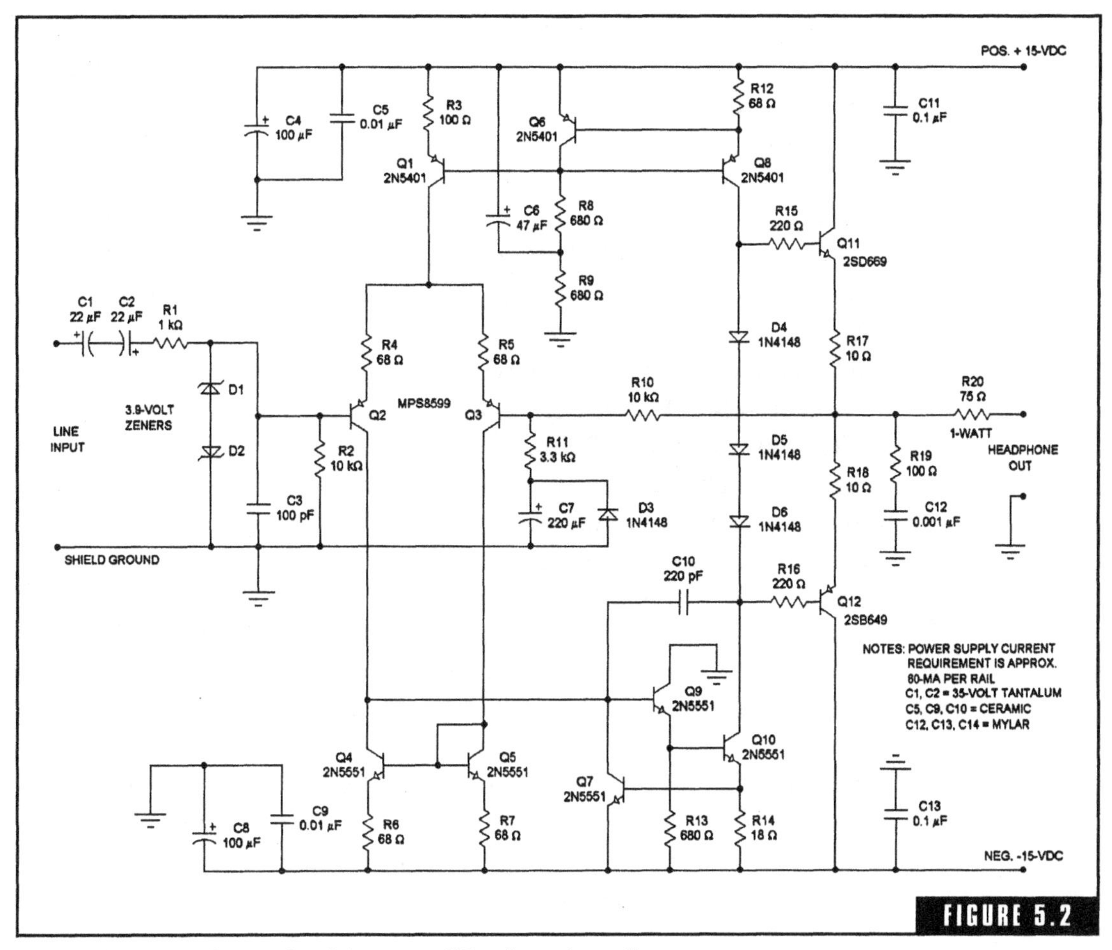

High-performance discrete headphone amplifier (one channel).

rent source is receiving its reference voltage from the constant current source of Q6, Q8, and their associated components). Q4, Q5, and their associated degeneration resistors, R6 and R7, form a current mirror circuit, which balances the differential amplifier's collector currents (reducing distortion) and provides active loading for the differential stage. Conventional power supply decoupling is provided by C4, C5, C8, C9, C11, and C13. Q9, Q10, R13, and R14 make up a beta-enhanced voltage amplifier stage that is actively loaded with the constant current source of Q8, R12, Q6, R8, R9, and C6. This amplifier incorporates the same output stage design as the previous Fig. 5-1 circuit, with the addition of the two base resistors, R15 and R16. These resistors provide a suitable impedance match between the voltage amplifier stage and the output stage.

Stabilizing high-frequency rolloff is provided by R19 and C12, while R20 sets the output impedance at 75 ohms. Global negative feedback is provided by R10, and the overall voltage gain is established by the ratio of R10/R11. As illustrated, the voltage gain is approximately 4, but if a higher gain is desired, the resistance value of R11 can be decreased accordingly. C7 provides 100% global DC feedback to the differential stage, while maintaining the AC feedback as a ratio of R10/R11 (i.e., the same as the signal voltage gain).

The design of Fig. 5-2 incorporates several interesting protection techniques. The back-to-back zeners, D1 and D2, protect the differential input transistors from input signals that could exceed their maximum 5-volt base-to-emitter parameters. Q7 monitors the emitter current of Q10 by means of the voltage drop across R14. If this voltage should try to increase above approximately 0.7 V (proportional to a 38-mA emitter current), Q7 will turn on and shunt the drive signal to the negative power supply rail. In this way, Q10 is protected from excessive currents. Note that Q8 is inherently protected from excessive current flow since it is part of the aforementioned constant current source. Q11 and Q12 are limited from excessive current or overload conditions by R20. D3 protects C7 from destruction due to voltage polarity reversal that could result from component failure or user error.

I purposely went into a more detailed circuit description for the Fig. 5-2 headphone amplifier because it is a very versatile circuit, and can be modified for a great variety of general "line amplifier" applications. The 2SD669/2SB649 output transistors are a good choice for the output stage, but the remaining small-signal device types are not critical. I used the 2N5551, 2N5401, and MPS8599 devices because I had plenty of them in stock. If you want to use different device types, there are only a few points to keep in mind. The differential transistors (Q2 and Q3) should be low-noise, high-gain types, and they should be matched (i.e., they should be of the same device type and the beta values should be within 10% of each other). The same holds true for the current mirror transistors, Q4 and Q5.

CHAPTER

6

Audio Power Amplifiers

Audio power amplifiers are very popular as construction projects among audiophiles and electronics hobbyists for two primary reasons. First, it is typical to achieve dramatically noticeable performance improvements, since the power amplifier is often a "weak link" in the audio chain. Second, the monetary savings incurred by personal construction can be enormous—much larger than can be realized with other types of audio equipment.

Unfortunately, audio power amplifiers are a little more difficult and involved in comparison to most audio construction projects. This should not be interpreted to mean that you will have great difficulty with your first few audio power amplifier projects. I am simply suggesting that you should put a little more thought and planning into your project, and, by all means, familiarize yourself with the fundamentals of power amplifiers before jumping headlong into an expensive project. Since the PC board fabrication and layout is a little more critical to the overall performance of audio power amplifiers, I have included the full-size PC board artwork and layout illustrations for most of the cookbook designs in this chapter. I have also been considerate of the PC board fabrication facilities available to most hobbyists and audiophiles, so all of the PC board artwork is for single-sided construction. (A few of the photographs in this chapter illustrate the double-sided commercial versions of these same designs, but the performance of the single-sided versions is identical.)

There is no question that a great deal of confusion regarding audio power amplifiers is present within the audiophile and hobbyist fields. I believe this is partially a result of misinformation pertaining to the literal *performance goals* associated with power amplifiers. I also recognize, through first-hand experience, that the basics of power amplifier operation are not widely understood, and this situation certainly adds to the overall confusion level. Therefore, before jumping into cookbook projects, we'll examine these subjects in a little greater detail.

Audio Power Amplifier Performance Goals

It is difficult to evaluate performance standards of audio power amplifiers if the performance goals are not clearly understood. There are valid complaints and concerns offered by the audiophile community, but they are sometimes overshadowed by an equal quantity of marketing hype and esoteric nonsense. In the effort to sift through all of the conflicting information, I believe it is wise to begin by considering how an *ideal* audio power amplifier would perform. If we can firmly establish the ideal performance goals, it then becomes much easier to evaluate a power amplifier's real-life performance by comparing how close it comes to the target.

1. *The ideal audio power amplifier would be perfectly linear.* This is synonymous with saying that the ideal amplifier would be capable of *zero percent distortion* performance. Such a hypothetical amplifier could be said to be *perfectly transparent.* The concept of perfect linearity would seem to be a rather obvious performance goal if sonic accuracy is desired, but this simple concept has been misunderstood and misinterpreted by many in past years. The confusion started decades ago, when the term *transparency* was applied to earlier-version solid-state amplifiers. In reality, such amplifiers were not even close to being sonically transparent, but the association of *transparency* with *solid-state sound* became inseparably linked, leading to the modern myth that *transparency* is a dirty word.

While on the subject of sonic accuracy (or transparency), I'd like to add a few related comments. Many audiophiles harbor personal preferences for certain types of distortion, which is certainly their privilege—I neither agree nor disagree with matters of taste. However, there is an old proverb that is particularly relevant to this discussion that states, "Don't put the cart before the horse." The point being, there is a correct order for the methodologies inherent to high-performance audio. Any type of sonic coloration should always be performed at the line level, with such coloration never existing as a function of the operational characteristics of the power amplifier. This is a matter of common sense. There is nothing wrong with adding some low-frequency harmonic content to certain types of music. I can think of a half-dozen outboard processors that are well suited to accomplish that goal, while simultaneously providing a great deal of user flexibility (such as the option to "turn it off"). In contrast, if the audio power amplifier is producing such distortion, you will never have the capability of listening to anything that isn't sonically colored, and you will be "locked in" to one characteristic sound. Sonic accuracy is the necessary foundation for any high-quality audio system. You can choose to go in a thousand different directions in terms of tonal settings, sound effects, or harmonic modifiers, but you will always need the capability of returning to the "baseline" of sonic accuracy for comparison purposes.

2. *The ideal audio power amplifier would be completely free of any internally generated noise artifacts.* The most severe problems typically associated with audio power amplifiers are those relating to distortion mechanisms, but it is also important to practice good noise reduction techniques, especially regarding the input stage design. Over half of the questions I receive from hobbyists experiencing problems with audio power amplifier construction relate to hum, and hum is considered part of the noise spectrum. Consequently, noise considerations also relate to wiring techniques and power supply considerations.

3. *The ideal audio power amplifier would be capable of infinite bandwidth and slew rate.* Bandwidth and slew rate are related characteristics important to an audio power amplifier's high-frequency distortion performance and its capability of processing high-level, high-frequency musical transients.

4. *The ideal audio power amplifier would look like a perfect voltage source to the speaker load.* In other words, the output impedance of a perfect power amplifier would be zero, resulting in bandwidth/distortion performance that would be unaffected by a variable speaker load, and an infinite damping factor (desirable for controlling speaker cone resonances).

5. *The ideal audio power amplifier would clip in a perfectly symmetrical fashion when overdriven, and the clipping characteristics would be "soft."* In my opinion, the importance of symmetrical clipping has been traditionally underemphasized. High-power professional applications require symmetrical clipping for the accurate operation of anticlip circuitry (important for the protection of horn and tweeter systems). Symmetrical clipping is desirable for domestic hi-fi applications because it will tend to produce a lower level of "nasty" harmonics if the amplifier is driven to clip levels during musical peaks. Also, since it is more or less inevitable that the power amplifier will be driven to clip levels occasionally, it is highly desirable that the characteristic of the clipping action is *soft.*

If you were to observe *hard clipping* action with an oscilloscope, relative to a pure sinewave signal input, it would appear as though someone took a pair a scissors and cleanly cut off the peak of the sinewave signal, leaving a perfectly flat plateau on the top and bottom, which abruptly begins and ends with the resumption of linear response. In contrast, *soft clipping* action will roll off into the flattened waveshape region, producing "rounded corners" at the beginning and end of the linear response. Soft clipping results in a very significant reduction of objectionable harmonics (roughly about five times less perceivable to human hearing), which is not only desirable relative to sonic quality but is also less damaging to sensitive tweeter systems.

It is important to recognize that the popularity of vacuum tube audio power amplifiers is largely due to their symmetrical, soft clipping characteristics. Vacuum tubes typically go into very nonlinear response regions at the ends of their transconductance curves. Due to this nonlinear transconductance response (coupled with the high-frequency limitations of most output transformers), the clipping characteristics of vacuum tube amplifiers tend to be very soft. From the perspective of sonic quality, this soft clip action (or "squashing" effect, as some refer to it) can actually produce a sense of "body" to musical passages, and produces the illusion of higher output power capabilities. For example, I often hear audiophiles comment that their 25-W vacuum tube amplifier sounds louder than a comparable 60-W solid-state amplifier. In reality, vacuum tube amplifiers are not louder than their solid-state counterparts, but they can be driven into relatively high levels of clipping without sounding distorted, so the illusion of higher output power is produced.

Out of all of the audio power amplifier designs offered in this chapter, only one (i.e., the Fig. 6-21 topology) is capable of symmetrical clipping together with soft clipping action. Soft clip action is very rare among solid-state audio power amplifiers. The harsh sonics created by hard clipping is one of the major objections that some audiophiles have toward solid-state designs.

6. *The ideal audio power amplifier would provide infinite peak power output capability for high-level signal transient processing.* During the early days of audio power amplifier development, it was discovered that power amplifiers could produce higher levels of output power for short intervals in comparison to their long-term, or continuous, power capabilities. The maximum short-term (i.e., about 20 ms) power output capability of an audio power amplifier is classified as its *music power* rating. In most conventional power amplifier designs, the music power rating is about double the RMS output power rating. The music power rating does not provide a very accurate picture of an amplifier's real-life transient capabilities, but it does illustrate the fact that there is a significant difference between the maximum continuous ratings and instantaneous capabilities. Some audiophiles measure the short-term power output capability and compare it to the maximum continuous output capability, terming the difference between the two as *headroom.* In conventional audio power amplifier designs, the difference between music power and RMS power is a result of *power supply droop* (i.e., the expected drop in power supply voltage levels when heavily loaded).

In a colloquial sense, many audiophiles use the term headroom to discuss the level of *excess power capability* that must be reserved for handling the normal dynamic range of the program material. For example, assume you connected a 100-W RMS power amplifier to a nominal audio load, applied a typical music program to the input, and adjusted the volume so that the power output was 20-W RMS. If you looked at this output with an oscilloscope, you would see a great variety of "peaks" and "voids" in the musical program. (This broad variation of musical amplitude levels is referred to as the *dynamic range* of the music.) You would probably observe some occasional high-level peaks in the music that would drive the amplifier up to its maximum 100-W rating (unless the program was compressed). It would also be readily apparent that if you were to increase the continuous power output up to 50-W RMS, many of the musical peaks would drive the amplifier into clipping levels. Therefore, the important question is, "What is the typical ratio between the required peak processing power and the continuous RMS power level?"

Back in the early '70s, the Heathkit Company recommended a peak power handling capability of 10 times the continuous RMS listening level. In other words, if you were listening to a musical program at a power level of about 10-W RMS, you would need a peak power output capability of about 100 W. If we assume that a conventional audio power amplifier rated for an output power of 100 W will provide about 200 W of "music power," this would mean that a 100-W RMS-rated amplifier could process most musical programs up to 20-W RMS levels without compromising the transient response. Personally, I think the 10-to-1 rule of thumb is a little extreme, and such decisions are based on numerous variables. One important variable is the clipping action of the power amplifier. If it goes into *asymmetrical hard clipping* when overdriven, the clipping distortion components will be a worst-case scenario, and be noticeable at relatively low levels. If, on the other hand, the power amplifier goes into *symmetrical soft clipping* when overdriven, you could achieve much higher volume levels before any clipping action would be perceived. Another important consideration is the dynamic range of the program material. For example, cassette tape has a dynamic range of approximately 80 dB in contrast to videotape, which boasts a dynamic range of 120 dB. Also, it is common for many types of musical programs to be *compressed,* which drastically reduces the dynamic range.

7. *The ideal audio power amplifier would be totally unaffected by the shortcomings of its associated power supply.* The *power supply rejection ratio* (PSRR) denotes the immunity level of a power amplifier to ripple content and signal components that may be superimposed on the power supply rails. Poor PSRR performance will manifest itself in the forms of hum, crosstalk (within stereo systems), and distortion.

8. *The ideal audio power amplifier would protect itself from any type of reasonable output short-circuit or overload condition.* I have a simple rule that I follow on all of my audio power amplifier designs—If any type of output short-circuit or overload condition will cause a rail fuse to blow, the design is poor. Well-designed protection circuitry should not generate any undesirable sonic effects during the normal operation of the associated power amplifier. Unfortunately, there are a lot of amplifiers in the world today suffering from poorly implemented protection circuits.

One of the common compromises inherent to many commercially manufactured audio power amplifiers is an insufficient number of output devices (usually bipolar power transistors) required to provide high performance at the specified output power levels. The commercial technique is to "squeeze" as much power as possible from each output transistor, which usually ends up in running the output devices on the "hairy edge" of their maximum secondary breakdown ratings. In the effort to provide a reasonable measure of reliability under these circumstances, the locus of the short-circuit protection network is typically adjusted so that it will not activate into purely resistive loading, but as soon as the real-world capacitive effects of most speaker systems is introduced at the output, false activation can become chronic and perceivable. The cure for this problem, of course, is to incorporate as many output devices as needed for high-performance operation into reactive loading without activating the protection circuits. This is easily accomplished, but the additional output devices represent an increased cost, and manufacturers are not always willing to sacrifice their competitive pricing edge for the sake of improved performance (especially when the *resistive loading* used to measure their product specifications won't reflect the improved performance).

The undeserved bad reputation that has been summarily attached to short-circuit protection networks has resulted in many audiophiles designing power amplifiers without short-circuit protection, or removing the short-circuit protection from existing amplifiers. I strongly advise against the practice of such techniques.

9. *The ideal audio power amplifier would provide optimum performance in conjunction with any practical speaker system impedance.* All modern solid-state audio power amplifiers should be capable of providing acceptable performance with speaker system impedances ranging from 2 ohms up to 16 ohms (or higher). I don't recommend operating any power amplifier into 2-ohm speaker loads, due to insurmountable drawbacks originating from the laws of physics. I recognize that 2-ohm loading is sometimes a necessary evil with automotive sound systems, but that still doesn't mean that I have to like it. I am often asked if a certain audio amplifier design is *stable* down to 2 ohms. A well-designed audio power amplifier should be stable down to a *dead short.*

10. *The ideal audio power amplifier would be 100% reliable and would not exhibit any performance changes throughout its operational lifetime.* All solid-state power components have a predictable life span based on the associated *thermal cyclic curve* for the particular device. Simply stated, in the case of virtually all solid-state components, as the power dissipation demand is increased, the reliability factors decrease (by logarithmic proportions). Reliability factors are frequently compromised in commercially manufactured power amplifiers.

Certain properties and/or specifications of electronic components will change with age. Such time-related component variations are usually called *drift.* Component drift, if severe enough, can result in virtually any type of distortion or operational problem incurred with a power amplifier.

I consider the previous list of ideal performance characteristics to be the most important issues to consider when evaluating an audio power amplifier. Some of the performance goals have been achieved in an (almost) ideal manner. The remainder have been achieved to a level that we could call "practical perfection," meaning that the performance is not perfect, but an improvement would probably not produce an audibly perceivable result.

In addition to the aforementioned performance characteristics, there are also some common "more-or-less standardized" specifications applicable to most audio power amplifiers. The *input sensitivity* will typically be about 1-V RMS, or lower, and the *input impedance* will probably be about 10 kohm (unbalanced) and about double the unbalanced input impedance for the balanced inputs.

Audio Power Amplifier Classes

Audio power amplifiers are categorized into a variety of groupings, or *classes,* depending on the operational principles and bias levels applied. The more common classes that you may have heard or read about (applicable to audio power amplifiers) are class A, AB, B, D, G, H, and S. I cannot begin to provide adequate detail in discussing all of these linear amplifier classes within this chapter, so I'm going to make this short and only focus on what is important to the context of this textbook.

If you want to construct excellent, high-performance audio power amplifiers, you really have only two choices at our present state of technological development—either *class A* or *class B*. It could be justifiably argued that class S is also a high-performance alternative, and I would agree. However, the problem with class S is that the materials required to do justice to the design are probably not going to be available to the typical hobbyist or audiophile. The performance levels of class S do not exceed the more common A and B types, nor do the efficiency levels, and its plunge into the realms of obscurity are not without good reason.

The popularity of class A designs within a minority of the esoteric audiophile community seems to be partially founded on several misguided concepts. One is that crossover distortion is a *constant,* meaning that its absolute level remains the same regardless of the output amplitude of the amplifier, and this is untrue. Another notion is that the linearity of class A designs is superior, and such a claim would surely show up in reduced THD levels. It doesn't, and ironically, most esoteric class A designs measure out with considerably more distortion than well-designed class B amplifiers. In any event, if you can achieve *worst-case* THD performance levels of 0.1% or better (which can be obtained with either class A or class B methodologies), any linearity differences are going to be far below human perception levels, so the whole issue appears to be a pointless argument.

I don't have anything against class A designs. They are impractical, but so are diamond rings and two-seater sports cars. However, I believe I owe a certain loyalty to the project desires of the majority of those reading this textbook, so my sole concentration will be in the realm of class B designs. Virtually all of the classic amplifiers of the past and present are class B designs, and 99% of all amplifiers currently being manufactured are class B.

Before leaving this topic, I'd like to clarify one area of confusion that many people have regarding the definition of class B designs.

Technically speaking, class B is any output stage design wherein the complementary output devices are biased into a precise 180-degree conduction angle. Class AB is often used in the colloquial sense to convey a concept of a complementary output stage with a slight forward bias added to reduce crossover distortion. The term "class AB" is technically incorrect unless the forward bias is adjusted so that the conduction angle of the output devices is in *excess of 180 degrees.* In other words, class AB defines an output stage that is overbiased, which results in a significant degradation of distortion performance characteristics. Therefore, when you read or hear someone defining the topology of an amplifier as being class AB, they are probably referring to class B.

Vacuum Tube Power Amplifiers versus Solid-State Power Amplifiers

My primary goal in writing this book, as it has been with all of the books I have written, is to distribute information. I'm not too keen about getting into debates over issues relating to taste, and especially when they apply to a hobby or pastime (both of which are intended to provide *joy* in the life of the participant, not *controversy*). However, it doesn't seem appropriate for me to cover the subject of audio power amplifiers, even at a casual level, without *distributing information* about the issues involved in the tube versus solid-state controversies.

Allow me to begin by directly quoting several statements from recent catalogs pertaining to vacuum tube amplifiers: *"The warm, rich, detailed sound of vacuum tubes has yet to be surpassed by either transistors or FETs,"* states one catalog. Another states, *"Bringing true valve quality to your budget!"* Such statements insinuate, or directly claim, that valve (vacuum tube) sound is superior to solid-state sound. Is this true?

In order to classify one item as better than another, we must first understand the criteria for evaluation. If we intend on using sonic accuracy (i.e., distortion performance) as a baseline for evaluation, then the bulk of the commercial vacuum tube amplifiers don't even come close to solid-state models. For example, the vacuum tube amplifier that was being advertised by the first of my aforementioned quotes provides a THD performance of 0.7% @ 1 kHz @ 0 dB, and I happen to know that its THD performance rises to over 2% @ 0 dB levels at both ends of the audio spectrum. Such THD performance would be considered extremely poor for a comparable solid-state amplifier.

Of the 10 performance goals established in the previous section, the only one that vacuum tube amplifiers inherently excel in is number 5. This does not mean that a vacuum tube amplifier of excellent linearity and performance characteristics cannot be designed. There have been classic designs of the past that were very competitive with solid-state designs in terms of performance. In truth, however, the audiophile enthusiasts who prefer vacuum tube equipment don't desire their equipment to perform equally with solid-state units, *because it would then sound like solid-state units.*

The modern "high-end" (synonymous with "high-cost") single-ended vacuum tube amplifiers will produce very high levels of harmonic distortion, sometimes exceeding 4%. I can most factually state that if you are serious about obtaining sonic accuracy from your audio system, vacuum tube amplifiers are not the place to look. It should be understood that the majority of vacuum tube enthusiasts are not after the ultimate in sonic accuracy, but rather, a unique type of sonic experience that they prefer. And that's OK! There is nothing wrong with individuals achieving their individual goals and enjoying something according to their personal tastes. However, there is something very wrong in advising an audio enthusiast on a budget that he or she must go out and pay $3000 for a vacuum tube system that will provide the so-called *ultimate sound experience.* Vacuum tube sound is *different,* but whether or not it is *better* is a matter of personal taste.

I don't want to beat this subject to death, but I do want to share one final bit of information. There has been a myth going around for a long time (thankfully, it's begun to subside) that there are mysterious, unmeasurable, analytically invisible attributes to music that vacuum tubes magically magnify into brilliant blossom, whereas solid-state amplifiers stomp them into the mud. Such claims are nonsense. Check out the professional recording studios and see what type of amplifiers they are using to create the music you are listening to.

Audio Power Amplifier Fundamentals

An exhaustive investigation of the operational physics relevant to audio power amplifiers can easily fill a rather large-sized book, so it is quite impossible to accomplish such a task in the space of a single chapter. If you would like to pursue the subject of audio power amplifiers in a more technical, design-oriented fashion, I recommend my 1999 publication entitled *High-Power Audio Amplifier Construction Manual,* available from McGraw-Hill Publications. Three other excellent reference books on this subject are

1. *Audio Power Amplifier Design Handbook,* by Douglas Self (Newnes Publications)
2. *High Performance Audio Power Amplifiers,* by Ben Duncan (Newnes Publications)
3. *Valve & Transistor Audio Amplifiers,* by John Linsley Hood (Newnes Publications)

I believe the easiest method of understanding the fundamental operational principles involved with audio power amplifiers is to begin with a very common "generic" audio amplifier design. After the operational principles of this generic amplifier are fully detailed, we will progress onward to more complex designs, continually building on the preestablished foundations. As we progress through a variety of amplifier projects, I will introduce additional concepts and terms as they become applicable to the design, and in this way, I hope to overcome much of the confusion that arises from discussing concepts without applications (we tend to forget things when we don't understand what they are good for).

Figure 6-1 illustrates a complete schematic diagram of a very conventional audio power amplifier design. Utilizing dual-polarity 38-V DC power supply rails, this design is capable of driving 8-ohm speaker loads at a maximum power output of approximately 60 W RMS. The THD performance is better than 0.01%, rising to levels of about 0.1% under worst-case conditions. I have purposefully left out the component values for the sake of clarity; they are provided later in this chapter.

C1 and C2 in Fig. 6-1 are input coupling capacitors. Their function is to block any DC levels while allowing the AC line level input signal to freely pass. C4 and C5 function as power supply decoupling capacitors (often called *rail decoupling capacitors*). R1 sets the input impedance, and C3 is used to shunt RF frequencies to signal common. Note

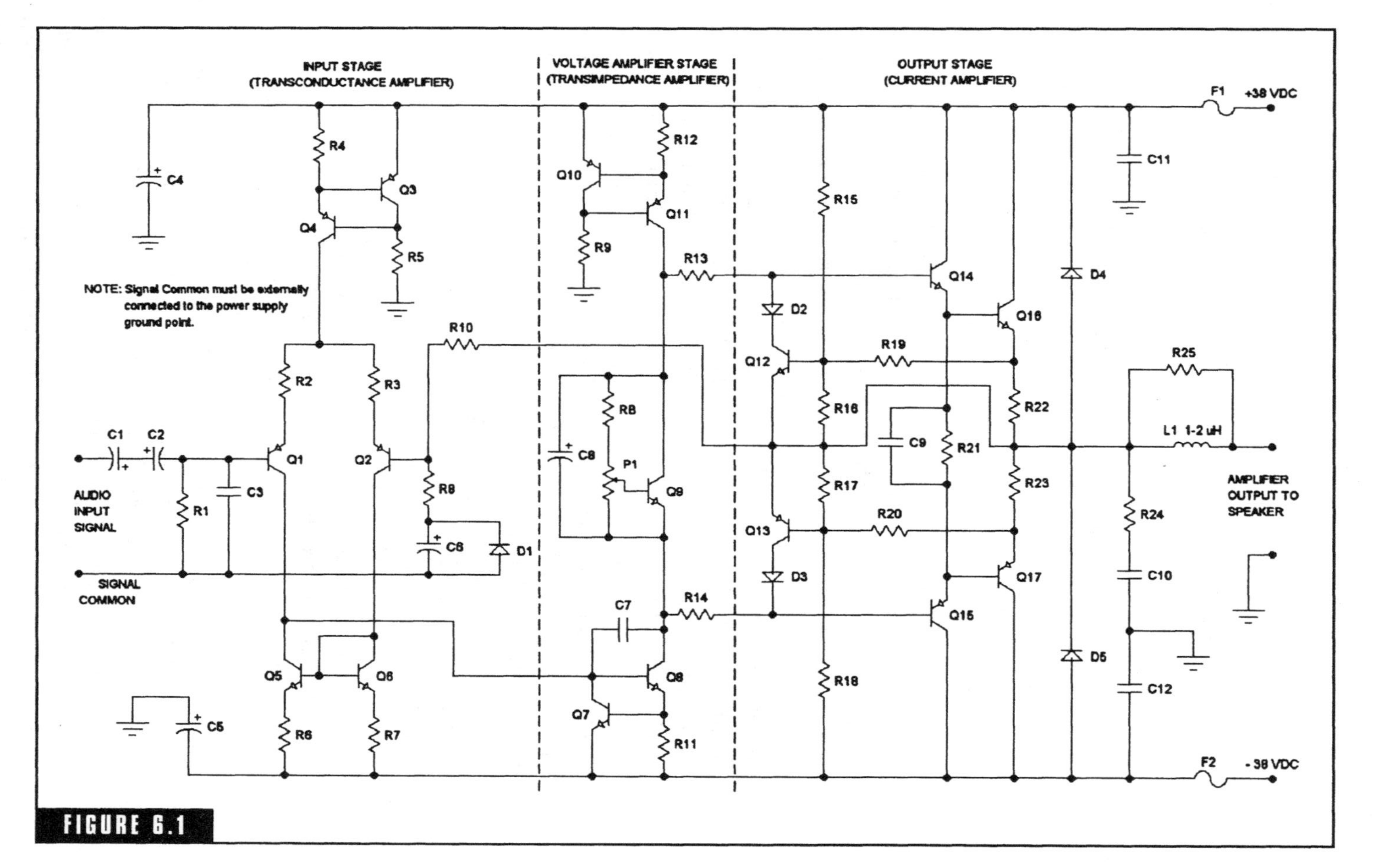

FIGURE 6.1

Complete schematic of the "Basic EF" (Pro-60) model amplifiers.

the *signal common* line connected to the bottom side of R1, C3, C6, and D1. This line must eventually be connected to the circuit common point, but it is isolated from any other circuit common points on the PC board to eliminate the effects of ground noise interference. Q3, Q4, R4, and R5 make up a constant current source to provide the tail current for the Q1/Q2 differential input amplifier. R2 and R3 provide linearizing degeneration for the differential transistor pair. Q5 and Q6, together with their associated degeneration resistors, R6 and R7, form a current mirror. Current mirrors provide two important functions in high-performance input stage design: (1) They force a current balance through the differential input stage, which reduces distortion, and (2) they provide active loading for the differential collectors, which essentially doubles the current output capability of the differential stage.

Continuing to refer to Fig. 6-1, global negative feedback is provided via R10, while the AC voltage gain of the amplifier is determined by the ratio of R10/R8. Note that the DC negative feedback is 100%, since C6 (in series with R8) will block any DC levels being fed back (via R10) from the output. This type of 100% DC feedback is desirable to keep the DC output offset voltage (i.e., the DC voltage continuously present at the speaker output terminals) to a minimum. In this type of design, the typical values of DC output offset will be somewhere between 10 and 30 mV. Diode D1 is installed to protect C6 in the event of a significant voltage polarity reversal.

The output of the input stage is applied to the input of the voltage amplifier stage, which is the base of transistor Q8. The collector of Q8 is actively loaded with the constant current source of Q10, Q11, R9, and R12, which establishes the quiescent collector current of the Q8 collector circuit. In series with Q8's quiescent current flow is the *amplified diode* circuit of Q9, P1, RB, and C8. Amplified diode circuits are used to provide the small forward bias voltage (i.e., typically about 2.8 V in emitter-follower amplifiers, such as this one) required to bring the class B output-stage transistors into an exact 180-degree conduction angle. The bias potentiometer, P1, allows precise adjustment of the bias voltage for the lowest possible crossover distortion artifacts. Q9 is typically mounted to the same main heatsink as the output devices (i.e., Q16, Q17) so that the bias voltage provided by the amplified diode circuit will track the temperature of the output devices. In this way, the output devices are protected from *thermal runaway* failures, and the optimum distortion performance is maintained regardless of temperature variations. Compensation is provided by C7. The function of Q7 is to protect Q8 from excessive current flow, by monitoring the emitter current flow through R11. If the voltage developed across R11 begins to exceed about 0.7 V, Q7 will turn on, diverting Q8's drive current, and limiting Q8's maximum collector current to safe levels. (Q8's collector current could increase to destructive levels if an output short-circuit condition occurred, saturating Q13, and effectively connecting the right side of R14 to the output rail.)

The output of the VA stage is applied to the predriver transistors, Q14 and Q15, through two impedance matching resistors, R13 and R14. The function of Q14 and Q15 is to provide beta enhancement (i.e., current amplification) required to drive the output transistors. R21 provides emitter stabilization for Q14 and Q15, while C9 reduces the effects of *switching distortion* from the output transistors. Switching distortion is a result of the sluggish nature of most high-power bipolar transistors. It occurs when one complementary device cannot turn off fast enough, causing it to continue to conduct while the other complementary device turns on. Thus, a condition of *cross-conduction* occurs (i.e., both output devices conducting simultaneously), creating distortion effects very similar to crossover distortion. The action of C9 helps to rapidly drain any stored charge carriers in the base junctions of the output devices, and greatly enhances their ability to turn off rapidly.

Resistors R22 and R23 are emitter resistors for the output transistors, Q16 and Q17. In this particular amplifier design, their only real function is to provide a *current feedback* signal to the protection circuitry. With higher-power amplifier designs incorporating multiple output pairs, such emitter resistors aid in promoting equal current sharing of the parallel output devices, and are often referred to as *ballast resistors*. Q12, Q13, D2, D3, R15, R16, R17, R18, R19, and R20 make up a single-slope overload protection circuit for the Fig. 6-1 amplifier. The voltages developed across R22 and R23, which will be proportional to the output current flows to the speaker system, are summed at the bases of Q12 and Q13, together with the power supply rail voltages. If the voltage drop across R22 or R23 exceeds the design limits, the associated protection transistors will turn on, diverting a portion of the drive current away from the predriver transistors, and safely limiting the maximum output current.

Diodes D4 and D5 are often called *catching diodes.* Their function is to protect the output transistors from reverse-polarity voltage spikes, which could result from inductive "kick-back" transients of highly reactive speaker loads. R24 and C10 make up a *zobel network.* Zobel networks compensate for the inductive-reactive nature of some speaker loads, and their design is fundamentally based on a "best-fit" estimation of the speaker load, not the amplifier design. Therefore, R24 will almost always be about 8 to 10 ohms, and C10 will be about 0.1 μF. C11 and C12 are conventional power supply decoupling capacitors. Finally, L1 is the output inductor for the power amplifier, and R25 is its associated *damping resistor.* A small measure of output inductance is beneficial to audio power amplifier designs to compensate for any capacitive-reactive nature the speaker system might present. Essentially, the output inductor improves the high-frequency stability characteristics. Again, the value of L1 is based on an estimate of the anticipated speaker loading effects, so its inductance value will almost always range from 1 to 10 microhenries (μH). Obviously, the gauge of the wire used to construct L1 will depend on the maximum output power capabilities of the audio power amplifier. The damping resistor, R25, is incorporated to reduce high-frequency "ringing" effects that could occur from the combination of L1 and any associated capacitance inherent to the speaker system.

Thus far, we have examined all of the functions of the components incorporated into the Fig. 6-1 audio power amplifier design. In addition to component operation, it is also necessary to understand how the individual *amplifier stages* work together for the best possible performance. Note that Fig. 6-1 illustrates three individual stages inherent to the amplifier architecture; the *input stage,* the *voltage amplifier stage,* and the *output stage.* Such a topology is commonly referred to as the *Lin three-stage topology,* after its developer, Mr. Lin of RCA (circa 1956). About 99% of all audio power amplifiers in the world are designed according to this fundamental three-stage architecture, and those that are not, for the most part, are outside of the classification of high-performance. In order to understand why such a topology is so important, we should probably begin by considering the enormous job that audio power amplifiers are called upon to accomplish.

Audio power amplifiers are required to perform in an *ultralinear* fashion while being designed to provide about 50 dB of power gain to an input signal exhibiting a dynamic range of up to 120 dB and a bandwidth of 10 octaves! Such a formidable task must be accomplished within an operational environment of 500% load impedance variations, 1000% load reactance variations, and 300% temperature climbs. Needless to say, a simple cascaded series of single-stage transistor amplifiers "ain't gonna make it!"

Referring back to Fig. 6-1, transistors Q1 and Q2 make up the input differential amplifier. Differential amplifiers are usually chosen as the first stage of an audio power amplifier because they provide a convenient point of applying global negative feedback (i.e., the base of Q2, which is the inverting input), they are capable of high current gain and high input impedance, and they are relatively insensitive to power supply fluctuations. The *input stage* functions as a *transconductance amplifier* (i.e., a voltage-to-current amplifier), providing a signal current output at the collectors of Q5/Q1. The next stage of the amplifier, usually called the *voltage amplifier stage,* is a *transimpedance amplifier* (i.e., a current-to-voltage amplifier), receiving the amplified signal current from the input stage, providing further amplification, and converting the signal back into a high-level signal voltage. In most conventional amplifier designs, all of the voltage gain is accomplished in the first two stages, so the output of the voltage amplifier stage represents the approximate maximum voltage output of the power amplifier.

The *output stage* of a conventional audio power amplifier design is a class B, unity voltage gain, current amplifier. It receives the high-level voltage output of the voltage amplifier stage and simply buffers it, providing the current amplification necessary to drive low-impedance, high-power speaker loads. Output stages can be configured in a variety of ways, as we shall detail in the upcoming cookbook amplifier designs. In this particular topology, the output stage design is called an *emitter-follower,* because the emitter voltage "follows" whatever is applied to the base (it is analogous to the fundamental *common-collector* transistor configuration). Note that the output signal at the junction of R22 and R23 (this is commonly called the *output rail* of the amplifier) is fed back to the input stage via R10. This feedback loop is referred to as the *global negative feedback loop.*

The combined transconductance and transimpedance functions of the first two stages provide very high levels of *open-loop gain* (i.e., the voltage gain before any negative feedback is applied) with a minimum of phase shift. High-performance power amplifiers require very high open-loop gains (in the realm of +100 dB) so that large quantities of negative feedback can be applied for optimum linearity response. In the general sense, the higher the open-loop gain, the better the THD performance will be. Capacitor C7 is the compensation capacitor, which is incorporated to achieve good Nyquist stability.

The interaction of the three stages within the Fig. 6-1 amplifier topology is rather fascinating when you begin to perceive the broadband physics involved with its operation. Imagine, for the sake of discussion, that you were to apply an input signal to the input of the amplifier, beginning with a frequency of about 20 Hz, and slowly increasing the frequency up to about 5 MHz (assuming all other operating parameters of the amplifier are at nominal levels). When you first apply the 20-Hz input signal, the *open-loop gain* of the amplifier will be very high (probably about +80 dB in this design) while the *closed-loop gain* will be at about +30 dB (established by the ratio of R10/R8). As you slowly go up in frequency, the open-loop and closed-loop characteristics stay flat until you get to about 1 kHz. At this frequency point (commonly called the *first-pole frequency,* or P1), the capacitive reactance of C7 is such that a small portion of the collector signal of Q8 will be capacitor-coupled back to its base (note this will be "negative" feedback, reducing the voltage gain of the voltage amplifier stage). Consequently, at this P1 point, the open-loop gain will begin to roll off, and since C7 basically represents a first-order filter, the roll off will be approximately −6 dB per octave. In addition, the input and output impedance of the voltage amplifier stage begins to decrease above the P1 frequency.

As you continue to increase the frequency above the P1 point, the open-loop voltage gain continues to drop, the input/output impedance of the voltage amplifier stage continues to drop, but the closed-loop voltage gain remains the same (because it is a function of the R10/R8 ratio). I should also mention that the output signal remains in *absolute phase* with the input signal during this entire process, so we don't have to concern ourselves with any possible phase distortion problems in regard to modern solid-state power amplifiers.

Again, as you continue to increase the input frequency, you will eventually reach the frequency point where the falling open-loop gain will intersect with the "flat" closed-loop gain. With most high-performance solid-state power amplifier designs, this frequency point will be somewhere in the 60-to 100-kHz range. Just prior to this intersection frequency point (maybe about 5 to 10 kHz below it), the *phase-quadrature action* of the input differential stage loses control, causing the output signal to begin to *lag* in its phase relationship to the input. At the actual intersection point, the open-loop and closed-loop gain responses merge into a single gain response primarily dictated by the compensation factors.

Again, as you continue to increase the input frequency, the overall voltage gain of the amplifier continues to fall (there is no longer any difference between the open-loop and closed-loop response), and the phase lag of the output signal continues to increase. If the value of the compensation capacitor has been properly chosen, the voltage gain should drop below unity before the output signal phase lag reaches −180 degrees. If this is accomplished, the amplifier is said to have *Nyquist stability* (i.e., the amplifier cannot degenerate into self-sustaining oscillation). With most solid-state amplifiers, this unity-gain point, or P2 point, will be at a frequency of between 1 and 3 MHz.

As mentioned earlier, when you exceeded the P1 frequency, the input and output impedance of the voltage amplifier stage began to drop. This is a good phenomenon because it minimizes the phase shifts due to semiconductor capacitances within the input and output stages. This technique of minimizing phase shifts with a "middle stage" is called *pole splitting.* Pole splitting is another advantage to the conventional three-stage topology.

You may remember that in the first chapter of this book I briefly commented on a matter of physics relating to audio power amplifiers, which would cause the distortion to rise as the fundamental frequency was increased. I was referring to the rolloff of the open-loop gain at frequencies above the P1 frequency (remember, this is typically about 1 kHz in most solid-state amplifiers). As the open-loop gain decreases, there is proportionally less global negative feedback available for linearizing purposes, so the THD levels begin to climb. However, this effect is not as drastic as you might think at first. Referring back to Fig. 6-1, the compensation capacitor, C7, begins to provide some level of negative feedback above the P1 frequency, reducing the voltage amplifier stage gain. Like most other forms of negative feedback, this has a linearizing effect upon the voltage amplifier stage. Therefore, as the amplifier distortion is rising at higher frequencies due to the loss of global negative feedback, the increasing *local negative feedback* within the voltage amplifier stage is counteracting the distortion rise. The end result is still rising distortion at higher frequencies, but the overall effect is not nearly as pronounced as it would be without the benefit of the feedback action within the voltage amplifier stage. The higher-frequency linearizing effect of the voltage amplifier stage is another advantage of the three-stage architecture.

If you have never been exposed to the operational physics of a conventional solid-state audio power amplifier before, your head may be spinning a little by now. You may want to go back and reread this section a few times. If anyone ever belittles your efforts in constructing an *ordinary old power amplifier,* and suggests that you ought to get into something *high-tech for a change,* you may want to let that person wade through this section of this book!

Audio Power Amplifier Cookbook Designs

Figure 6-2 illustrates the amplifier topology of Fig. 6-1 (with the component values inserted), so the technical aspects of its functional description have already been covered in some detail. However, from the perspective of a "cookbook" project, there are a few construction details that should be mentioned.

The input coupling capacitors, C1 and C2, are 22-μF @ 35-V "tantalum" types, mounted back-to-back, forming a nonpolarized input coupling capacitor. Many audiophiles criticize the use of tantalum coupling capacitors due to a type of rectifier action that can occur if the values are significantly mismatched. In my experience, I haven't seen this phenomenon in conjunction with the typical line-level signals and currents involved. I believe the overall THD analysis provides further evidence against any need for concern in this regard. However, I do have one fault with the use of tantalum (or electrolytic types) for input coupling applications. It is not unusual to develop a fault condition from a preceding audio stage (such as a preamplifier) that will apply a significant DC level to the input coupling capacitors. In such a case, tantalum capacitors will fail very rapidly (usually shorting in the process), which will then apply the DC level to the remaining "DC-coupled" portion of the amplifier circuitry, sometimes

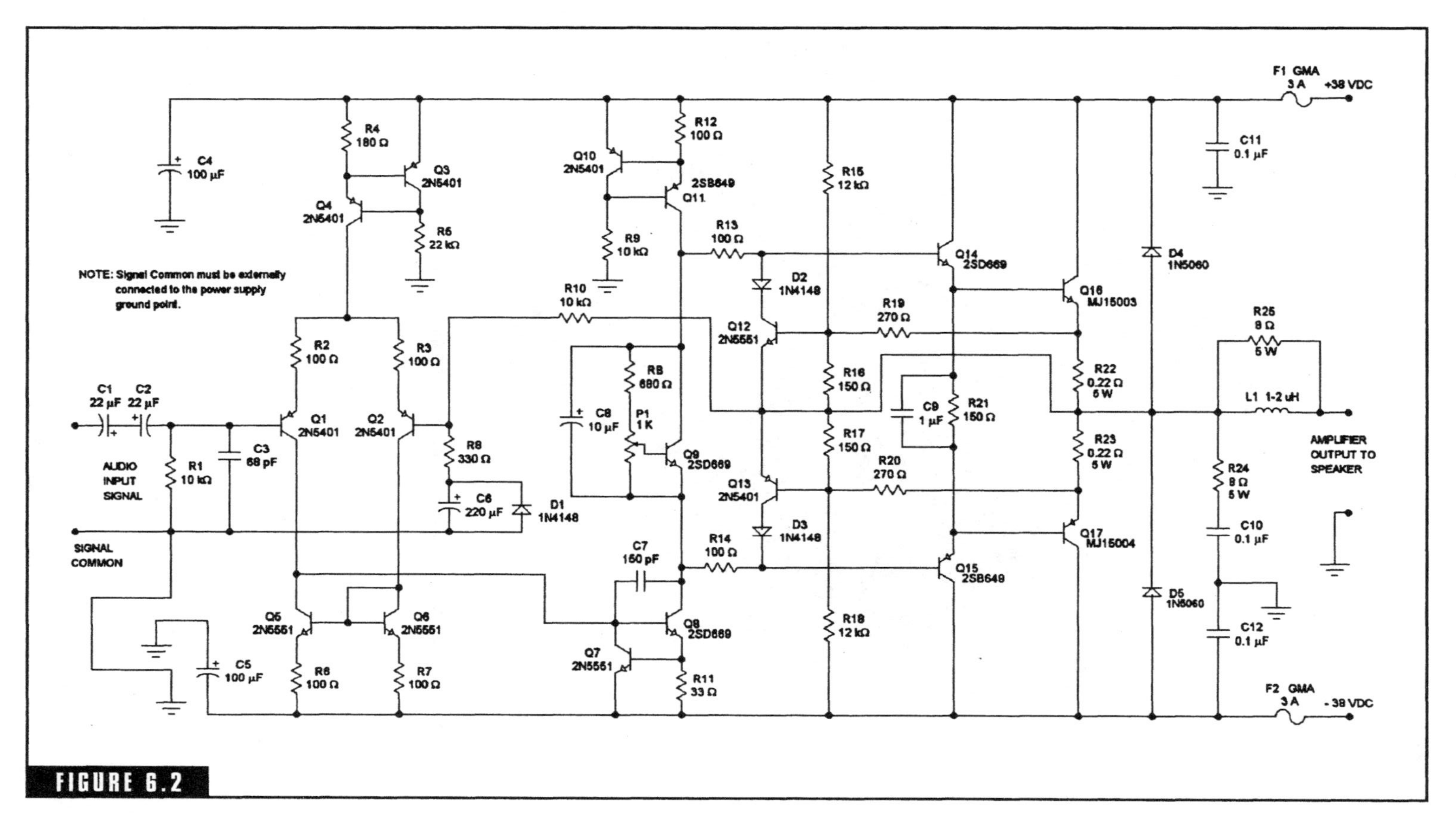

FIGURE 6.2

Complete schematic of the "Basic EF" (Pro-60) model amplifiers.

The Fig. 6-2 emitter-follower audio power amplifier.

resulting in catastrophic component failure. If you really want to go first class in every respect, you could replace the input coupling capacitors with a 10- to 15-μF low-loss type of capacitor, rated for much higher voltages, such as aluminized polypropylene. These types of capacitors are available in smaller sizes now, and they can be installed in the PC board designs included in this chapter by installing them in a vertical position. In all of the following amplifier designs, the specified input coupling capacitors are tantalum types.

Referring to Fig. 6-2, you'll note that I isolated the signal common (i.e., the "shield" connection from the audio signal input) and feedback common (the signal common for the global negative feedback signal). These points are the most susceptible to signal injection problems from other sources (such as rail decoupling capacitors), so it is best to run a dedicated ground wire from this signal common point directly to the *high-quality ground* (HQG) point.

As stated previously, transistors Q5 and Q6 are *current mirror* transistors, and as such, they should be matched fairly well. If the transistors are the same device types, and the beta values are within 10% of each other, this degree of matching should be sufficient (although 5% beta matching is preferable). R6 and R7 primarily serve to compensate for differing V_{BE} characteristics.

The differential input transistors, Q1 and Q2, should also be matched *to some degree,* but their similarity is not as critical as in the case of the current mirror transistors. Significant V_{BE} differences in Q1 and Q2 will manifest themselves in slightly higher DC offsets at the amplifier's output.

The heatsink I use for the Fig. 6-2 amplifier design is a common type, predrilled to accept two TO-3 case-style output devices (such as the MJ15003/MJ15004 pairs illustrated) and thermally rated for approximately 0.8°C/W. The predriver transistors, Q14 and Q15, require a small amount of heatsinking. Almost any size of small TO-220 type heatsink should be sufficient. The output inductor is an "air-core" type (as are all of the other amplifier output coils in this chapter), made by winding 20 turns of 16-AWG magnet wire around a ½-in form (such as a wooden dowel or old ink pen). Don't forget to mount the bias transistor (Q9) to the main heatsink for thermal tracking purposes.

The top-view composite, top-view layout, and bottom-view reflected PC board illustrations are provided in Figs. 6-3, 6-4, and 6-5, respectively. If you decide to construct this PC board design (or any other PC board design in this book), be sure to follow the *reflected artwork* illustrations.

Figure 6-6 illustrates a high-performance 80-W RMS audio power amplifier design, which is very similar to the previous project, incorporating the same fundamental topology and protection circuitry. However, there are a few refinements. You'll notice that the configuration of the two constant-current sources for the input stage and voltage amplifier stage have been changed. Q6 and Q8 make up a conventional

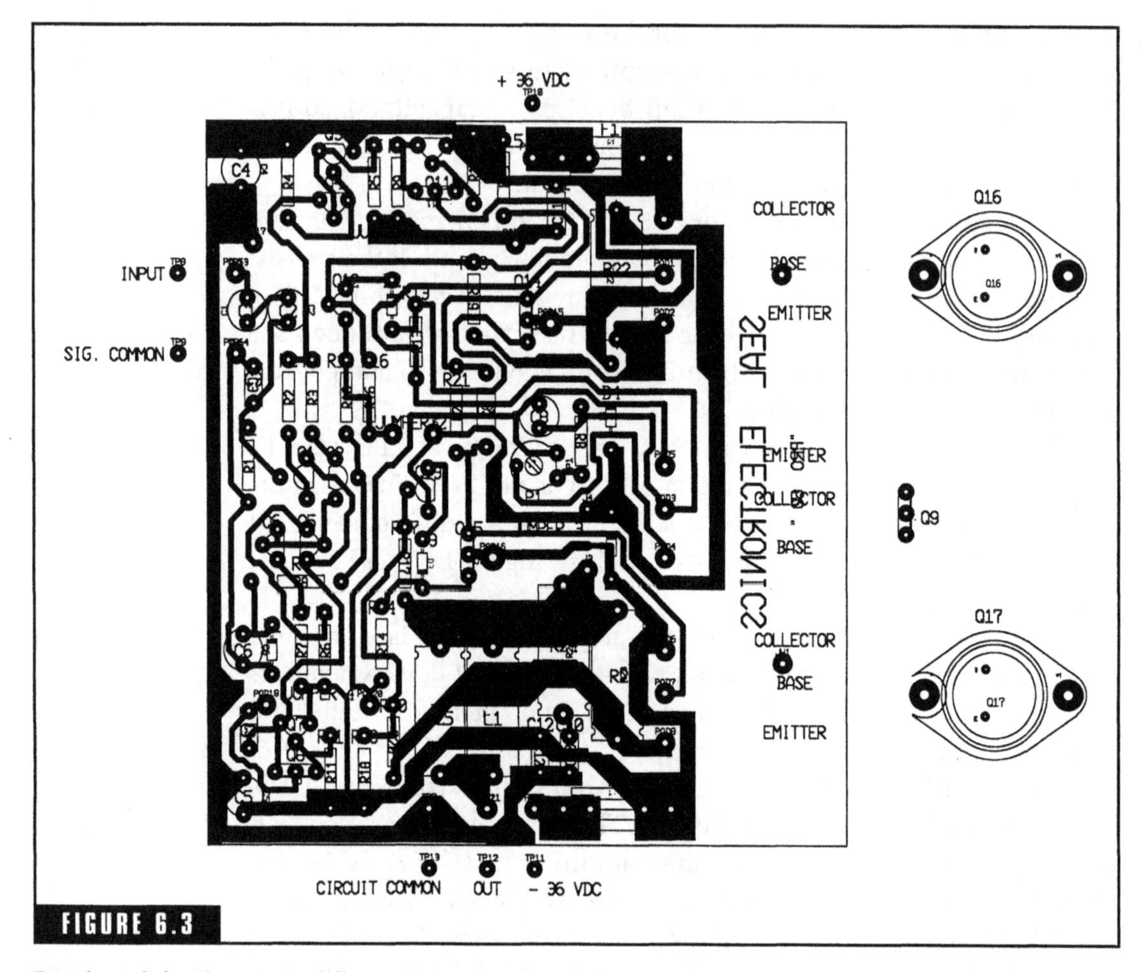

FIGURE 6.3

Top view of the Fig. 6-2 amplifier.

constant-current source, while Q3 receives its base reference voltage from the same source as Q8 (i.e., the collector of Q6). This is essentially a method to eliminate a needless transistor. C6, R7, and R8 make up a very effective RC network to stabilize the collector reference voltage of Q6, and to isolate this reference from any ground interference.

Resistor R14 has been placed in series with the amplified diode bias generator (Q9 and the associated circuitry). R14 helps to regulate the precise V_{BIAS} setting by compensating for current variations through Q9, resulting from power supply rail and temperature variations. For example, if the collector current of Q8 increased slightly, there would be the tendency for the collector-emitter voltage of Q9 to

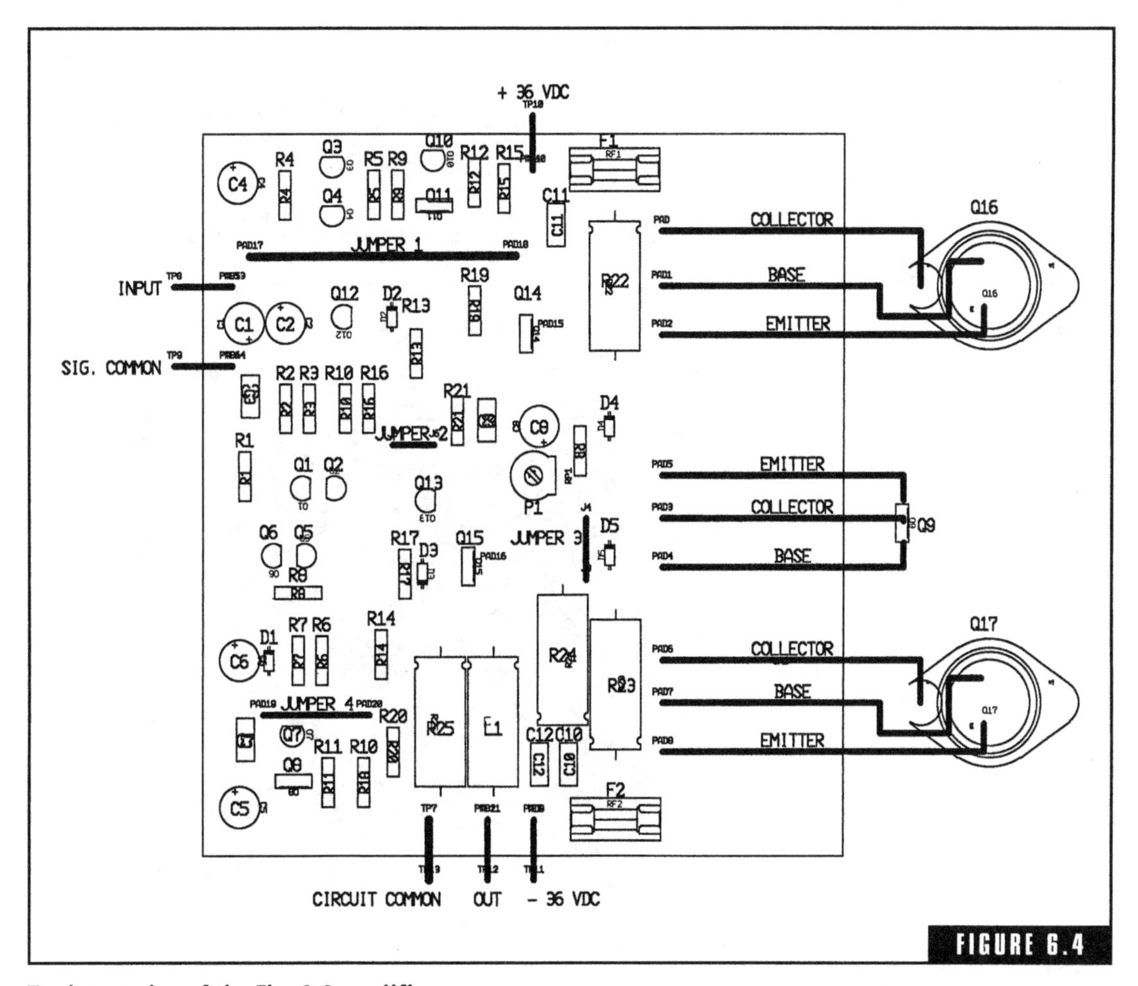

FIGURE 6.4

Top layout view of the Fig. 6-2 amplifier.

increase slightly, but the current increase of Q8 will be dropped by R14, keeping the collector-emitter voltage of Q9 stabilized.

Q10 is incorporated as a beta-multiplier for Q11 (a beta enhancement, or Darlington, technique). This increases the open-loop gain characteristic of the amplifier, which results in improved distortion performance. C8, C9, and R17 form a type of compensation network referred to as *two-pole compensation.* As you may recall from the previous discussion of the Fig. 6-1 amplifier, the single compensation capacitor provided the same basic response as a first-order filter, causing the open-loop gain to decrease at a rate of approximately 6 dB per octave above the P1 frequency. C8, C9, and R17 are simply a second-

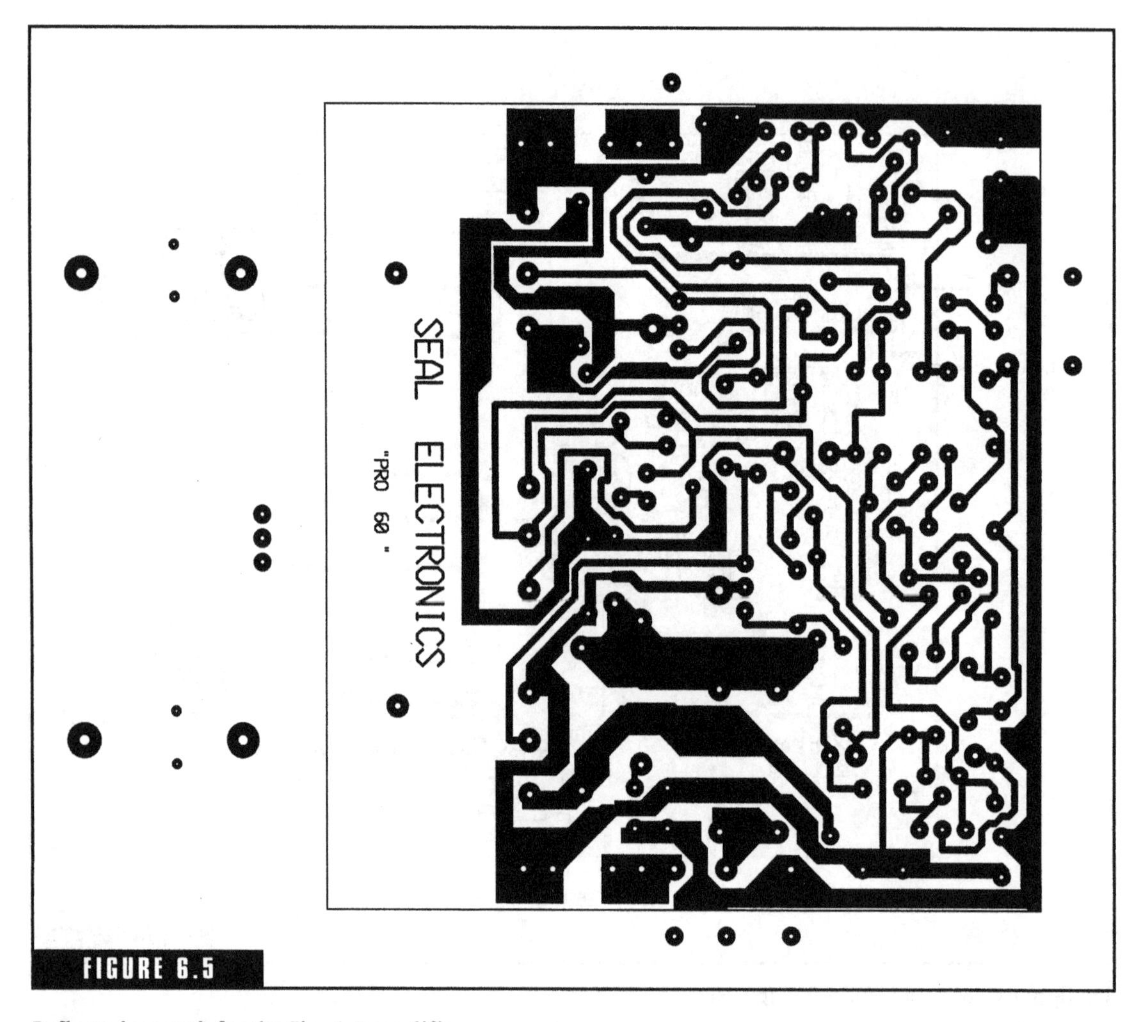

FIGURE 6.5

Reflected artwork for the Fig. 6-2 amplifier.

order filter, providing a 12 dB per octave decrease in open-loop gain above the P1 frequency. Since the open-loop gain rolls off at a steeper rate, the P1 frequency can be *moved up,* typically to about 10 kHz, while still achieving a unity gain response at the same P2 frequency. (The technicalities of two-pole compensation are a little more involved than my description, but the principle is accurate.) Moving the P1 point up to about 10 kHz keeps the beginning of the THD rise (due to loss of open-loop gain) from occurring at much lower frequencies, which keeps it from getting as "high" by the time the end of the

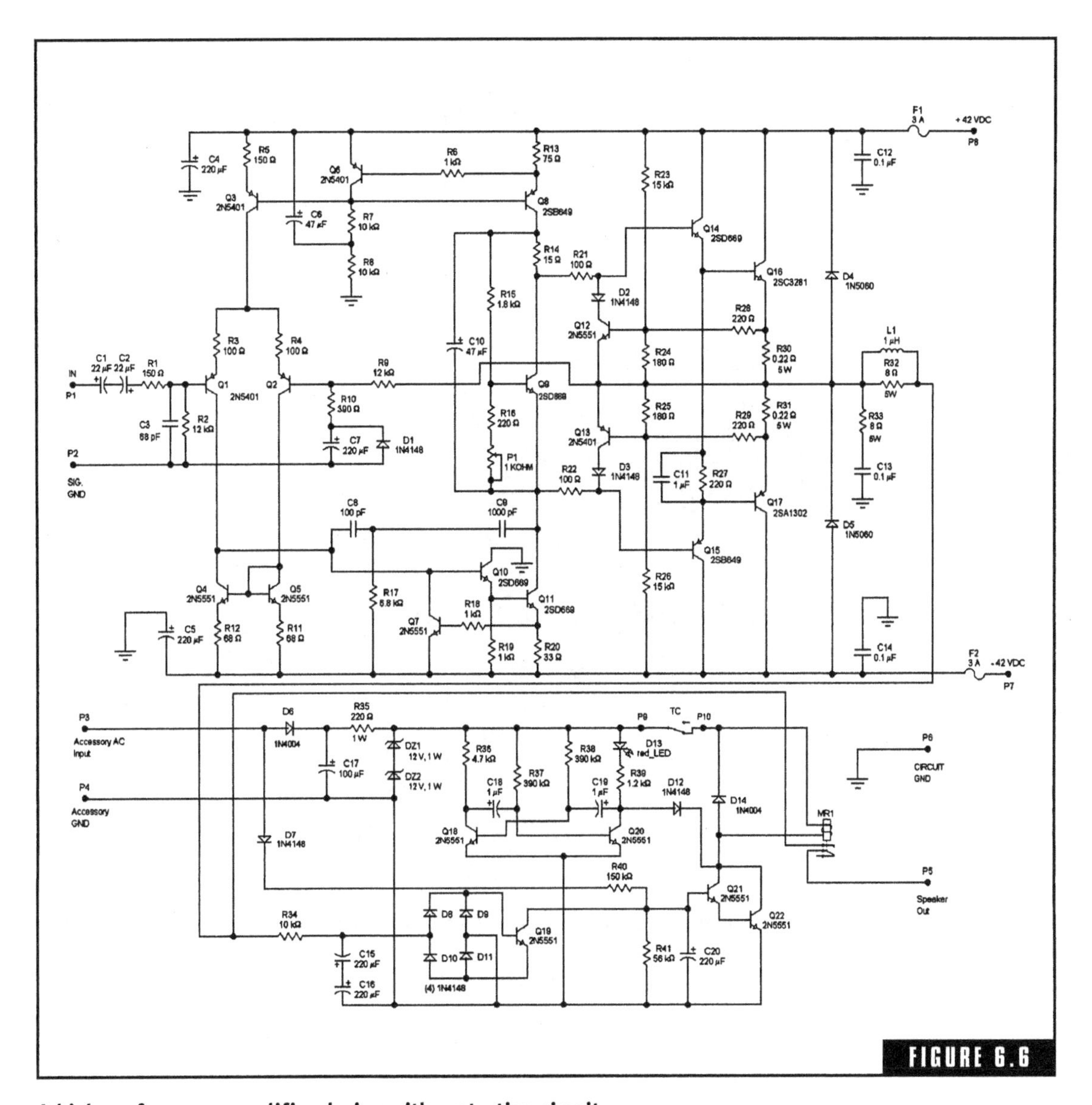

A high-performance amplifier design with protection circuit.

audio spectrum is reached. In short, two-pole compensation improves the high-frequency THD performance.

I chose the "flat-pack" 2SC3281/2SA1302 output devices for this amplifier design, so you can use any type of "flat mounting face" heatsink with similar thermal ratings as used in the Fig. 6-2 design. L1 is formed by winding 18 turns of #16 AWG magnet wire around a

The Fig. 6-6 audio power amplifier with protection circuit. Note the thermal bar mounted to the front surface of the output transistors, and the thermal switch mounted to the main heatsink (above the thermal bar) that provides automatic thermal shutdown.

1/2-in diameter form. Note that transistors Q8, Q11, and Q10 should be mounted to small TO-220 heatsinks for heat dissipation, even though this fact is not shown in the layout illustrations (also note that Q10 and Q11 are mounted back to back on the same small heatsink). Most other construction details are the same as for the Fig. 6-2 design.

The lower circuit in the Fig. 6-6 illustration is an *on-delay/DC speaker protection circuit* for this amplifier design, which is also included in the PC board design. The functional details of this protection circuit will be covered in Chap. 8 of this textbook.

The PC board designs for the Fig. 6-6 amplifier are included in Figs. 6-7, 6-8, and 6-9. Referring to Fig. 6-8, note that the predriver transistors (Q14 and Q15) are mounted to the same main heatsink as the output devices, so it isn't necessary to incorporate separate small heatsinks for these devices. Also, note that transistor Q9 (the V_{BIAS} transistor) is positioned a small distance back from the main heatsink. When Q16 and Q17 are mounted to the heatsink, you must fabricate a small bar of aluminum that will be held to the exterior "face" of the output transistors by their mounting screws. Q9 should be mounted to this small bar (we'll call it a *thermal bar*). In reality, the V_{BIAS} transistor needs to track the *junction temperature* of the output devices, but it takes a significant amount of time for the main heatsink temperature to "catch up" with the junction temperature of the output transistors (due to the *thermal mass* of the heatsink). The face of the output transistors responds to the junction temperature much faster than the heatsink, and by mounting a small aluminum bar (of low thermal mass) to the face of the output transistors, the thermal bar will be a more accurate temperature indicator for the V_{BIAS} transistor. The motivation of this technique is to reduce *thermal delays.*

The overall performance of the Fig. 6-6 amplifier is really quite exceptional. It is rated for 80 W RMS into 8-ohm speaker loads, but will function well with other conventional speaker impedances (at increased distortion levels). The THD performance is about 0.004% @ 1 kHz @ 0 dB, and only rises to a slight fraction above 0.01% under *worst-case conditions* (i.e., approximately 500 mW @ 20 kHz). If good construction techniques are used, the SNR should be considerably better than −100 dB, and the bandwidth is 3 Hz to 80 kHz. This is a true audiophile-quality amplifier and a very impressive project.

The Fig. 6-10 amplifier design (which I affectionately refer to as the "Micro-Mite" design) actually began as a "joke" project, instigated by an audiophile friend of mine from Denmark. The goal was to come up with a "cheap" high-performance audio power amplifier with an RMS output power of about 12 W. My friend said he couldn't stand the quality of his multimedia computer speakers any longer! The entire amplifier, including power supply, is illustrated in Fig. 6-10. The power transformer is a 24-V.C.T., 2-A model that I had lying around (a very common unshielded, laminated type—I believe it came from Radio Shack). All of the small-signal transistors are 2N3904/2N3906 types, and the output transistors are TIP31/TIP32 devices. All of the resistors are the garden-variety 5% carbon-film types, and the output coil is formed by winding 20 turns of 18-AWG magnet wire around a pencil.

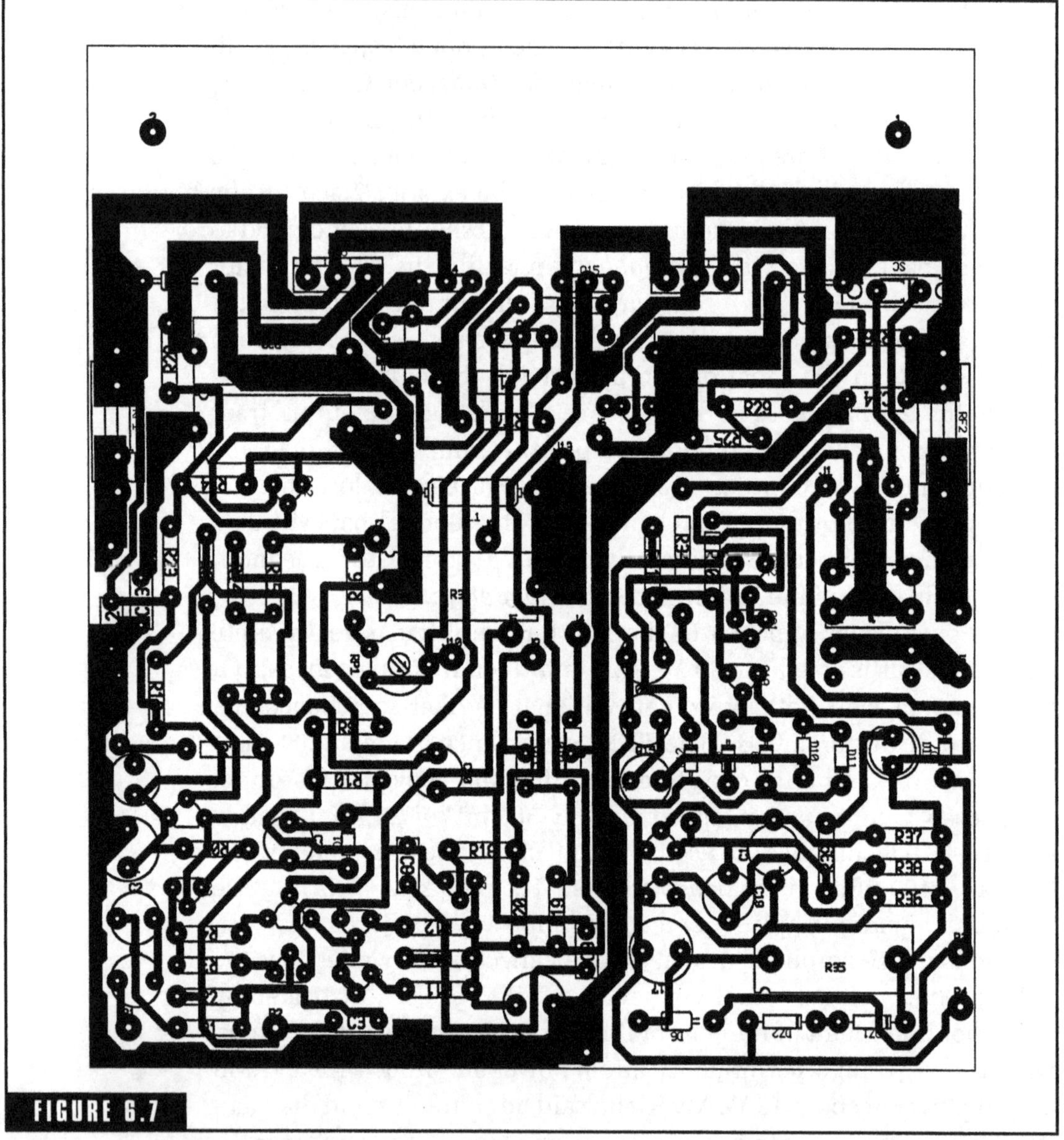

FIGURE 6.7

Top view of the Fig. 6-6 amplifier.

The main heatsink was only a scrap piece of aluminum plate, and if you were to mount this amplifier in a metal enclosure, the output transistors could be mounted to the enclosure, which would provide more than adequate heatsinking. The only devices I splurged on were the 2SD669/2SB649 devices, and they only cost about 75 cents from most retail distributors.

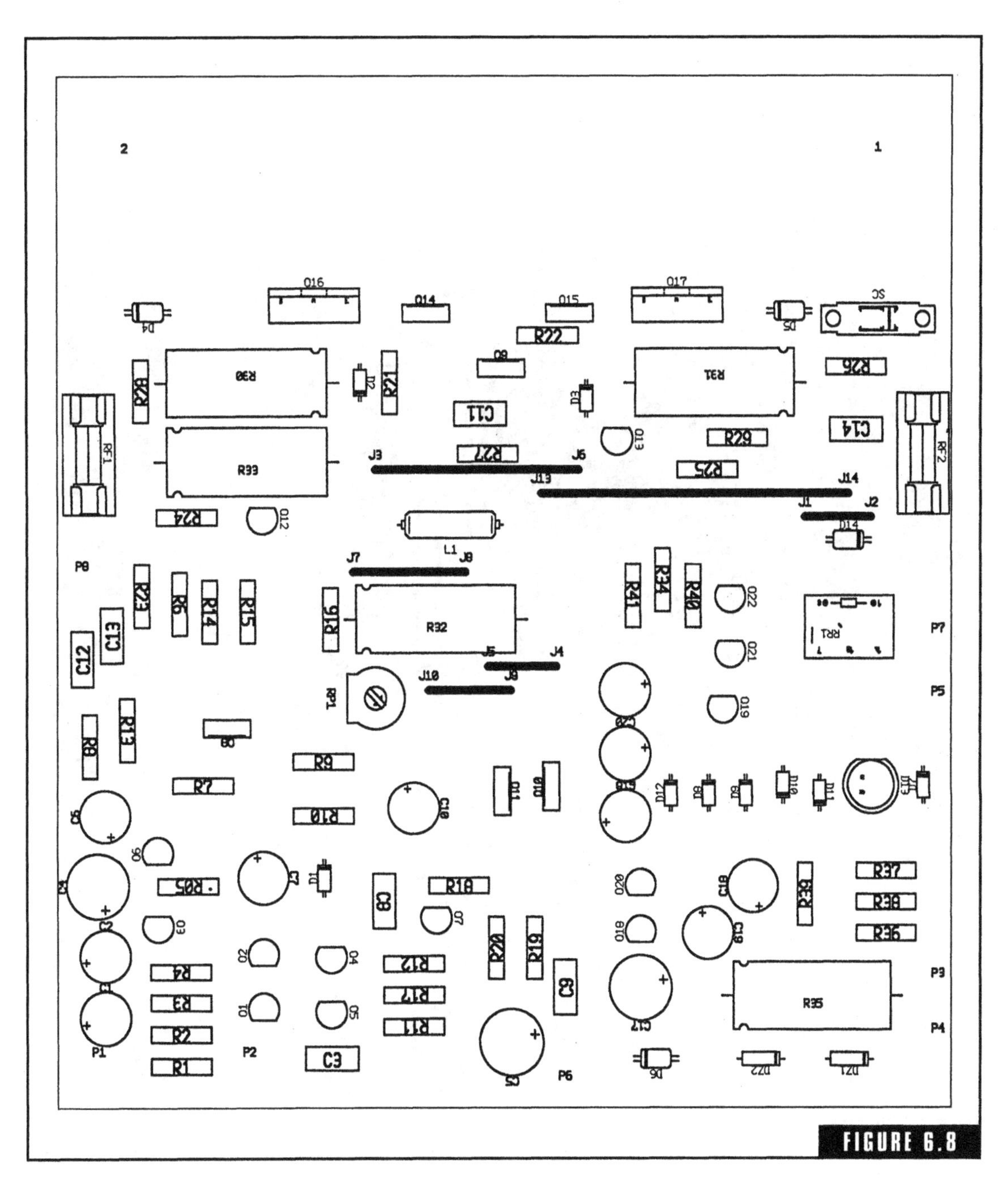

FIGURE 6.8

Top layout view of the Fig. 6-6 amplifier.

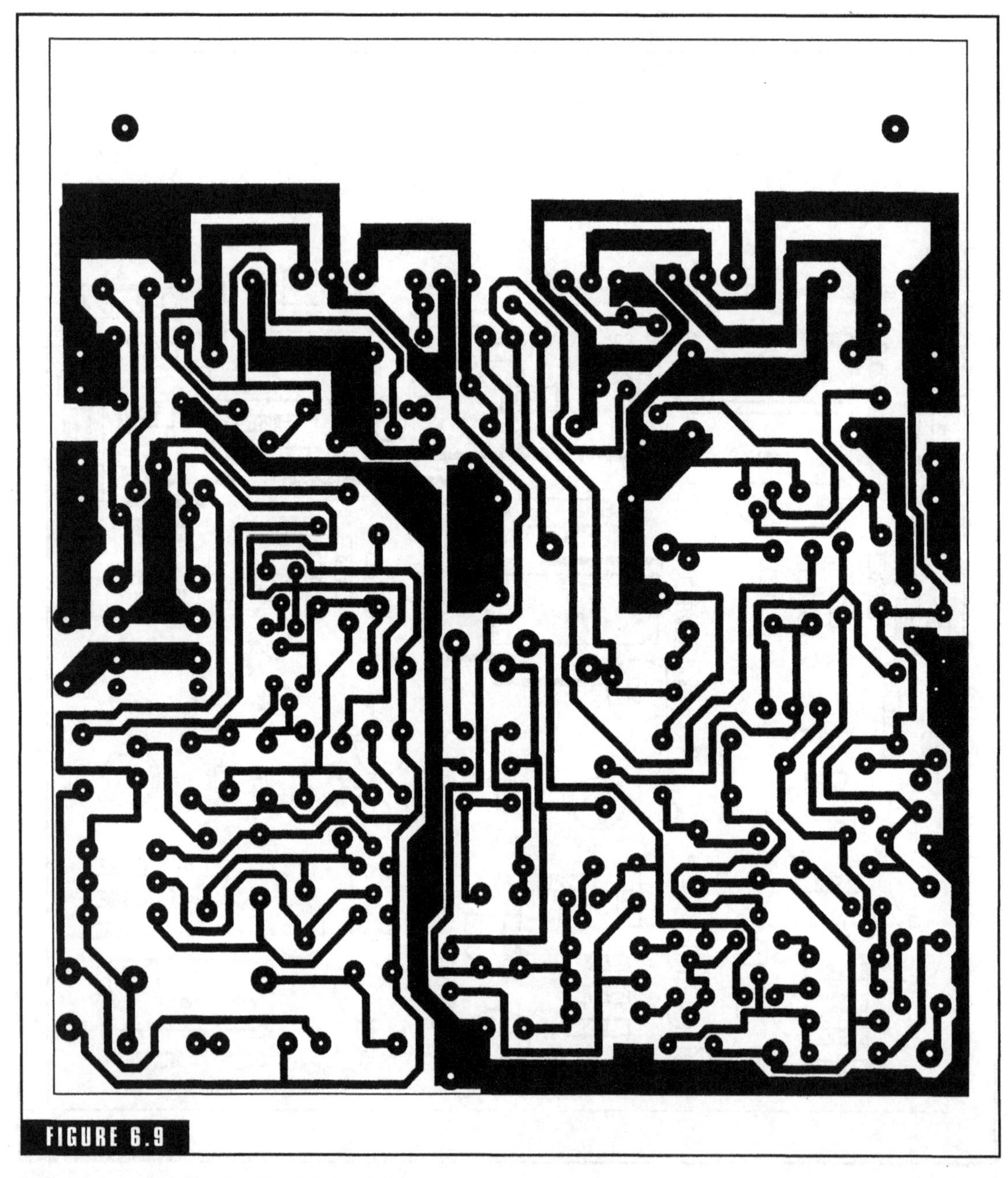

FIGURE 6.9

Reflected artwork for the Fig. 6-6 amplifier.

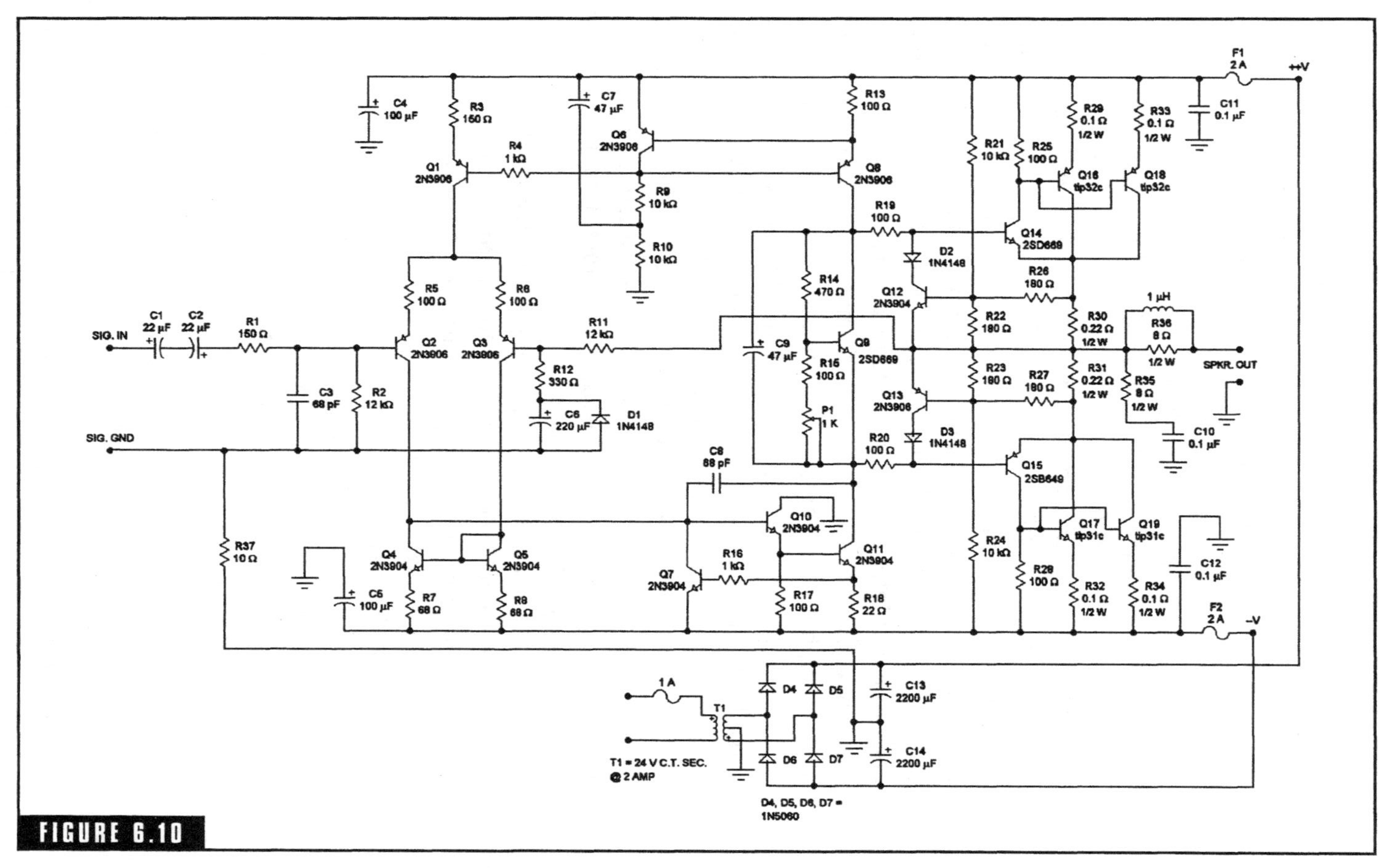

FIGURE 6.10

A high-performance 12-watt RMS amplifier design.

Believe it or not, this design will perform very well, which is a testament to the fact that audio power amplifier performance is more a matter of design techniques than component quality (please don't take that statement to mean that I'm against using good-quality components). I didn't spend a great deal of time analyzing all of the quality aspects of this amplifier, but the THD performance was better than 0.01% @ 10 kHz and the high-frequency bandwidth was in excess of 120 kHz. The only less-than-optimum performance parameter was the SNR, which came out to about −83 dB. However, I'm not surprised, because I mounted the unshielded power transformer and power supply directly to the same PC board containing the amplifier electronics (only a few inches away). In any event, I sent the completed prototype and design information to my friend in Denmark, and he was very pleased with it. Of course, his multimedia speakers were no longer good enough for the power amplifier, but that's another story.

I included the Fig. 6-10 design in this textbook for several reasons. First, there is the obvious chance that someone may have an application for a high-performing 12-W amplifier. Secondly, it occurred to me that some of my inexperienced readers might want to try this design as a "first project." (There sure wouldn't be much of a monetary risk involved.) And finally, this design incorporates two design techniques that have not yet been discussed.

You'll note that the topology of the Fig. 6-10 amplifier is very similar to the previous amplifier designs except for the output stage. First, note that the output device polarities (i.e., NPN or PNP) have been reversed, with PNP output devices amplifying the "positive" half-cycles, and NPN devices amplifying the "negative" half-cycles. In addition, the orientation of the output devices has been reversed, with the emitters going to the power supply rails, and the speaker load being driven by the collectors. This output stage configuration is called a *complementary-feedback* design, or *Sziklai design* (or *Sziklai pairs,* after the inventor). Complementary-feedback designs have been very popular for a long time, so let us take a few moments to discuss them.

Referring to Fig. 6-10, note that Q14 and Q15 (i.e., the predriver transistors) are configured in basically the same manner as in the previous *emitter-follower* output stage designs. However, they incorporate collector resistors (R25, R28), and the output signal is taken from the collectors instead of the emitters. This means their outputs are inverted. Likewise, the output transistors (Q16, Q17, Q18, and Q19) receive their inputs from their bases, but the output (i.e., the speaker output) is taken from their collectors, so the signal is inverted again. Therefore, since the signal is inverted *twice,* it is applied to the speaker in its original noninverted form. Going back to Q14 and Q15, their emitters are tied directly to the collectors of the output transistors, which means that the *total output of the output transistors will be applied as negative feedback to the predriver transistors.* In other words, a complementary-feedback output stage design provides the benefit of a 100% *local negative feedback loop* for linearization purposes. This all boils down to an overall THD performance improvement. In very rough terms, a high-quality amplifier with an emitter-follower output stage may measure out to 0.005% THD @ 1 kHz. If this same amplifier incorporated a well-designed complementary-feedback output stage, the THD performance may drop to 0.001% under identical conditions. In amplifier topologies of somewhat inferior design, the improvement may be more significant.

Another interesting feature of complementary-feedback designs (which is really an advantage) is the operation of the V_{BIAS} circuit relative to temperature tracking. Since there is a 100% local negative feedback loop inherent to the output stage design, this feedback also applies to the variance of leakage currents in the output transistors as their temperature varies. Consequently, the temperature characteristics of the output transistors are automatically compensated for (by the negative feedback loop). Temperature compensation is required for the predrivers, however, since their temperature variations are essentially amplified by the loop characteristics. Thus, the V_{BIAS} transistor must thermally track the temperature of one (or both) of the predriver transistors. In the Fig. 6-10 design, I physically mounted Q9 in a "back-to-back" configuration with Q14, assuring accurate thermal conductivity and thermal tracking. The temperature tracking method of complementary-feedback designs is advantageous because the temperature variations of the predriver transistors are not as extreme as those inherent to the output devices. Unfortunately, this advantage is offset by the disadvantage of the V_{BIAS} adjustment being much more critical and "touchy."

Occasionally, you may come across an audio power amplifier design incorporating some degree of voltage gain as a function of the output stage. In such cases, the output stage design is always a complementary-feedback arrangement, with a voltage divider network inserted between the emitters of the predrivers and the collectors of the output devices. In other words, the local negative feedback loop is adjusted so that the percentage of negative feedback is *less than* 100%, thus providing a voltage gain greater than unity. Personally, in regard to conventional discrete-component power amplifier design, I have never found a good reason to incorporate voltage gain in the output stage.

So why don't all audio power amplifiers incorporate complementary-feedback output stages? One reason relates to stability concerns. Complementary-feedback designs are more prone to instability problems (it is critical to ensure that the predriver transistors incorporated into complementary-feedback designs are very low-capacitance types).

Another disadvantage is their susceptibility to rapid self-destruction when driven at high frequencies (around 80 kHz). A third fault is their tendency to produce higher levels of crossover distortion (in comparison to emitter-follower designs) during low-volume conditions. Actually, there is a fairly long list of advantages and disadvantages associated with both emitter-follower and complementary-feedback designs, but it all boils down to a matter of personal choice of the designer. Both output stage designs are capable of excellent *audiophile-quality* performance if well designed and skillfully constructed.

A final point I wanted to mention regarding the Fig. 6-10 design relates to the function of resistor R37. Note how this resistor has a low resistance value (i.e., 10 ohms), and it is used to connect the *signal ground* to the high-quality ground point (i.e., the junction of the power supply filter capacitors). This is a hum reduction technique used in many amplifier designs. The low resistance value of R37 is not enough to affect the signal grounding properties significantly, but it is high enough to isolate most of the ripple and injection signals that could be riding on the circuit grounds from mixing with the signal ground, thus reducing hum and noise problems.

Figure 6-11 is a "big-brother" version of the previously discussed Fig. 6-10 amplifier. It is obviously a complementary-feedback design of the same fundamental topology. However, I added a few refinements, such as two-pole compensation (the C8, C9, R13 network) and the V_{BIAS} stabilizing resistor (i.e., R16). Both of these refinements have been discussed in previous emitter-follower designs. The only remaining topics to discuss are performance issues and a few of the physical construction details.

If you're into the "numbers game," the Fig. 6-11 amplifier is capable of extraordinarily low distortion performance, typically in the range of 0.002% THD (or even a little lower). The worst-case THD performance is about par with the previously detailed Fig. 6-6 emitter-follower topology (about 0.01%). I might mention at this point that you will definitely not be capable of perceiving any sonic differences between the Fig. 6-6 design and the Fig. 6-11 design—their worst-case distortion performance is about 30 times lower than human perception levels. The Fig. 6-11 design will provide better "overall" distortion performance, but the Fig. 6-6 design is a little more stable and forgiving of construction shortcomings and component types.

The Fig. 6-11 amplifier is rated for 120-W RMS output power into 8-ohm loads, and it is stable under all conventional speaker loading. It is designed to operate from dual-polarity rail voltages between 40 and 55 V. The input sensitivity is approximately 800 mV, and the input impedance is 12 kohm. If good construction techniques are followed, the SNR performance should be better than −100 dB.

The V_{BIAS} transistor, Q9, is mounted back to back on the same small TO-220 heatsink as used for cooling the Q14 predriver transistor. Refer to the Fig. 6-12 layout illustration to see how this is accomplished. L1 is formed by winding 16 turns of #14-AWG magnet wire around a 1/2-in form. Two main heatsinks are required for this amplifier project, with the same basic design and thermal ratings as used in the Fig. 6-2 amplifier (the Fig. 6-2 amplifier only required "one" such heatsink). Complementary-feedback amplifier designs are the only type of *bipolar design* wherein it is acceptable to mount the output transistors on separate heatsinks. This is due to the fact that we monitor the temperature of the predriver transistors for temperature tracking purposes, while the local negative feedback loop inherent to the output stage maintains the proper quiescent levels for the output transistors. You'll note that transistors Q15, Q11, Q10, Q8, and Q1 also require small TO-220 heatsinks for dissipation purposes.

Referring to the layout illustration of Fig. 6-12, observe how the "flat faces" of transistors Q2 and Q3 are opposite to each other, and the same holds true for transistors Q4 and Q5. Audiophiles traditionally mount differential transistor pairs and current mirror pairs in such a manner that their cases can be tied together, face to face, and held in place with a small tie wrap or crimped metal ring. The motivation is to force the temperature of the devices to track each other, so that a temperature imbalance can never occur between the matched pairs. The PC board design for the Fig. 6-11 amplifier allows this to be easily accomplished, as shown in the Fig. 6-12 illustration. In reality, the

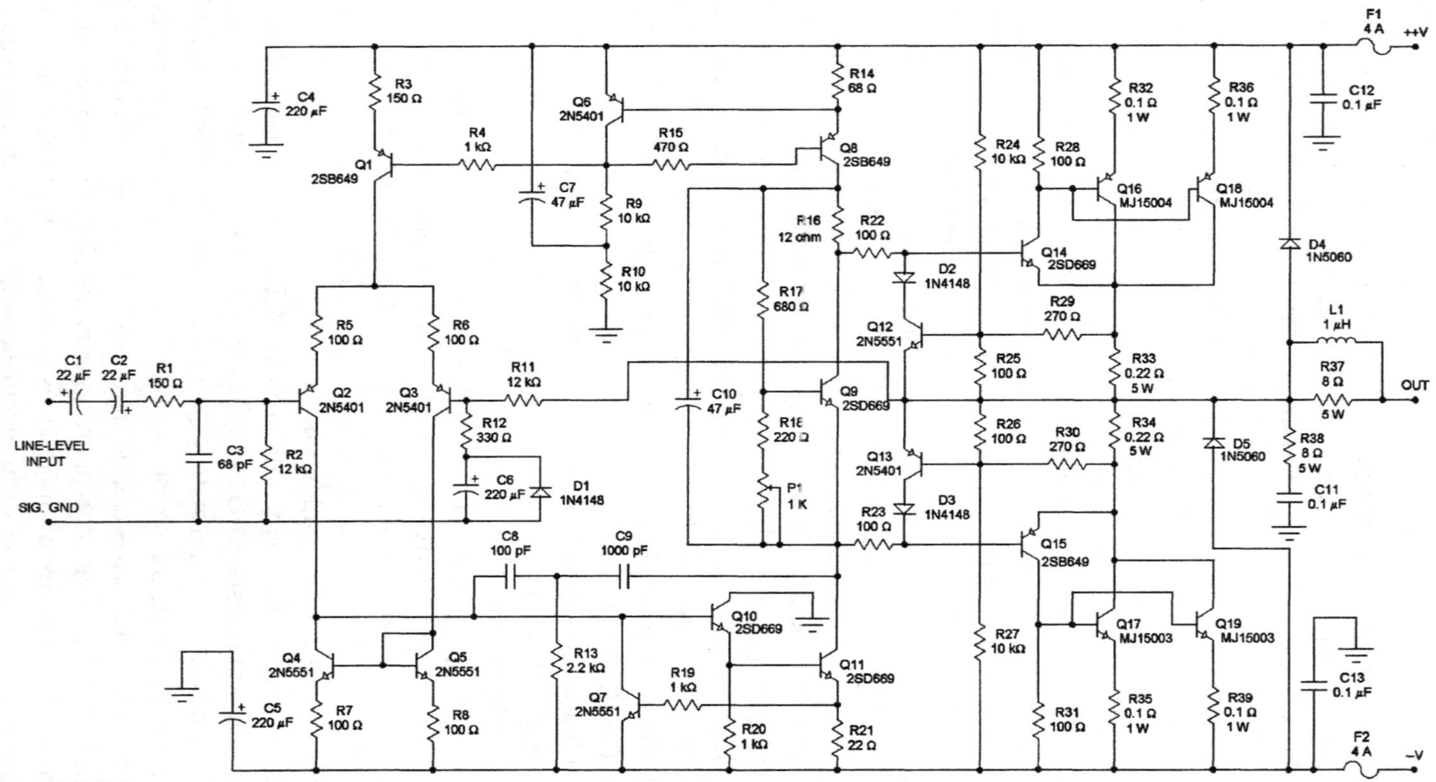

FIGURE 6.11 A very high-performance complementary-feedback BJT audio power amplifier.

The Completed Fig. 6-11 complementary-feedback audio power amplifier. The small rectangle of PC board material mounted to the top of the dual heatsinks is for mechanical support.

An "inside" view of the Fig. 6-11 audio power amplifier."

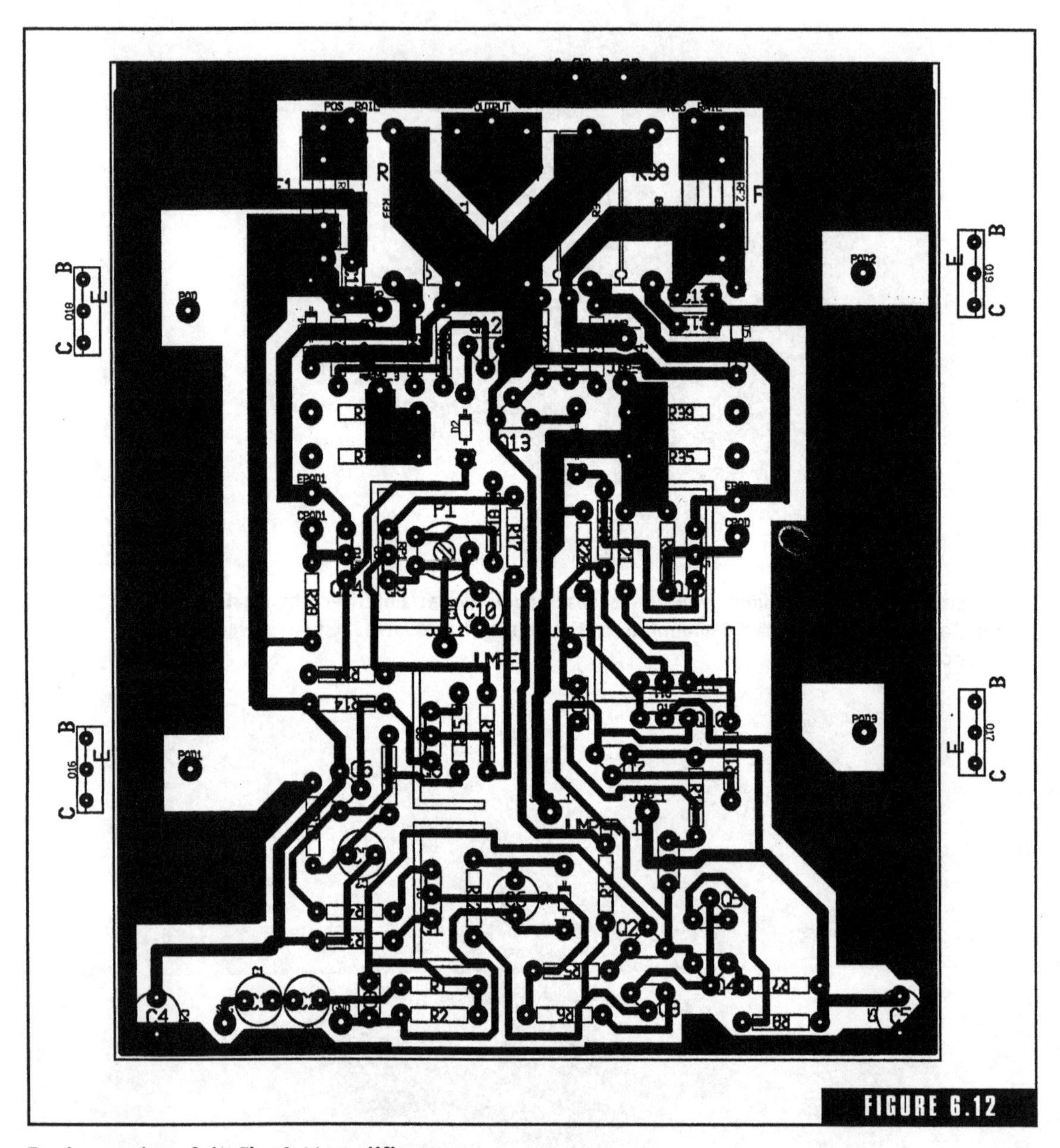

FIGURE 6.12

Top layout view of the Fig. 6-11 amplifier.

self-correcting characteristics of the Fig. 6-11 design make such a technique unnecessary, but it "looks nice," and "audiophiles like to see it done that way."

The remaining construction details for the Fig. 6-11 amplifier are fairly self-explanatory by studying the layout and artwork illustrations of Figs. 6-12, 6-13, and 6-14. The layout illustrations show the output devices as having "flat-pack" case styles, but the specified output devices in the Fig. 6-11 schematic are TO-3 packages. I inserted the flat-pack case styles as a means of showing the lead connections, so the illustrated case styles for the output transistors can be disregarded.

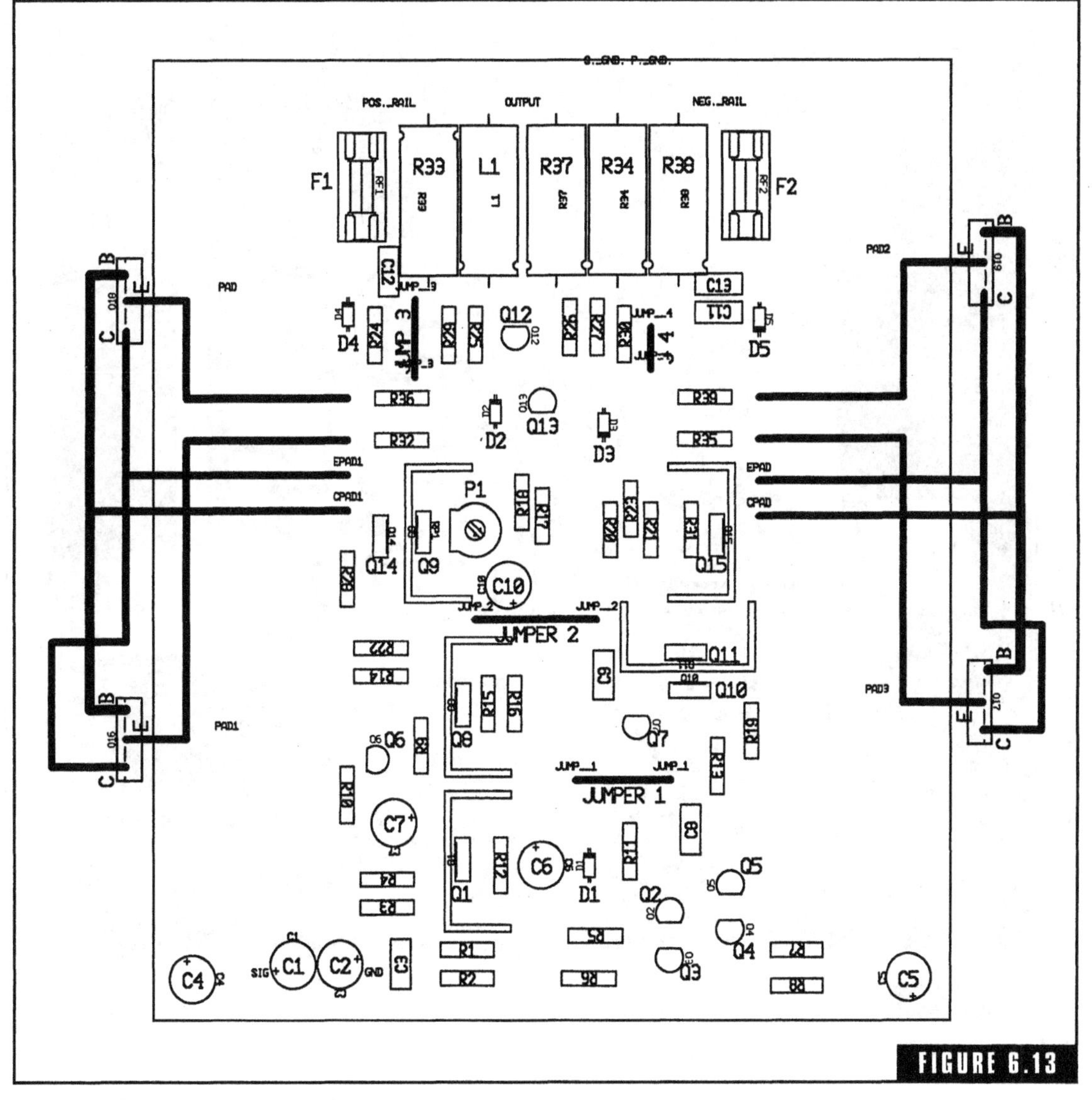

Top view of the Fig. 6-11 amplifier.

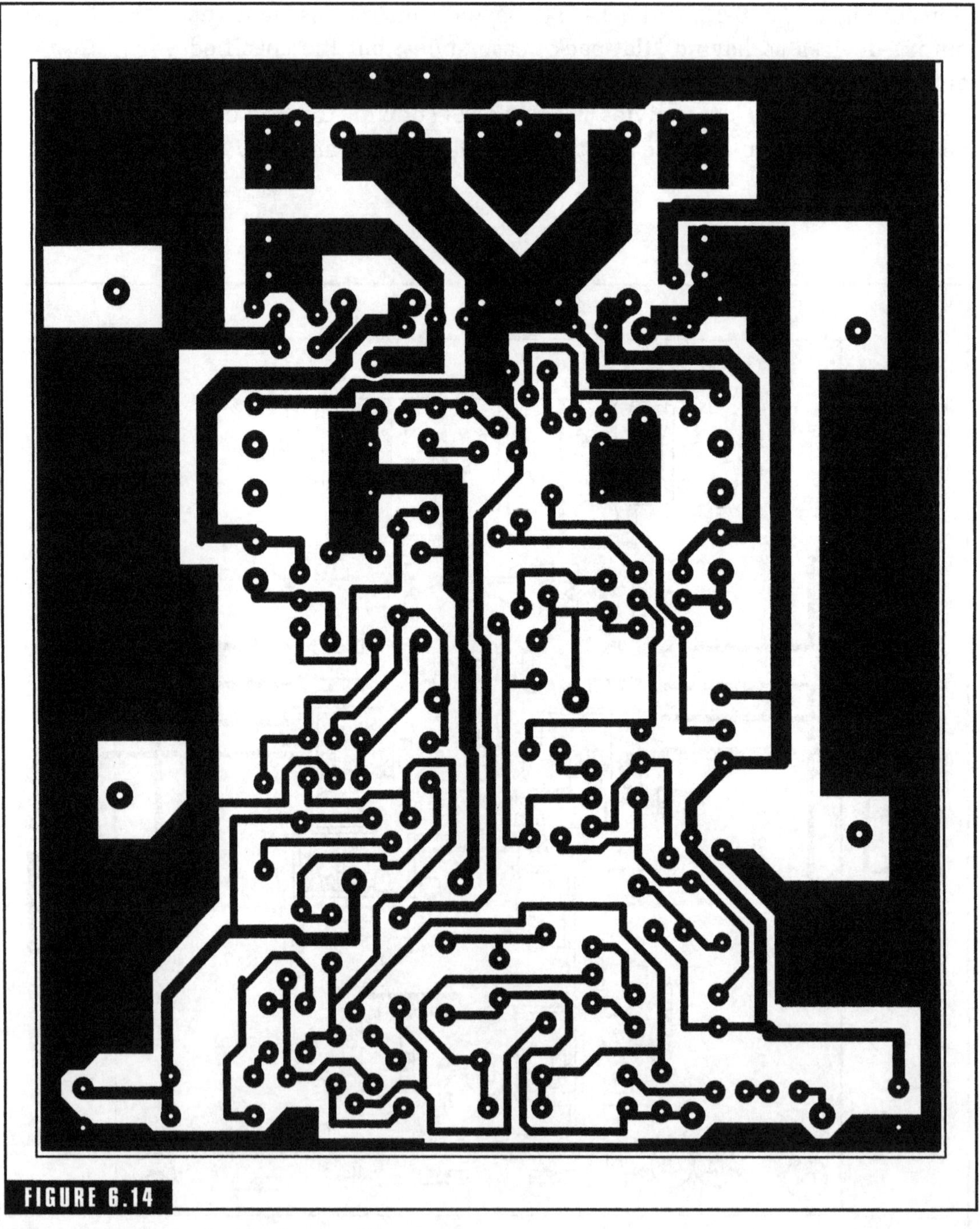

FIGURE 6.14

Reflected artwork for the Fig. 6-11 amplifier.

Figure 6-15 is the schematic illustration for my favorite BJT (bipolar junction transistor) audio power amplifier design. It is obvious from the increased complexity that there are a few more refinements added. In terms of fundamental architecture, it is a *fairly conventional* emitter follower topology.

Beginning with the back-to-back zeners incorporated into the input stage, these are protection diodes, which keep the input voltage from exceeding the maximum 5-V V_{BE} parameter of the input differential transistors (Q2 and Q3).

Another addition is the input cascode stage, consisting of Q4, Q5, R6, D3, and C5. There are two purposes for incorporating this cascode stage. First, it isolates the input differential transistors (Q2 and Q3) from the negative power supply rail. Thus, the voltage drop across the differential input transistors is only about 25 V, allowing *low-noise, high-gain* differential input transistors to be used. The end result is an overall improvement in the SNR of the amplifier (approximately −110 dB or better). The second improvement relates to the PSRR (power supply rejection ratio) of the amplifier. The PSRR determines how much of an effect an adverse power supply anomaly will have on the performance of the amplifier. The most sensitive stage relative to power supply variances is the input stage. Referring to Fig. 6-15, note how the differential input stage is effectively isolated from the positive power supply rail by the conventional constant current source. The cascode stage effectively isolates the input stage from the negative power supply rail as well, so the input stage is totally isolated from both power supply rails, and PSRR is significantly improved. There is also an increase in the open-loop gain provided by the cascode stage, but this is of relatively little importance.

The overload and short-circuit protection circuit used in the Fig. 6-15 design is also a different type than previously detailed. It is commonly referred to as a *multislope protection circuit.* Note that there is no longer any power supply reference provided to the protection circuitry. A multislope protection circuit utilizes a resistor/diode network (referenced to circuit common) to sense the output current flow (via the voltage drop across R34, R35, R36, R37) relative to the output rail voltage. As the output rail voltage rises, the allowable output current flow increases (high current output at low output rail voltage is indicative of a short-circuit condition, and such current/voltage conditions would

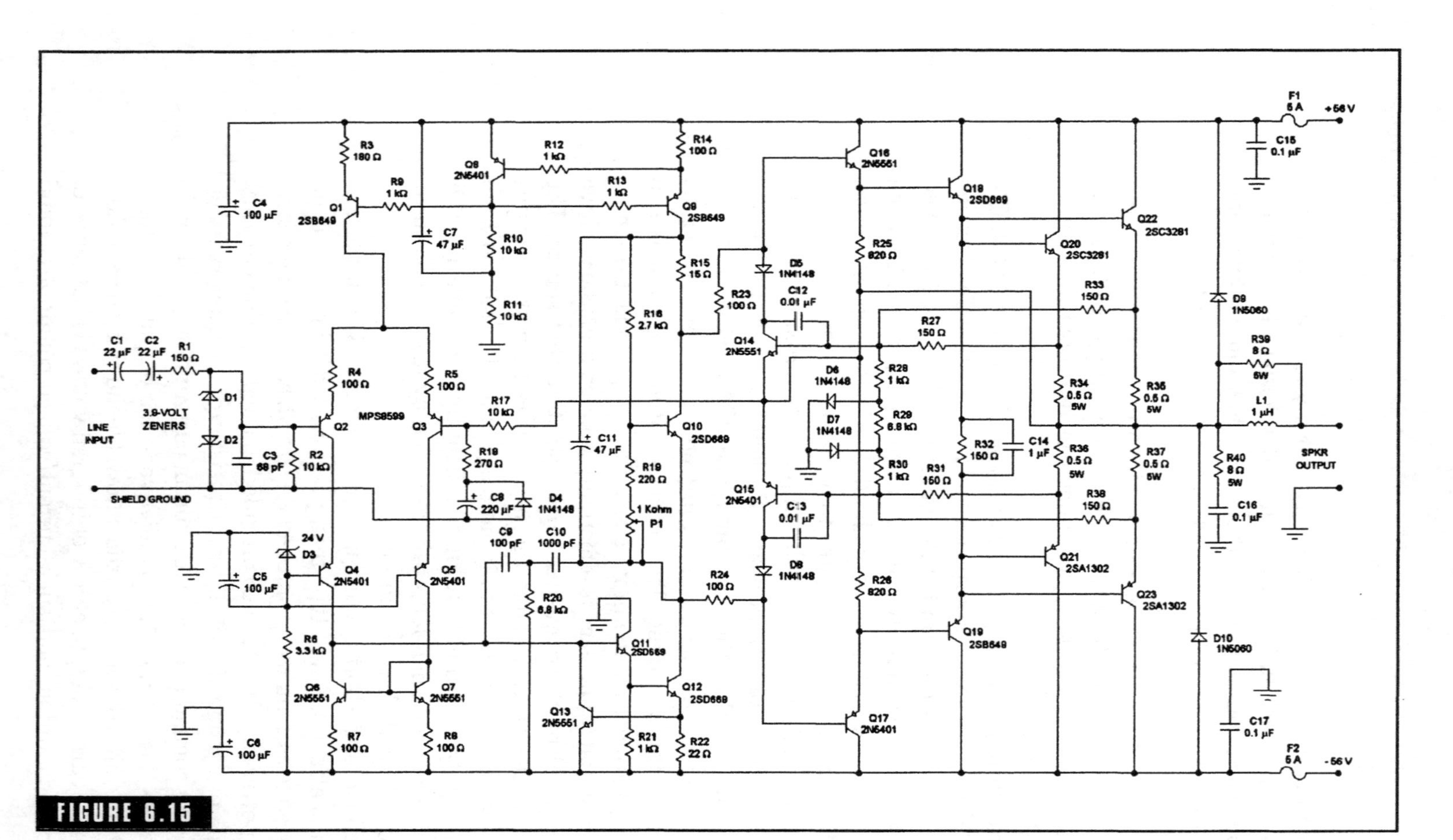

FIGURE 6.15

My favorite BJT audio power amplifier design.

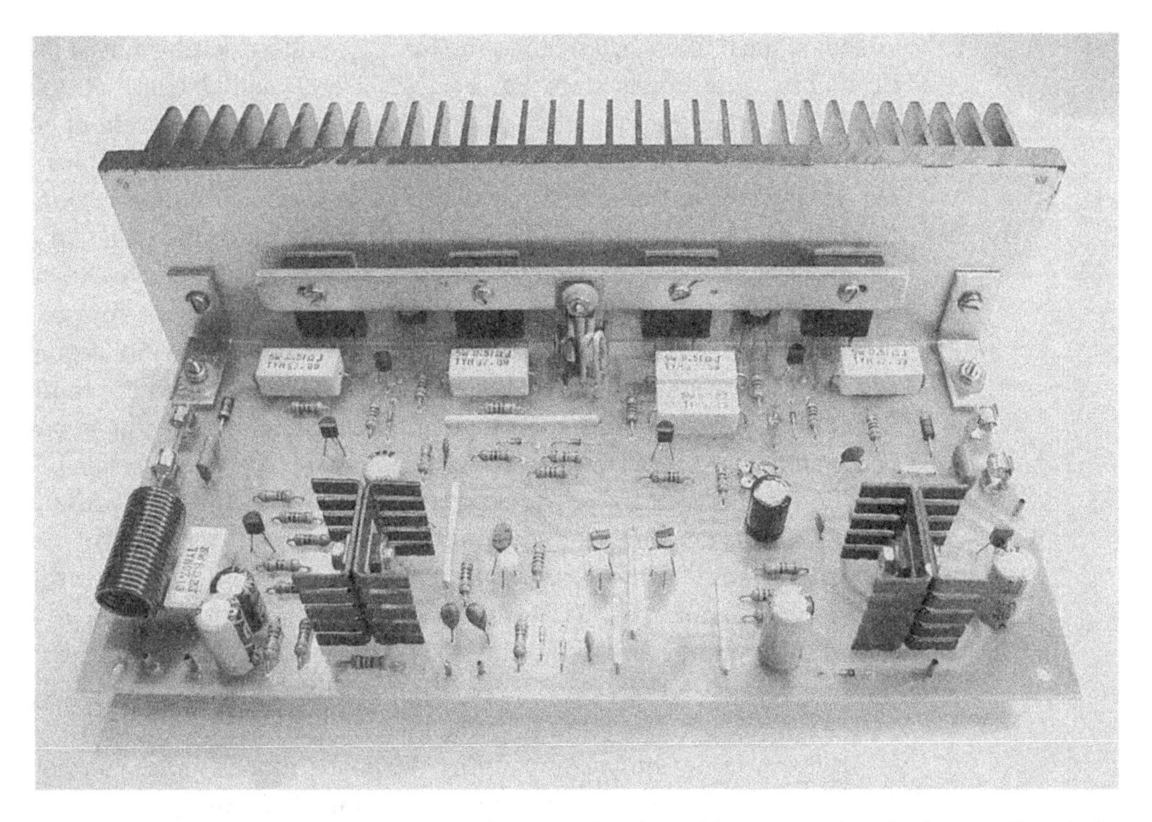

The completed Fig. 6-15 audio power amplifier. Note the thermal bar mounted to the front surface of the output transistors, with the bias transistor (Q10) mounted to the center of the thermal bar.

exceed the secondary breakdown parameters of the output devices). Such a circuit provides a fairly accurate reproduction of the output transistors' SOA (safe operating area) curves. In contrast to many higher-powered BJT amplifier designs, the Fig. 6-15 design monitors the output current flow from *all four* output devices. This is a preferable technique, since fairly high transient currents can exist during power-on/power-off sequences, and an accurate current balance between the paralleled output devices cannot be depended on during such "settling" periods.

The final addition to the Fig. 6-15 design is the buffer stage (i.e., Q16, Q17, R25, and R26) providing a higher input impedance for the voltage amplifier stage to interface with (reducing "loading distortion" on the voltage amplifier stage) and additional drive current for the predriver transistors (Q18 and Q19). The additional semiconductor junctions added by the buffer stage will require the V_{BIAS} voltage setting to be a little higher. This buffer stage has very little detrimental effect on the *phase margin* of the overall amplifier circuit. (The phase margin of an amplifier is the difference between the high-frequency unity-gain phase lag, in degrees, and the required phase lag to cause sustained oscillations due to Nyquist instability factors. As a rule-of-thumb generality, we never want the unity-gain phase lag to exceed −180 degrees. Since the unity-gain phase lag of the Fig. 6-15 amplifier design is approximately −150 degrees, subtracting this figure from the −180 degree *assumed maximum* renders a phase margin of 30 degrees. You can think of the term *phase margin* as being synonymous with the term *phase shift safety margin,* because, in reality, this is exactly what it means.)

The Fig. 6-15 audio power amplifier is rated for 150-W RMS into 8-ohm loads, operating from nominal dual-polarity 56-V rails. You can "push" these rail voltages a little without any danger of exceeding the component parameters. THD performance comes out to 0.004% @ 1 kHz @ 0 dB, increasing to about 0.008% under worst-case conditions. Bandwidth is 5 Hz to 80 kHz, the input sensitivity is approximately 800 mV RMS, and the input impedance is 10 kohm.

By examining the accompanying PC board illustrations, you can see that the overall construction of the Fig. 6-15 design is pretty straightforward. The output coil, L1, is formed by winding 18 turns of #14-AWG magnet wire around a 1/2-in-diameter form. As in the case of the Fig. 6-6 amplifier design, the V_{BIAS} transistor, Q10, is physically mounted to a *thermal bar,* which is mounted to the external faces of the output transistors. If you don't want to fabricate a thermal bar to fit over the output transistors, Q9 can be mounted to the main heatsink, but there will be some thermal delays created. As illustrated in Fig. 6-16, transistors Q1 and Q9 are mounted back to back on two back-to-back TO-220 heatsinks. This makes a very thermally stable, physically strong configuration. Q11 and Q12 are mounted in the same fashion. The predriver transistors, Q18 and Q19, mount directly to the flat facing of the main heatsink, as do the output transistors.

For the main heatsink of the Fig. 6-15 amplifier, I used a flat-faced 10 in (W) × 3 1/4 in (H) × 1 1/2 in (D) vertically finned heatsink, with a thermal rating of 0.4°C/W. The PC board is designed so that the differential, cascode, and current mirror transistors can be squeezed together (i.e., face to face) and securely held with a small tie wrap. The remaining construction details are illustrated in Figs. 6-16, 6-17, and 6-18.

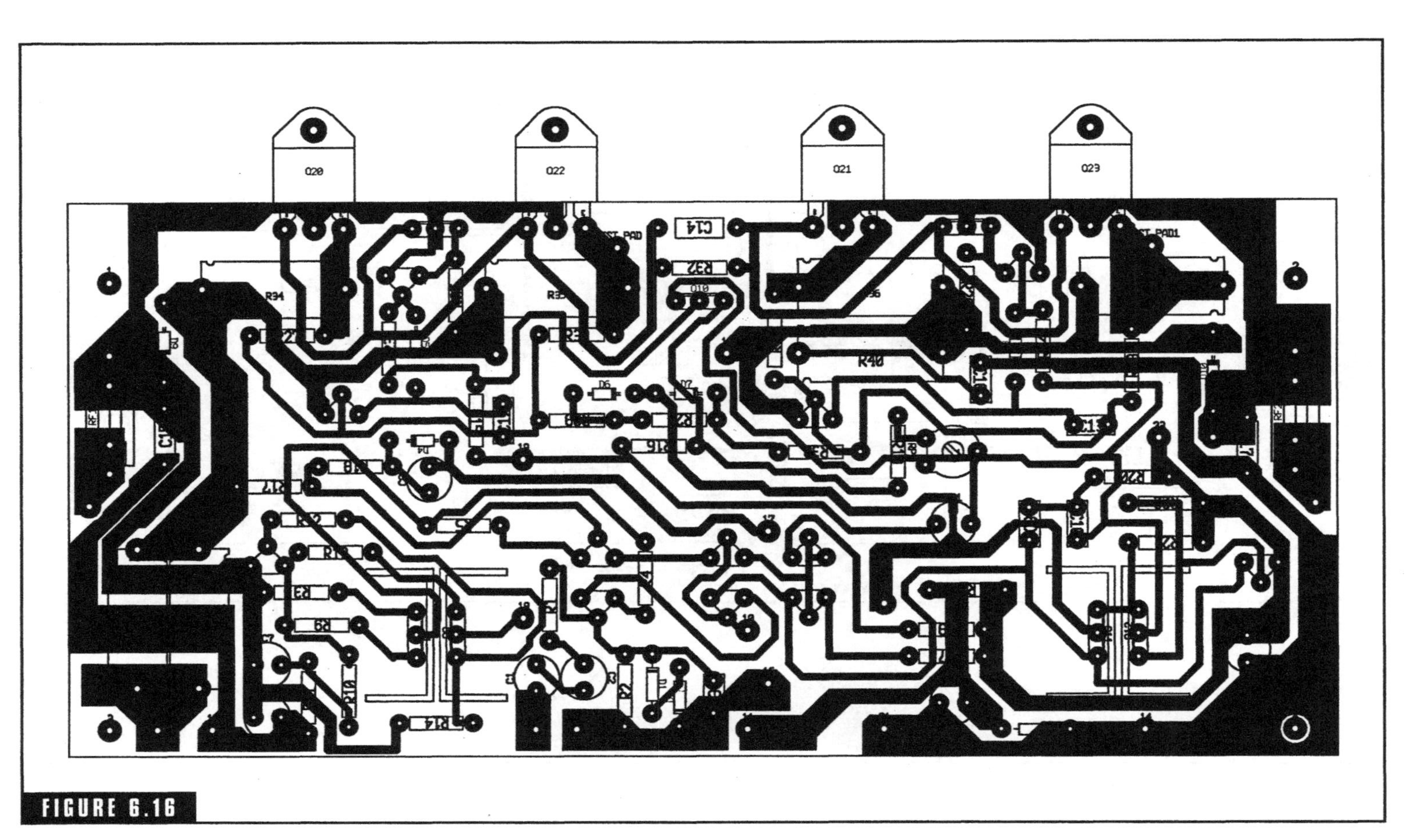

FIGURE 6.16

Top layout view of the Fig. 6-15 amplifier.

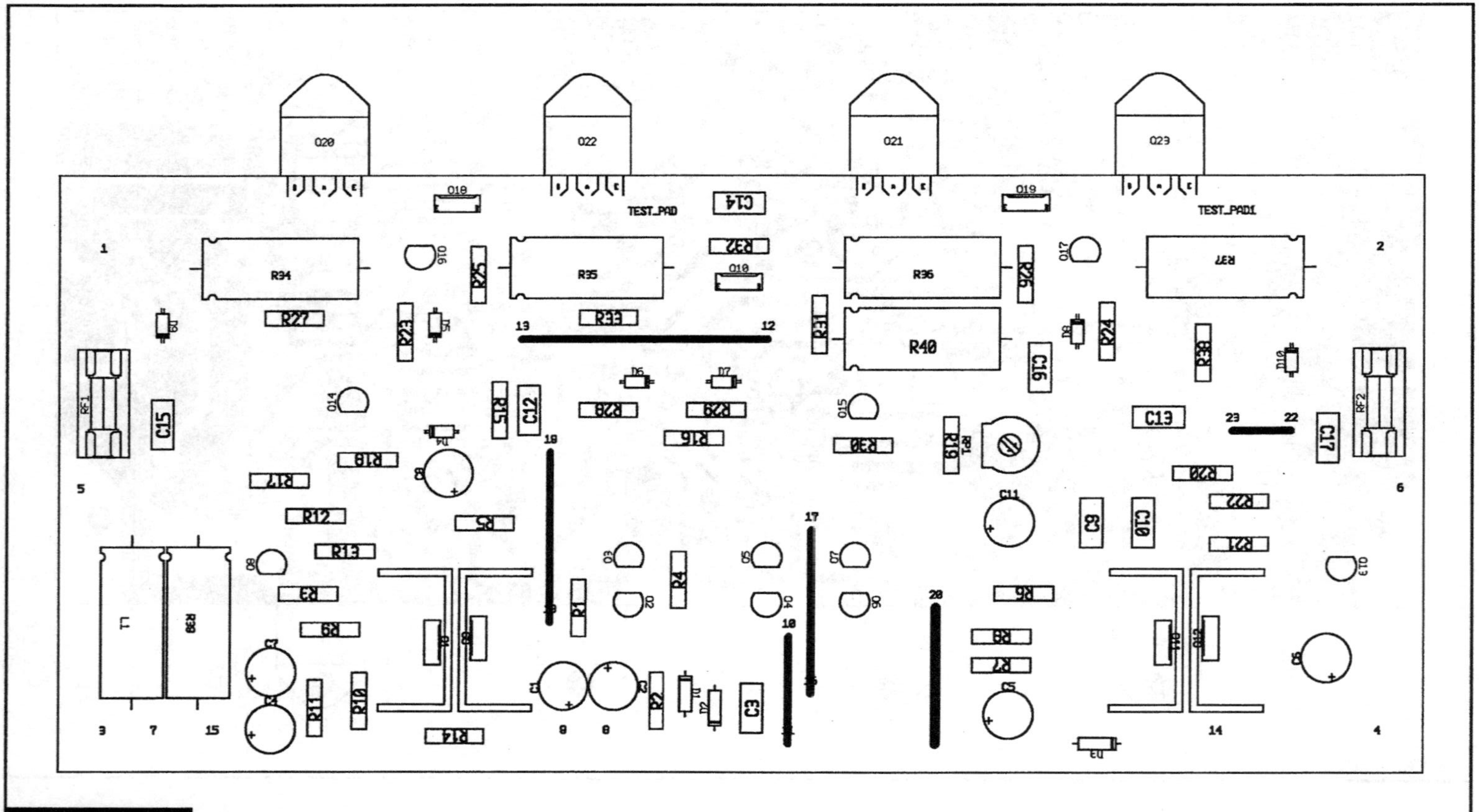

FIGURE 6.17

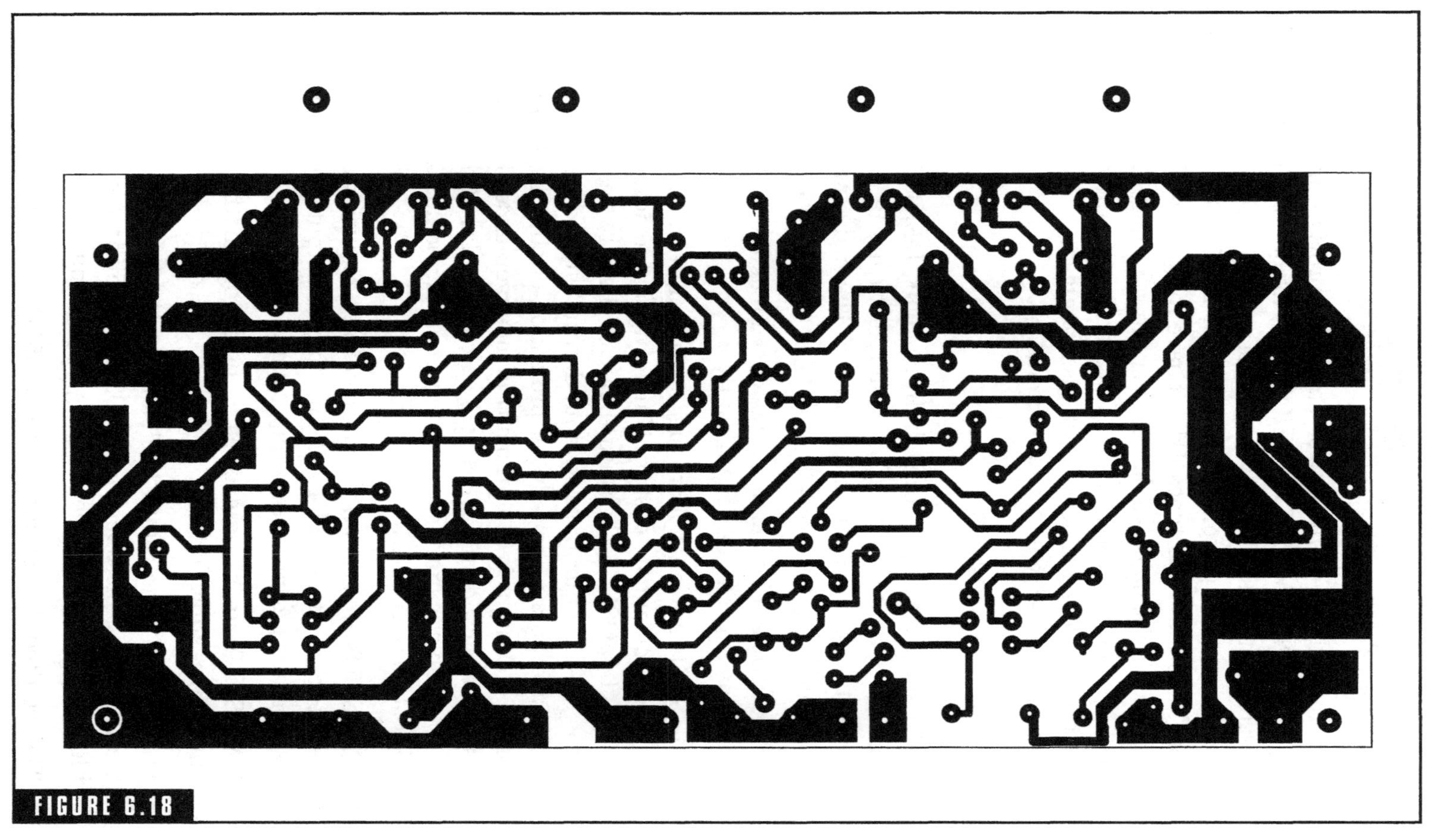

FIGURE 6.18

Reflected artwork for the Fig. 6-15 amplifier.

The final three audio power amplifier projects in this chapter incorporate *lateral MOSFETs* (L-MOSFETs) for output devices. I make it no secret that L-MOSFETs are my favorite output device types, and I believe the overall performance results of L-MOSFET power amplifiers are superior to BJT types. I don't have room in this single chapter to delve into an exhaustive debate over the advantages and disadvantages of L-MOSFETs versus BJTs. However, I will provide a short listing of the *major* advantages and disadvantages of L-MOSFETs, as compared to power-type BJT output devices.

Advantages of L-MOSFETs

1. When supplying significant drain currents, L-MOSFETs take on a *negative temperature coefficient,* automatically limiting drain current as the case temperature increases. Thermal runaway is impossible, and V_{BIAS} temperature tracking is not necessary.
2. L-MOSFETs are immune to secondary breakdown, thus increasing reliability expectations significantly.
3. L-MOSFETs are much more rugged and forgiving; they are very difficult to accidentally destroy. Unlike bipolar devices, rail fuses will protect L-MOSFETs, because they can literally pass hundreds of amps for short durations without destructive effects. The only reason I incorporate short-circuit protection circuits in the following amplifier designs is to eliminate the inconvenience of replacing rail fuses if an accidental short or overload condition develops (after all, consumers aren't supposed to open the enclosures of modern solid-state equipment).
4. L-MOSFETs are primarily *voltage devices,* requiring almost negligible gate currents, which results in simpler output stage designs and less loading effects on the voltage amplifier stage.
5. L-MOSFETs are immune to switching distortion (there are no inherent charge-storage effects).
6. In the event of an L-MOSFET failure (which is rare), the failure is usually *soft.* Consequently, L-MOSFET amplifiers are less susceptible to *collateral damage* resulting from output device failure.
7. The required V_{BIAS} accuracy for L-MOSFET amplifiers is less critical.
8. L-MOSFETs have a much wider bandwidth capability.

Disadvantages of L-MOSFETs

1. L-MOSFETs require higher quiescent bias currents for optimum performance, resulting is slightly poorer efficiency. Also, the $R_{DS(ON)}$ resistance is higher than their bipolar counterparts, adding an additional measure of inefficiency.
2. The V_{GS} parameter is slightly higher for L-MOSFETs than the comparable V_{BE} parameter for bipolar transistors, which places a little more of the signal within the crossover region, adding to overall distortion levels.
3. The transconductance (i.e., gain factor) of L-MOSFETs is low, resulting in poorer inherent linearity.
4. The cost of L-MOSFETs is about two to three times higher than comparable BJT devices.

With all variables being equal, you can expect a well-designed L-MOSFET audio power amplifier to be a few millipercent higher in THD performance. For example, a high-performance L-MOSFET amplifier might measure 0.006% THD, in contrast to an equivalent emitter-follower BJT amplifier that measures 0.004% THD, or an equivalent complementary-feedback BJT amplifier that measures 0.002% THD (all evaluated at 1-kHz fundamentals and 0-dB levels). However, the MTBF (mean time between failure) rate of an audio power amplifier with an L-MOSFET output stage design will easily be *five times* better than BJT equivalents (probably closer to ten times better than complementary-feedback designs). This is assuming that the amplifier is treated in a typical manner (i.e., not overly abused, but implemented in a variety of applications and exposed to a few user errors throughout its lifetime). The MTBF rate will also vary depending on the rated output power of the amplifier. For medium-power applications, say up to 100-W RMS, BJT output stage designs are very reliable, but as you increase the power rating (automatically requiring numerous parallel BJT output pairs), the MTBF rate drops off radically. Therefore, in the cases of low- to medium-power applications, BJT output devices may be a better choice, but for high-output power requirements, L-MOSFETs are definitely a superior option. I might also mention that bipolar devices have a few inherent ills that surface in high-power, multiparalleled applications, which act to degrade the optimum THD performance. L-MOSFETs excel in high-power environments, so the overall THD performance of BJTs and L-MOSFETs in high-power applications are almost identical.

Figure 6-19 illustrates a high-performance subwoofer amplifier design. Actually, this amplifier will perform well in virtually any audio power amplifier application. The only reason I specify it for subwoofer applications is because there is a "better" L-MOSFET design choice for general-purpose applications (i.e., the Fig. 6-21 design, which we'll discuss shortly).

The overall THD performance of the Fig. 6-19 topology is very good, measuring out to 0.006% THD @ 1 kHz @ 0 dB. The worst-case THD performance is a little higher than the other designs in this book, measuring out to about 0.08%, but this is still well below human perception levels. Conventional L-MOSFET topologies of the past (both single-differential and fully complementary designs) have provided 1-kHz THD performance of about 0.03%, with worst-case performance close to 0.1% or 0.2%. The superior input stage and voltage amplifier stage designs of this Fig. 6-19 topology, coupled with higher open-loop gain values, provide significant improvements in THD performance. SNR should come out better than −110 dB if good construction techniques are implemented. Bandwidth is 5 Hz to 80 kHz, with an input sensitivity of 900-mV RMS and input impedance of 10 kohm. Utilizing the dual-polarity 56-V power supply rails, as illustrated, the Fig. 6-19 design is capable of about 135-W RMS into 8-ohm loads. It is safe to operate this design with up to 62-V rails for higher output power capability.

If you compare Fig. 6-19 with the Fig. 6-15 amplifier topology, you'll discover that the input and voltage amplifier stages are almost identical. I added capacitors C4 and C7 to the Fig. 6-19 design to further reduce any RF components from the DC power supply rails (RF interference signals are more of a problem with L-MOSFETs than BJTs, due to the much higher frequency response of L-MOSFETs). Also, you'll note that the V_{BIAS} transistor and its associated circuitry have been deleted, with a simple trimpot (P1) and stabilizing capacitor (C13) providing the required bias adjustment. No form of temperature compensation is required for L-MOSFET output stages. Also note that the same multislope protection circuitry is incorporated. The more conventional method of limiting excessive gate voltages in L-MOSFET designs is to incorporate a simple dual-zener, dual-diode clamping cir-

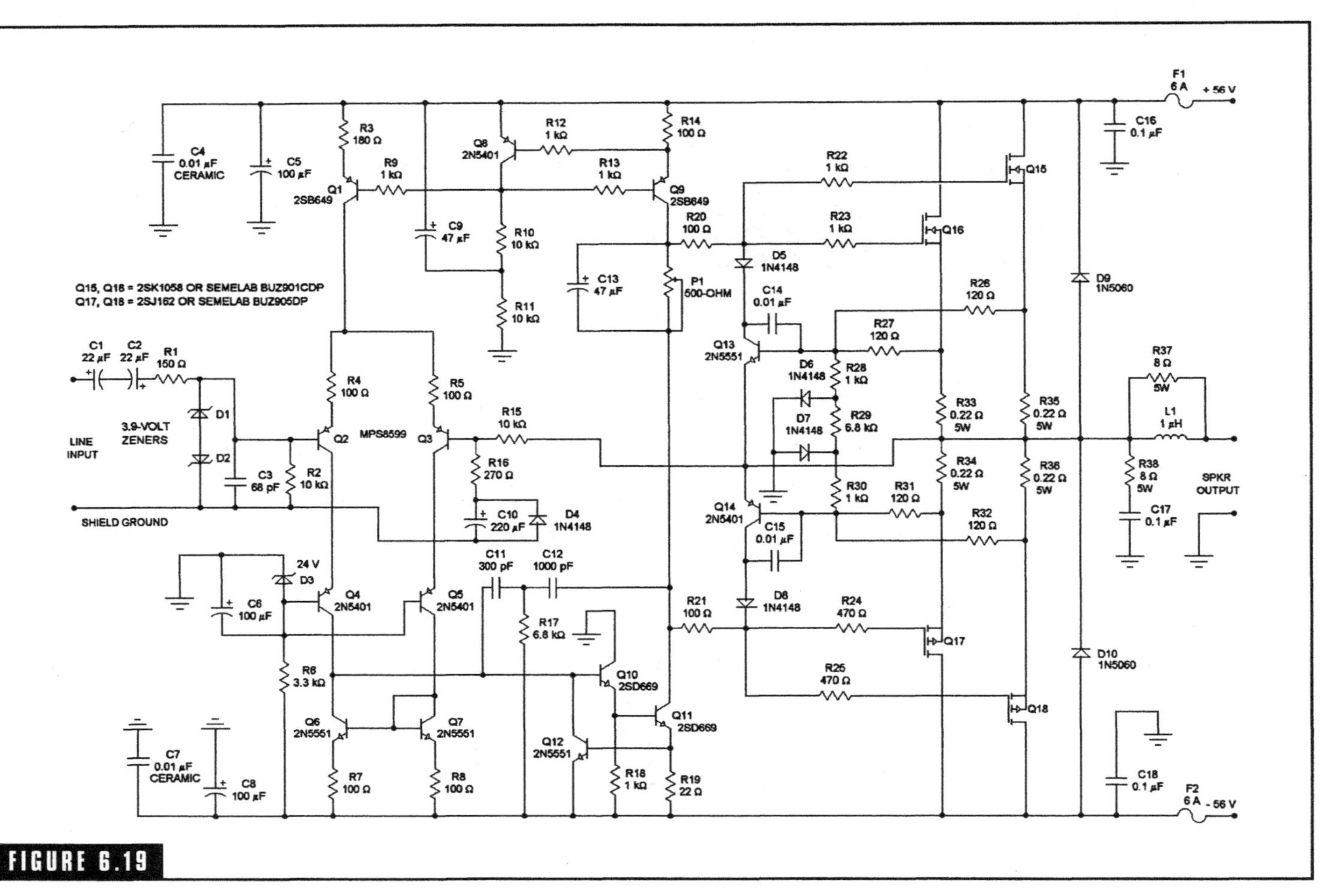

FIGURE 6.19

A nice L-MOSFET subwoofer power amplifier.

cuit. However, such circuits do not limit excessive output currents, which will result in blowing the rail fuses.

The L-MOSFET output stage design of Fig. 6-19 is commonly called a *source-follower* output stage design, which is analogous to the BJT *emitter-follower* equivalent. Note that the gate resistors for the N-channel devices (Q15 and Q16) are a different value than for the P-channel complements (Q17 and Q18). This is because the gate capacitance of the P-channel devices (i.e., about 900 pF) is higher than the gate capacitance of the N-channel devices (i.e., about 600 pF).

Regarding construction details, L1 is formed by winding 18 turns of #14-AWG magnet wire around a ½-in form. The heatsinking requirement for the L-MOSFET output devices is about the same as for the previous Fig. 6-15 design. Transistors Q1, Q9, Q10 and Q11 should be mounted to small TO-220 heatsinks. The length of the connection wiring to the L-MOSFET gate leads should be kept to minimum. If lead lengths longer than about 6" to 9" are required, it is best to solder the gate resistors directly to the L-MOSFET leads, and run the longer connection wiring from the opposite end of the gate resistors. The remaining construction requirements are the same as for any of the other previously discussed amplifier designs.

Figure 6-20 illustrates a lower-power L-MOSFET amplifier design, incorporating the L-MOSFET output devices in a complementary-feedback configuration with two predriver bipolar transistors. Note that the L-MOSFET device types are reversed, and they are placed "upside-down" in the circuit (i.e., the sources connect to the power supply rails). Such output stage configurations are often called *hybrid* designs, because they incorporate both bipolar and MOSFET devices within the stage design. As in the case of other complementary-feedback designs, the local feedback loop inherent to the output stage design provides for improved linearity over more conventional source-follower topologies.

You may notice that capacitors C18 and C17 have been added to the single-slope protection circuit. They are connected from the collector to the base of the protection transistors, Q12 and Q13. These should be ceramic disc types, and their function is to reduce instability effects during overload conditions. The installation of such capacitors within protection networks is a good technique to follow in almost all protection circuitry designs.

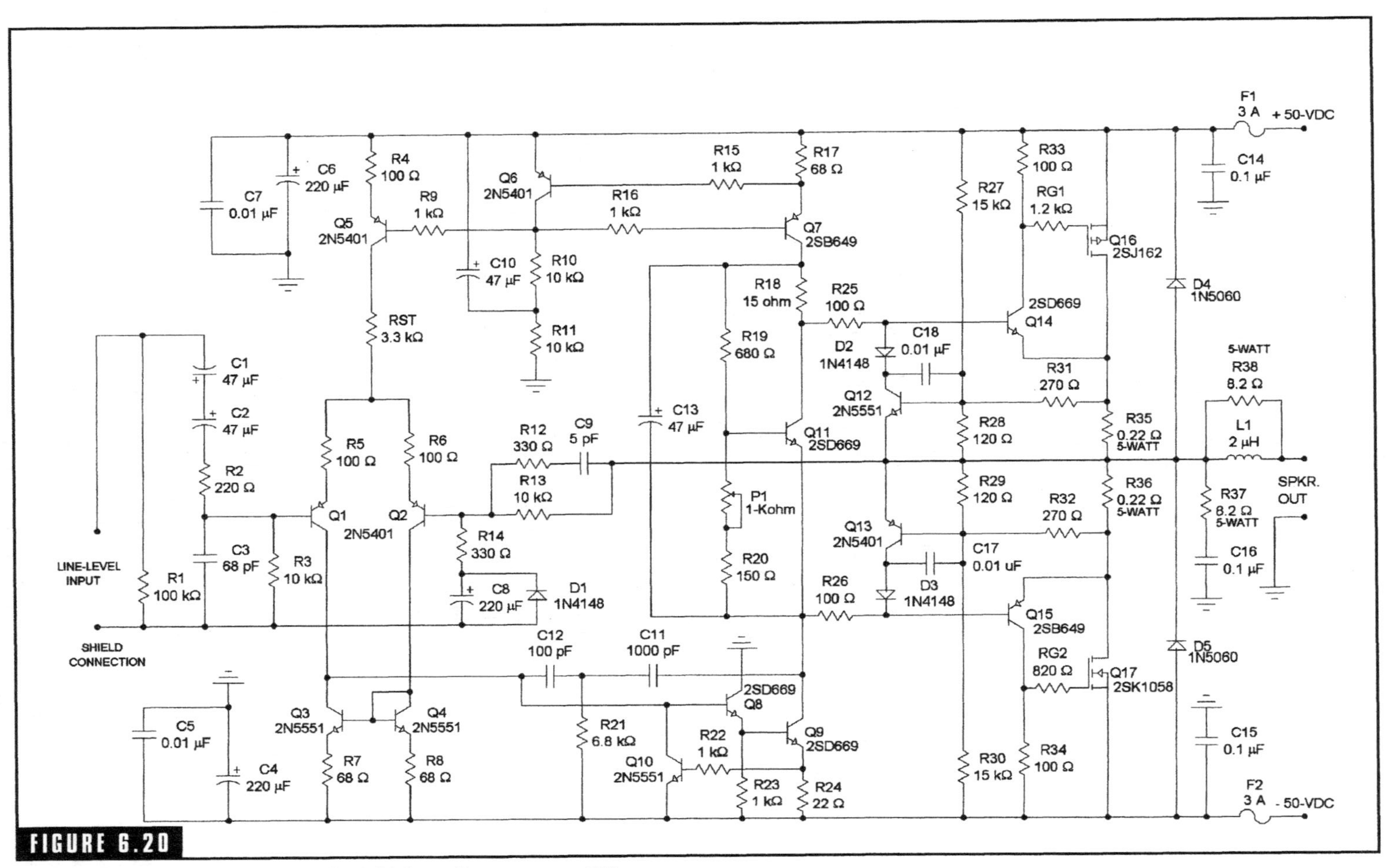

FIGURE 6.20

A complementary-feedback L-MOSFET high-performance amplifier.

The THD performance of this design is comparable to the best bipolar emitter-follower designs. The THD performance measured at 1 kHz @ 0-dB levels comes out to 0.003%, with worst-case performance at about 0.01%. The maximum output power is just slightly over 90-W RMS into 8-ohm loads. The SNR performance should be better than −100 dB (this is a simpler design, without the use of the input cascode stage and low-noise differential input transistors). The bandwidth is 5 Hz to 100 kHz, with an 800-mV input sensitivity and 10-kohm input impedance.

The construction details for the Fig. 6-20 amplifier are similar to the previous designs. Transistors Q7, Q8, Q9, Q14, and Q15 should be mounted to small TO-220 style heatsinks. Since bipolar transistors are used as predriver units, it is necessary to provide temperature compensation for the predriver transistors (this is to maintain good stability of the quiescent bias currents of the L-MOSFET output devices). Therefore, Q11 should be thermally coupled to one of the predriver transistors (either Q14 or Q15, or both). The output coil, L1, is formed by winding 18 turns of #16-AWG magnet wire around a 1/2-in-diameter form.

Finally, we come to Fig. 6-21, which is my favorite audio power amplifier design of *any type*. When you consider the overall performance characteristics of this particular topology, it becomes very difficult to surpass. The THD performance is excellent, measuring out to 0.005% @ 1 kHz @ 0 dB, with a worst-case THD performance of about 0.04%. Bandwidth is 3 Hz to 120 kHz, with an input sensitivity of 800 mV and input impedance of 10 kohm. The SNR can exceed −115 dB with careful attention to low-noise construction techniques. The topology, as illustrated, is capable of up to 200-W RMS output levels into 8 ohms (using dual-polarity 68-V power supply rails). With the rail voltages shown, the output power is about 140-W RMS into 8-ohm loads.

Referring to Fig. 6-21, the input stage and voltage amplifier stage designs are commonly referred to as *fully complementary*, or *mirror-image* topologies. The dual differential input stage (Q1, Q2, Q3, Q4, and associated components) is supplied tail currents from dual constant current sources (Q5, Q6, Q9, Q10, and associated components), with the input stage being totally isolated from the power supply rails by means of the dual cascode stages (Q7, Q8, Q11, Q12, and associated components). The complementary outputs from the input stage are fed

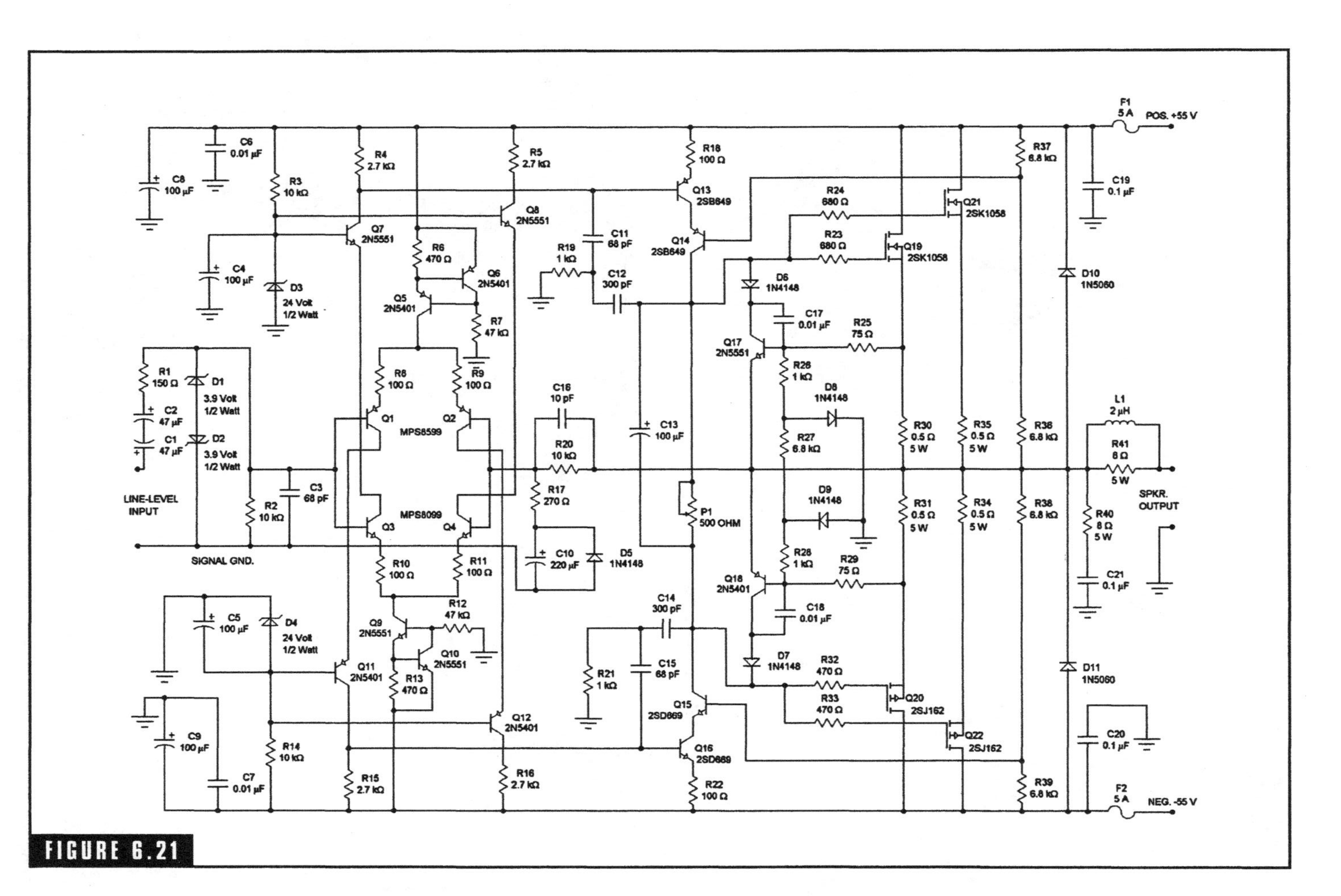

FIGURE 6.21

My favorite audio power amplifier design.

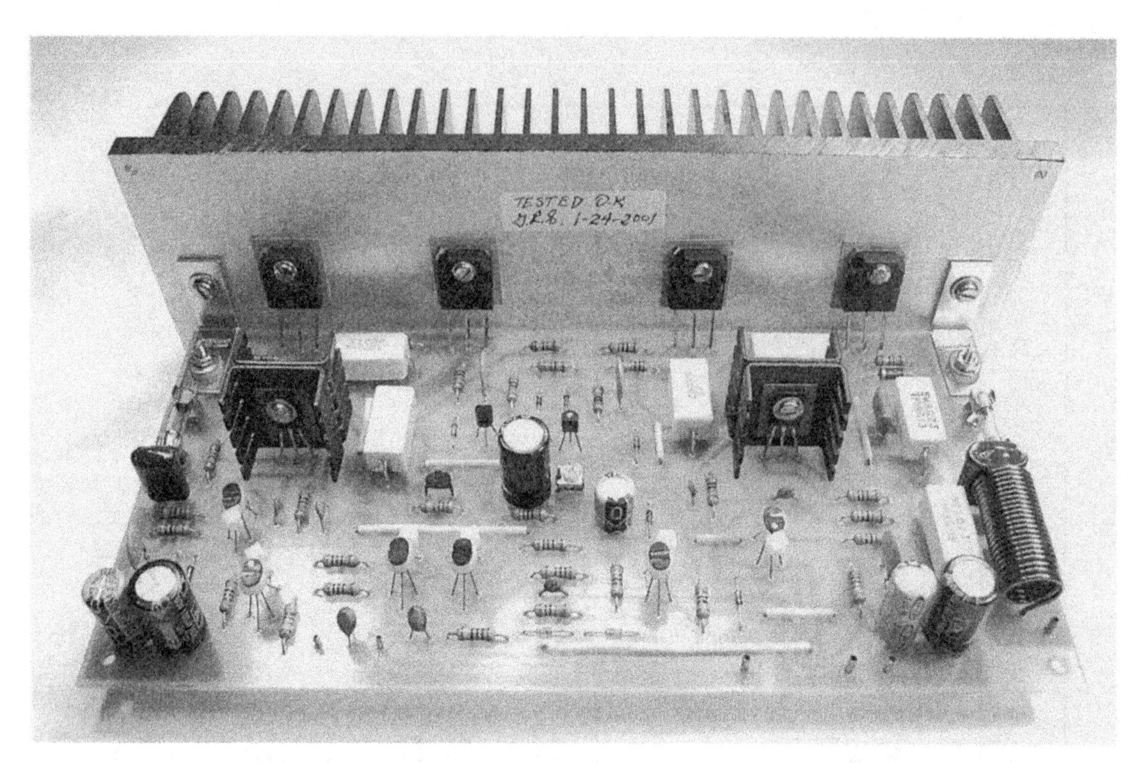

The completed Fig. 6-21 L-MOSFET audio power amplifier.

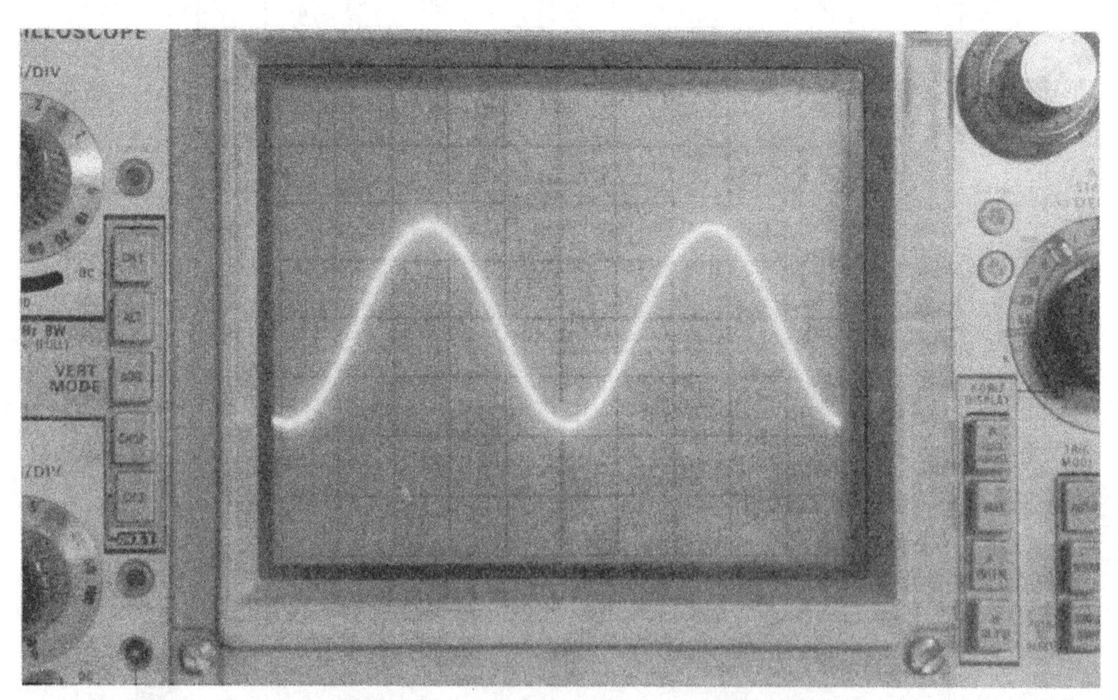

An oscillogram showing an amplified, nondistorted sinewave output from a typical solid-state audio power amplifier.

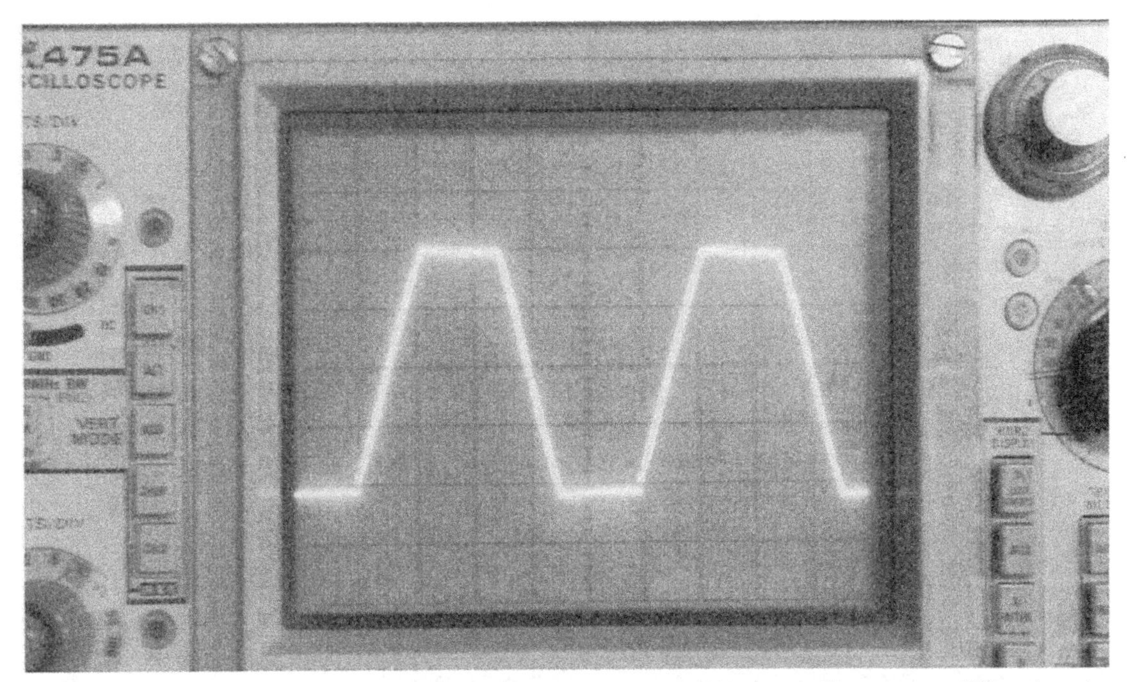

An oscillogram showing the hard-clipping action from a typical solid-state audio power amplifier when the amplifier is overdriven.

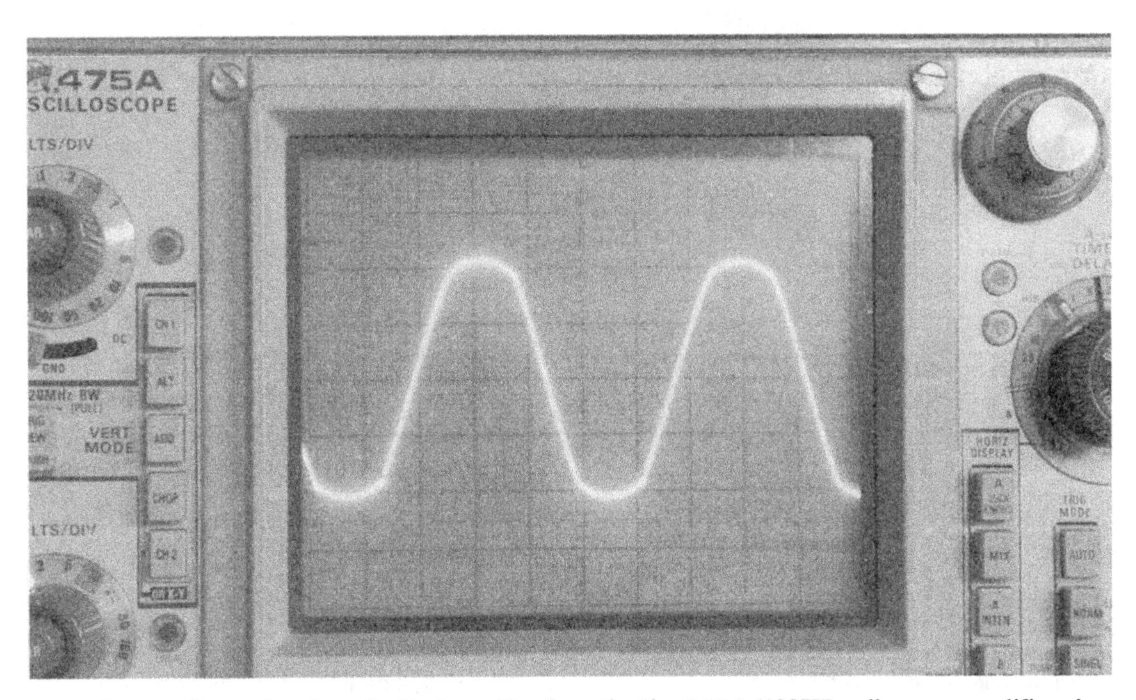

An oscillogram illustrating the soft-clipping action from the Fig. 6-21 L-MOSFET audio power amplifier when overdriven.

The commercial version of the Fig. 6-21 L-MOSFET audio power amplifier.

A 300-watt version of the Fig. 6-21 L-MOSFET audio power amplifier. The L-MOSFETs incorporated are SemeLab's double-die L-MOSFETs. A second heatsink is mounted directly under the top main heatsink for additional dissipation capability. Note the 75-degree C. thermal switch mounted to the aluminum plate between the two middle L-MOSFETs.

A 400-watt RMS version of the Fig. 6-21 audio power amplifier. This version also incorporates SemeLab's double-die L-MOSFETs.

to a class-A complementary voltage amplifier stage, with both signal amplification transistors (Q13 and Q16) being collector-loaded with two cascode stages (Q14, Q15, and associated components). The amplifier is compensated with dual two-pole compensation networks (C11, C12, R19, C14, C15, and R21). If you compare Fig. 6-21 with the Fig. 6-19 amplifier design, you'll note that the output stage design, including the protection circuitry, is very close to being identical.

Fully-complementary topologies provide several sonic advantages over the other topology types that we have examined. One advantage is symmetrical clipping, which has been previously discussed within this chapter. Another advantage is a *self-correcting tendency* that tends to cancel out various anomalies relating to power supply variations, component tolerances, and noise sources.

One of the more unique aspects of the Fig. 6-21 topology relates to the bias supplied to the two cascode stages incorporated into the voltage amplifier stage. Note that the bases of Q14 and Q15 receive a *dynamic bias* from the voltage divider network of R36, R37, R38, and R39. This bias comes out to half of the value of the difference between the output rail voltage and the corresponding power supply rail voltages (you might have to think about that statement for a few minutes, while reviewing the Fig. 6-21 schematic). For discussion purposes, refer to Q14 and its associated base resistors R37 and R36. During a positive "peak" output signal voltage, the output rail voltage will approach the value of the positive power supply rail voltage. If the signal peak is high enough in amplitude, half of the difference between the output rail voltage and the positive power supply rail voltage (which is Q14's base voltage) will become small enough to cause Q14 to enter into its nonlinear "knee" region of conduction. This condition will "round off" the peak output signals, and if clipping occurs, the clipping action will be *soft*. The same conditions apply to the complementary action of Q15, R39, and R38 as well. Note that the soft clipping action is a function of the ratio of the output rail and the power supply rails, so the absolute value of rail voltages have no effect on the soft clipping response. The benefits of soft clipping have been discussed previously in this chapter.

Although the complexity of the Fig. 6-21 audio power amplifier may seem somewhat formidable, the construction is actually quite straightforward and the overall design is forgiving of correctable mistakes (the rail fuses will protect the output devices in the event of a major error). Transistors Q13, Q14, Q15, and Q16 need to be mounted back-to-back on small back-to-back TO-220 heatsinks, as illustrated in the Fig. 6-22 layout diagram. The main heatsink should have about the same thermal characteristics and physical size as used in the Fig. 6-15 amplifier design. L1 is formed by winding 18 turns of #14-AWG magnet wire around a 1/2-in form. The PC board is designed so that the small-signal transistor pairs can be physically tied together for accurate thermal tracking (even though the importance of the technique is highly debatable). I believe the remaining construction details are self-explanatory by examining the PC board illustrations of Figs. 6-22, 6-23, and 6-24.

I might mention that I have constructed numerous 400-W RMS versions of the Fig. 6-21 topology, with the same outstanding performance results. The modifications for the higher-powered version are as follows:

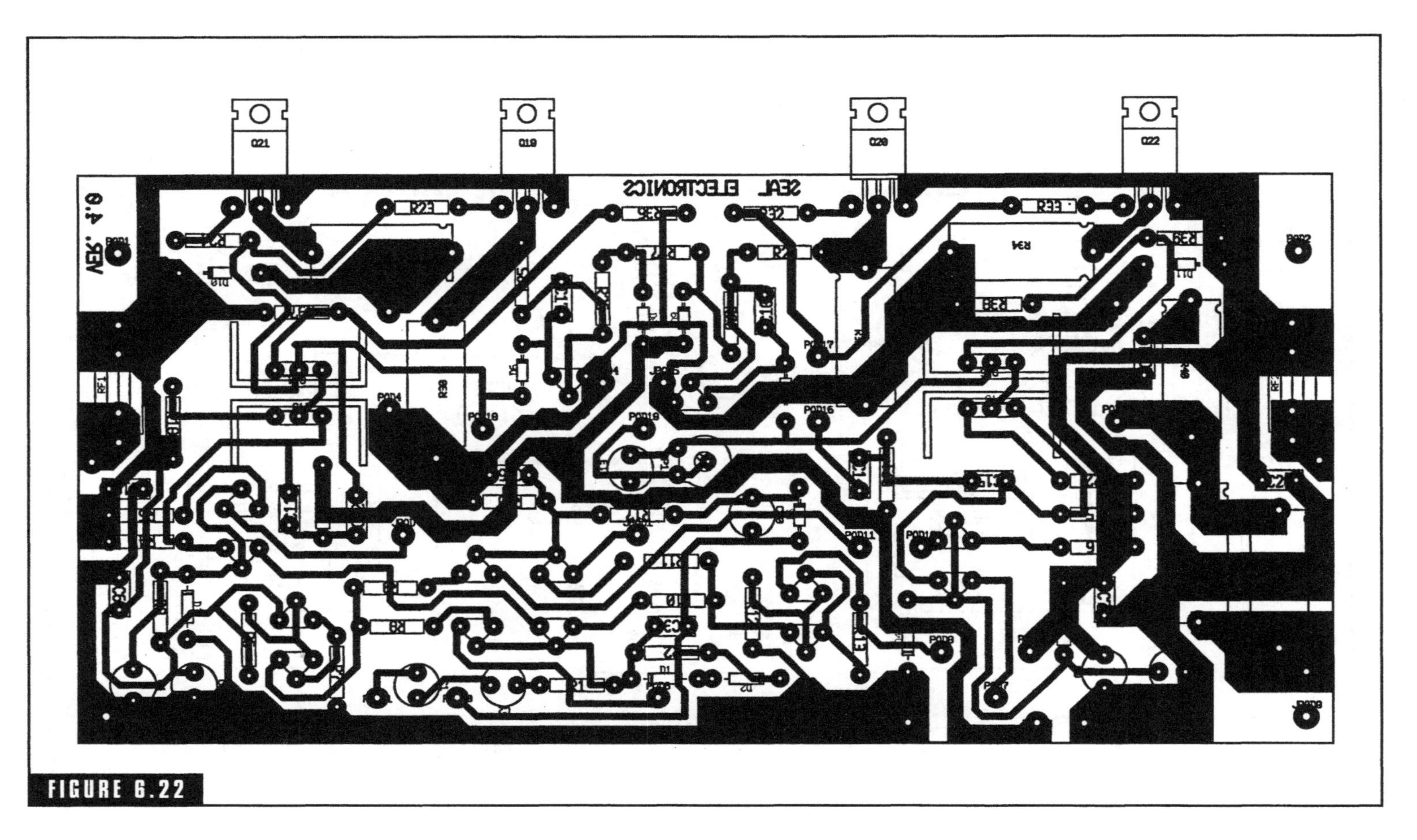

FIGURE 6.22

Top layout view of the Fig. 6-21 amplifier.

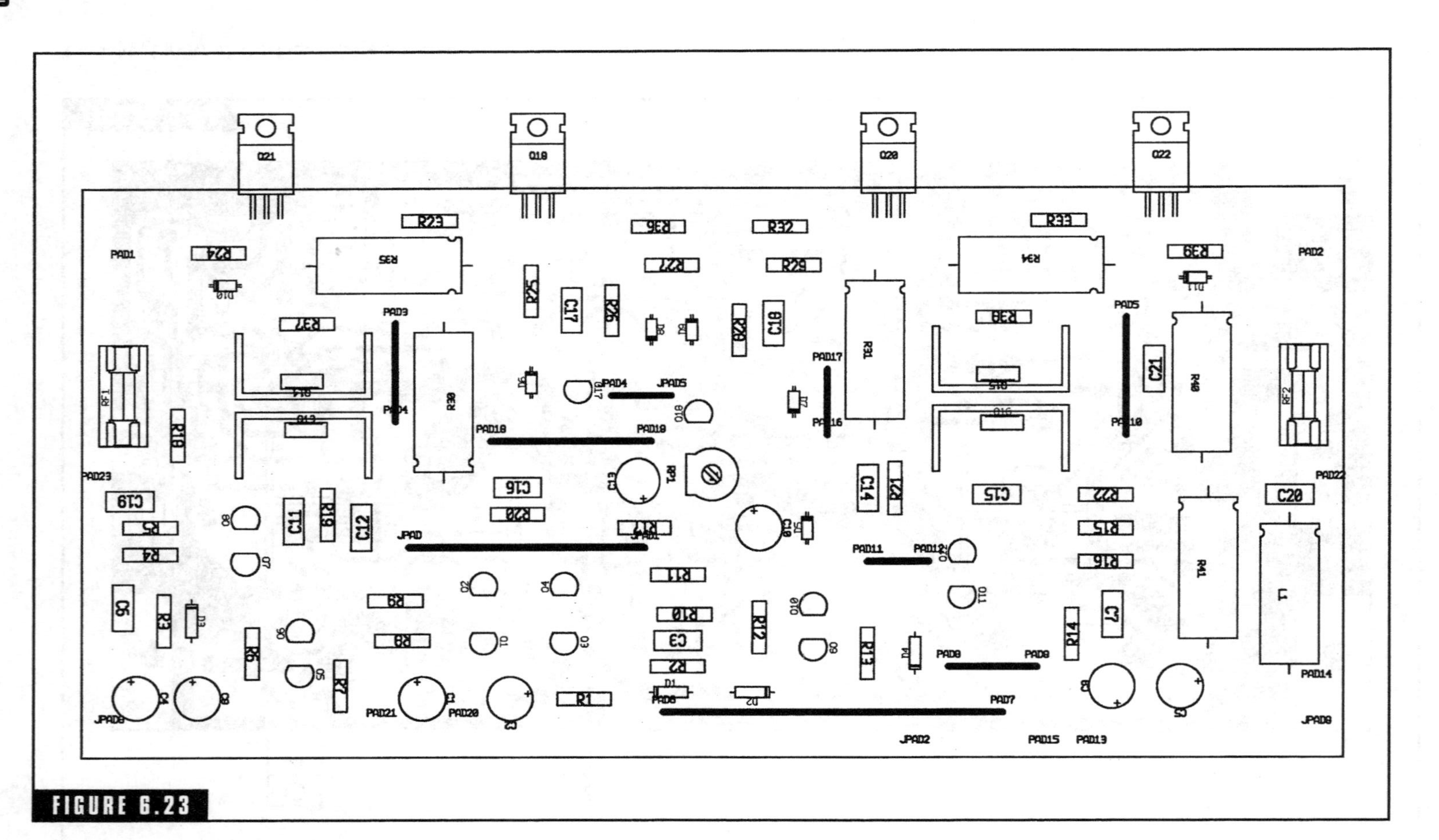

FIGURE 6.23

Top view of the Fig. 6-21 amplifier.

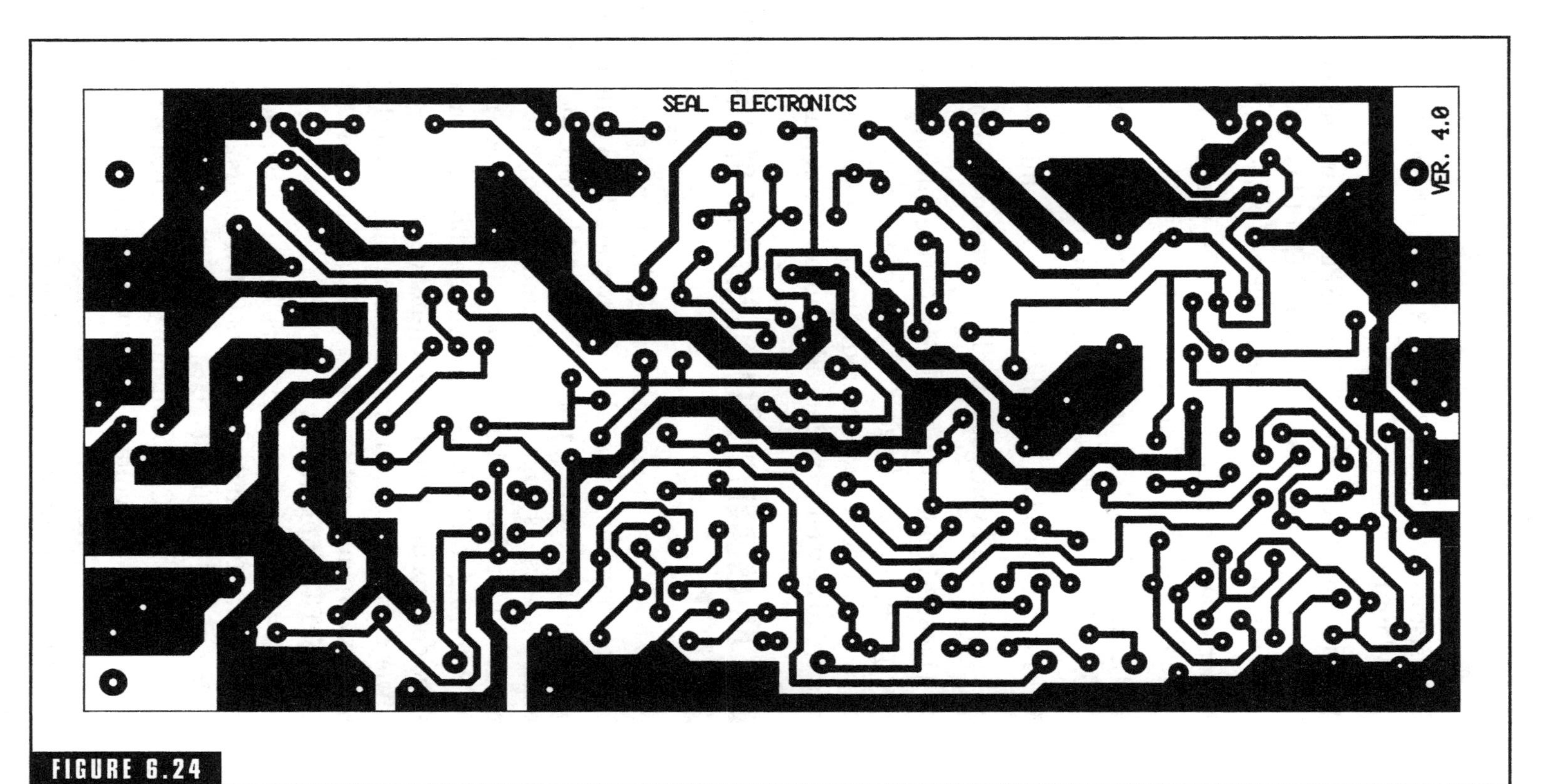

FIGURE 6.24 Reflected artwork for the Fig. 6-21 amplifier.

1. Resistors R30, R31, R34, and R35 need to be changed to 0.22-ohm, 5-W units.
2. SemeLab's "double-die" L-MOSFETs, part number BUZ901CDP, need to be substituted for Q19 and Q21 (this is equivalent to placing two "standard" L-MOSFETs in place of each "standard" L-MOSFET).
3. SemeLab's "double-die" L-MOSFETs, part # BUZ905DP, need to be substituted for Q20 and Q22.
4. The main heatsink should have a thermal rating of approximately 0.2°C/W (this will be a "large" heatsink).
5. F1 and F2 need to be replaced with 7-A units.
6. The dual rail voltages must be increased up to approximately 85 V.

The 400-W versions of the Fig. 6-21 amplifier are large and expensive projects (don't forget the cost of the power supply that will be involved!). I don't recommend attempting such a project unless you happen to be well-experienced in audio power amplifier construction.

Bias Adjustments Without the Use of a Distortion Analyzer

Not all hobbyists or audiophiles will have an expensive distortion analyzer or computerized analysis system available to them for the precise adjustment of the V_{BIAS} potentiometer. Although the use of a good distortion analysis system is still recommended, it is possible to achieve near-optimum bias adjustment based on a voltage measurement.

Referring to Fig. 6-25, observe the three common types of output stage designs illustrated. The left-hand illustration is a BJT emitter-follower configuration, the middle illustration is a BJT complementary-feedback configuration, and the right-hand illustration is an L-MOSFET configuration.

BJT amplifiers are most sensitive to the actual voltage drop between the two emitter resistors (RE resistors). In the case of emitter-follower designs, this voltage differential between the two emitters should be 47 mV. Complementary-feedback designs provide optimum results when the voltage differential between the collectors (i.e., the voltage drop across the two RC resistors) is 8 mV.

Optimum performance from L-MOSFET designs is based on the quiescent current rather than voltage differentials between the source

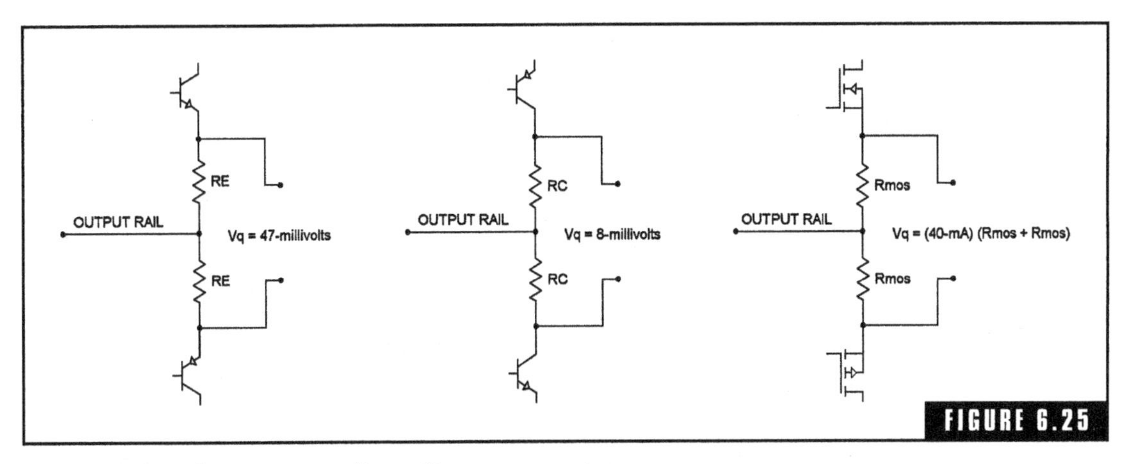

FIGURE 6.25

Setting of bias adjustment according to Vq measurements.

or drain leads. The optimum quiescent current should be very close to 40 mA. Therefore, the differential voltage between the source leads (or drain leads if the design is a hybrid complementary-feedback type) can be determined with a simple Ohm's law calculation. You want the quiescent current to be 40 mA, and you should know the resistance value of the two source (or drain) resistors, so simply multiply 40 mA by the total resistance between the MOSFET leads, and the resulting voltage will be the optimum adjustment point for the bias potentiometer.

For best results, the aforementioned bias adjustments should be initially performed with the amplifier operating from the intended power supply rail voltages, with no input signal applied, and without any type of dummy load connected to the output. After the initial adjustment, an input signal should be applied with a dummy load connected to the output, and the amplifier should be allowed to run until nominally "warm." The dummy load and input signal should then be removed, and the bias adjustment should be performed again (as in the first case) with the amplifier warm.

Some Final Thoughts on Audio Power Amplifier Construction

Anyone who concludes that all audiophiles and electronics hobbyists acquire audio power amplifiers for purely utilitarian purposes is naïve to the true *heart* of the audiophile community. Some audiophiles (like myself, I must admit) tend to view the performance and sonic accuracy of audio power amplifiers as their most beautiful characteristic. Other audiophiles and enthusiasts will spend enormous assets to acquire a certain "name brand" amplifier (some costing over $15,000), placing a great importance on the prestige and *sonic character* of a certain design philosophy or manufacturer. Other audiophiles develop a firm mental image of their ideal system, and nothing will satisfy them short of its literal fulfillment. Such dream systems may include the visualization of 14 warm-glowing vacuum tubes with the flicker of a crackling fireplace reflecting off the glass envelopes, and the loyal family dog residing at the feet of his or her master (direct from a Thomas Kinkade painting). Others imagine a towering inferno of raw power, tearing its way through the listener's head like aural missiles targeted at blowing the mind, and delving out sound pressure levels on the verge of sonic demolition.

Whatever your dream audio system may include, I sincerely hope you achieve it.

CHAPTER 7

Power Supplies

Virtually all of the audio projects contained within this textbook require some type of power supply to provide the operational power. The importance of a well-designed power supply to circuit performance cannot be overemphasized. Throughout the years, I have helped hundreds of hobbyists with attempting to isolate a problematic area of a new construction project, only to discover that the fault originated in the power supply (and the power supply is often overlooked as a source of problems). In some cases, the fault may be due to a defect in the power supply, such as a defective component or a wiring or construction error. But in other cases, the fault may be due to the *suitability* of the power supply for its intended application, and this seems to be an area of confusion for many audiophiles and hobbyists.

When we look at an overview of all of the power supply types incorporated into audio equipment designs, there emerges two main classifications: (1) the large, heavy, high-current, high-voltage power supplies required for audio power amplifier operation, and (2) the highly regulated, low-noise power supply designs required for signal-processing applications. Most do-it-yourselfers have little difficulty with regulated power supplies used in signal-processing applications, since they are usually designed around the multitude of modern integrated-circuit-type regulators. There is a great abundance of information available in almost every electronics textbook regarding the construction of such supplies, and the complexity is typically minimal. However, there seems to be considerable misinformation regarding audio power amplifier power supplies, and I can only conjecture that the reason for such confusion is the scarcity of design information available in the textbook marketplace.

This chapter will be divided into two sections. The first section will detail the considerations and design information applicable to power amplifier power supplies, while the second section will examine a good variety of signal-processing power supplies together with several regulation techniques commonly used in conjunction with audio equipment.

Power Supplies for Audio Power Amplifiers

The subject of audio power amplifier power supplies, like most areas of audio technology, is surrounded in controversy. Throughout the short history of audio power amplifier development, there have been shifts of professional opinions, usually prompted by new developments that made older technologies obsolete. Oftentimes, older technologies will be brought back into the forefront, claiming long-lost benefits in sonic performance or reliability factors.

The earliest solid-state amplifier power supplies were simple unregulated (raw) DC supplies, consisting primarily of a power transformer, rectifier, and filter capacitor(s) (often called *reservoir capacitors*). Most of the early power amplifier designs were either capacitor coupled or transformer coupled to the speaker load, so many of the early power supplies were *single-ended* (i.e., they only provided one polarity of DC output voltage).

Shortly afterward, it was discovered that a *linear regulated* power supply could add significant improvement in the performance of many solid-state amplifier designs of that time. Hum and intermodulation distortion were the performance parameters benefiting the most from this modification.

As power amplifier designs improved, along with improvements in solid-state components, the high-power boom of the 1970s began. Linear regulated power supply designs were all but abandoned due to the additional inefficiencies involved, and the vast majority of power supply designs became the *dual-polarity unregulated* variety that is still, by far, the most popular type today. (Single-ended power supplies cannot be used with direct-coupled amplifier designs.)

During the past few decades, there has been the occasional use of linear regulated power supplies in some esoteric amplifier designs, with a few prominent audiophiles claiming significant improvements in amplifier performance. Throughout this same time period, several power amplifier manufacturers began incorporating *switch-mode power supplies* (abbreviated SMPS) into their modern amplifier designs. (Switch-mode power supplies are often called *switching power supplies* or *switching regulators.*) The advantages to switch-mode power supplies are less weight, smaller size, and less cost (providing they are manufactured in quantity).

At present, there are strong advocates of all three power supply types (i.e., unregulated, linear regulated, and switch-mode). Obviously, before we can delve into power supply construction, we must decide on the best type of power supply to incorporate. In my opinion, the majority of the controversy in this area is unwarranted, as the facts are clear and easily interpreted.

An Evaluation of the Three Main Types of Power Amplifier Power Supplies

A dual-polarity unregulated DC power supply is very reliable and simple. In terms of cost, it is much less expensive than regulated designs, but can be more expensive than switch-mode power supplies if such power supplies are manufactured in large quantities. Unregulated power supplies provide an almost ideal environment for high-power audio amplifiers, delivering very high current peaks as needed by musical transients and wide dynamic ranges.

Unregulated power supplies have two disadvantages from the perspective of amplifier performance: the ripple content and the possibility of signal injection onto the power supply rails. Very large reservoir capacitors can reduce ripple to low values, but ultimately, ripple must be dealt with in terms of the amplifier's PSRR (power supply rejection ratio). Signal injection problems can be eliminated by the correct implementation of rail bypass capacitors and the simultaneous benefit of a high PSRR characteristic inherent to the power amplifier design. It should also be understood that rail injection problems are equally problematic within all three of the major types of power supply design, so unregulated power supplies should not be singled out as especially prone to this problem. From a convenience standpoint, unregulated supplies suffer from the heavy, bulky power transformer required for moderate to high-power amplifiers, but this factor has been improved greatly of late with the easy availability of toroidal designs.

The large power transformer will be a source of abundant electromagnetic interference (EMI) radiation. Unshielded E+I laminated transformers are the worst culprits in this respect, with shielded E+I transformers coming in second, and toroidal transformers producing the least magnetic field leakage. The high-current bridge rectifier is also a source of radio frequency (RF) emissions, caused by the abrupt turnoff of the high-current diodes. This condition usually worsens as current demands are increased, but can be substantially improved with the incorporation of *snubbing capacitors.*

The only advantage offered by linear regulated power supplies is the ability to reduce ripple content to the practical level of nonexistence. The price that must be paid for this luxury is an approximate 30% reduction in power supply efficiency, a much higher cost involved with the additional components and heatsinking required for the regulation devices, a decrease in overall reliability due to the additional components and circuitry, and the obvious increase in weight and bulk. Also, regulated power supplies suffer from all of the ills previously detailed in reference to unregulated supplies. In fact, these problems are actually increased in regulated supplies, because regulated supplies must be constructed on the foundation of an unregulated supply of greater capacity to compensate for the reduced efficiencies involved with effective voltage regulation.

The additional claimed benefits of regulated power supplies are mostly illusory. A modern well-designed power amplifier will probably reject ripple components as well as the active devices in a regulator circuit, so to include both in a complete power amplifier is an exercise in redundancy. While it is true that a regulated power supply will provide absolute consistency of output power levels in the face of line voltage reductions, it must be remembered that the unregulated portion of the power supply has to provide an output voltage about 20% to 30% higher, being regulated down to a lower level so that a level consistency can be maintained. An amplifier of the same size and weight without a regulator could theoretically start out with power output capabilities 20% to 30% higher, so a line voltage reduction of equal proportion would simply bring both amplifiers down to the same power output level. This does not turn out to be a 50-50 proposition, because the amplifier with the regulated power supply will still cost more. In addition, the amplifier without the regulator circuit will be greatly superior in processing high-level transients, since its *music power* rating will be higher.

There is still another, more serious, disadvantage with utilizing regulated power supplies in conjunction with audio power amplifiers. Regulated power supplies must incorporate some form of feedback loop to maintain regulation, and the power supply currents affected by this regulation effort are going to approximately equal the output currents of the amplifier section. Since we are already pushing amplifier components to their limits in terms of large signal swings and slew rates, the regulator section of the power supply has only a slight advantage over the amplifier devices in terms of speed associated with large current variations. In simple terms, both the power supply regulator and power amplifier must contain a global negative feedback loop, with the two loops operating in concurrent fashion, and each experiencing finite limitations regarding speed.

Two bad effects can develop from an amplifier loop trying to work with a power supply loop. First, as the amplifier loop is calling for a fast risetime from the output stage, the regulator loop has to be fast enough to provide the sharp current increase. At higher frequencies, the two response times become additive (i.e., the response time of the regulator must be added to the response time of the amplifier), and this can easily have a detrimental effect on slew rate performance (in fact, with high-power amplifiers, this will almost certainly be the result). Secondly, the global feedback loop of the regulator circuit will cause a rise in power supply output impedance with increases in frequency. While this effect can easily be negligible at audio frequencies, it must be remembered that the open-loop gain of a typical amplifier will not fall below unity until it reaches the megahertz region. The increase in high-frequency power supply output impedance automatically introduces a destabilizing effect within the amplifier. This situation amounts to a very unhappy state of stability performance for any high-performance amplifier, and becomes increasingly worse as amplifier output power increases.

The controversy over the use of linear regulated power supplies in high-performance audio power amplifiers boils down to a simple, bottom-line statement—it's a bad idea! Some audiophiles may consider this last statement to be overly blunt, but I cannot honestly conceive of how any other assessment can be made.

The decision to incorporate switch-mode power supplies in some commercial amplifiers has never been an issue of performance. In fact, there are serious shortcomings involved with switch-mode power supplies that must be resolved before they can even be used with amplifier circuits. The advantages of switch-mode power supplies are weight, size, and cost. Naturally, these attributes raise dollar signs in the eyes of manufacturers, so there has been a modern competitive push toward switch-mode power supply use within manufacturing circles. To fully appreciate the decision to utilize switch-mode power supplies in audio amplifiers, the hobbyist must try to put himself or herself in the shoes of the competitive manufacturer. If you examine a well-designed switch-mode power supply, it is immediately obvious that it is much more complex than an unregulated DC power supply. The quantity and diversity of the switch-mode power supply's electronic components is 10 to 1, or greater, in comparison to unregulated supplies. The first impression is that it must be much more expensive to develop and construct; therefore, it must also be superior in performance. All of these assumptions are not necessarily true.

The R&D costs of switch-mode power supplies are much higher than unregulated power supplies. However, R&D expenses are a one-time expenditure. Once a suitable design has been developed, it can simply be copied and manufactured for as many years as it remains competitive. In the long run, development costs can become quite insubstantial.

Manufacturers pay shockingly low prices for small electronic components when purchased in mass quantities. For example, a transistor that costs you about a dollar in your local electronics supply store will probably cost a large manufacturer about 3 cents! Electronic component manufacturers can afford to sell components so cheaply because they can be manufactured at very high rates with automated machinery. However, when it comes to the lowly power transformer, there is not much that can be done to effectively reduce costs. A large power transformer takes a lot of metal laminations and heavy-gauge copper wire, and the costs of these metals and the manufacturing processes cannot be reduced to microminiature levels. Or, in the case of toroidal designs, the installation of the windings is primarily a manual job. Consequently, manufacturers may have to pay as much for a large power transformer as they would have to pay for a thousand smaller electronic components.

This situation all boils down to the simple fact that a large manufacturer can probably manufacture a complete switch-mode power supply for less money than the original equipment manufacturer (OEM) cost of a substantial transformer. The switch-mode power supply may be more complex and contain a large quantity of components, but this all means little to a manufacturer that has set up a production line utilizing automated insertion (AI) machinery and wave soldering processes.

Switch-mode power supplies offer some significant advantages from a marketing perspective. Marketing personnel love to be able to put fancy phrases like "digitally stabilized" or "advanced digital technology" on the front silkscreen of commercial amplifiers. The insinuation is that some new and improved technology has been incorporated into their "new and improved" line of amplifiers. Such terminology may mean nothing more than the use of a switch-mode power supply (crudely defined as a digital apparatus). The term *digital* is especially meaningful in today's marketplace, because almost everyone associates digital with home computers, and the majority of consumers consider computers to be very reliable items.

It is ironic that the original motivation for going to switch-mode power supplies has not materialized as hoped for by the early proponents. In theory, switch-mode power supplies offered the capability of producing high-power amplifiers in much smaller enclosures, which would be advantageous from a marketing and cost perspective. Unfortunately, although some progress was made toward this goal, the attempt at significant size reduction has not been very successful. In fact, there are some manufacturing firms in the United Kingdom that are producing amplifiers utilizing conventional power supply design and class B output stages with *better* power density than competitive designs incorporating switch-mode power supplies and class G output stages! The primary size-determining factor in high-power audio amplifiers is the heatsinking and ventilation space required for dependable operation. About the only practical methodology for developing small, light, high-powered audio amplifiers is a major breakthrough in class D technology, but I hold out little hope for this materializing (at least in terms of sonically excellent, high-performance models).

Along this same line of discussion, switch-mode power supplies do provide the advantage of lower overall weight. If you have ever tried to install a 2-kW power amplifier into a 6-foot equipment rack, you already know how important that advantage can be!

Thus far, I have been discussing switch-mode power supplies on the basis of present concerns, but there are future considerations that are already starting to emerge in large professional amplifier systems. High-power audio amplifiers, having RMS output capabilities of 1 kW and higher, have already reached the practical limits of AC power requirements. Professional groups and entertainers may use several dozen such amplifiers when performing outside or in large auditoriums, with the combined power load resulting in a variety of AC power (mains) problems.

A new generation of smart switch-mode power supplies is being developed by numerous manufacturers, providing automatic *power factor correction* (PFC) for the reduction of problems associated with the AC mains power. Switch-mode power supplies are especially suited for the incorporation of PFC, so I expect to see more smart switch-mode power supplies incorporated into future professional audio power amplifiers.

There are several disadvantages associated with switch-mode power supplies. The first and most obvious disadvantage is their reliability. They are more complex, contain more active components, and are much more likely to fail (especially from the effects of power line transients and other AC power line problems). My personal experience in this respect is not very favorable. In virtually every instance of switch-mode power supply failure that I have been associated with, the manufacturer has been reluctant to supply parts or documentation for servicing. Manufacturers recognize that switch-mode power supplies are rather temperamental, and they are additionally concerned that a user could modify the power supply so that its RFI emissions would be adversely affected (possibly to the point of rendering the associated equipment illegal). Several amplifier manufacturers have introduced switch-mode power supplies and later discontinued their use due to reliability and servicing problems. It is not my intention to insinuate that all manufacturers are experiencing severe problems with their switch-mode power supply designs, but there is no doubt that the reliability of switch-mode power supplies will always be very poor in comparison to unregulated designs.

Switch-mode power supplies are notorious generators of RFI emissions because they function on the principle of high-frequency switching of the nonisolated AC power line. To put it mildly, their generation of high-frequency interference is prolific (i.e., in both directions—outgoing onto the AC line and internal to the amplifier circuitry). Many of the designs I have encountered appear to take up as much space with shielding protection as is needed for the power supply itself. Generally speaking, the problem is almost impossible to eradicate completely.

Switch-mode power supplies are likely to be slow in providing the transient current demands of the amplifier's output stage. In this respect, they suffer from the same problem as linear regulated power supplies, and this characteristic requires very close scrutiny to avoid slew distortion problems. Ripple content in switch-mode power supplies is an in-between parameter; it is improved over unregulated designs, but inferior to linear regulated designs.

Choosing a Power Supply Type for an Audio Power Amplifier Application

Looking at the facts in a condensed form, the following statements can be made:

1. Linear regulated power supplies are a bad choice from virtually every perspective. They are more expensive, more prone to failure, and require a larger unregulated DC power supply to start with. They can seriously degrade the performance of the amplifier in the areas of transient processing, slew rate, and high-frequency stability. Their only redeeming feature is the practical elimination of ripple, but this can be effectively accomplished within well-designed amplifier circuitry.
2. Switch-mode power supplies offer the advantages of lower weight and reduced ripple content in comparison to unregulated designs

(also, reduced cost if you happen to be a large manufacturer, or intend on buying OEM versions). Their disadvantages are very high levels of RFI emissions, complexity, reliability, and poor transient current response (in comparison to unregulated designs).

3. Unregulated DC power supplies have the disadvantages of the weight and cost of the power transformer, and worrisome levels of EMI radiation. In every other respect, their performance (as applicable to audio power amplifiers) is far superior to the other choices, definitely providing the opportunity for the best performance from a conventional audio power amplifier.

Power Transformer Considerations

The heart of a power amplifier's power supply is the power transformer. This is also likely to be the most expensive component in the entire amplifier. It suffers the disadvantages of being heavy, bulky, and expensive, but it also probably represents the most reliable device in the entire amplifier (providing appropriate secondary fuses are installed). Power transformers operate on a relatively simple principle of electromagnetic induction, they have no moving parts, and there is nothing internal to them that can wear out in the practical sense. I have scavenged power transformers for use in audio amplifiers that were in excess of 50 years old, with little concern that reliability would be compromised. This is an area where the hobbyist can really save a lot of money by scavenging old equipment!

The most important electrical specification of a transformer is its volt-ampere (VA) rating. The VA rating of a power transformer is simply the maximum specified secondary current(s) multiplied by the secondary voltage. For example, a power transformer with two secondaries, each providing 30 V @ 3 A, would have a *load VA rating* of 180 VA (3 A × 30 V × 2 secondaries = 180 VA). In many cases, the primary voltage, secondary voltage(s), secondary current(s), and VA rating are the only electrical specifications provided by the transformer manufacturer. This is unfortunate, because there are other important considerations.

Audiophile connoisseurs often jump to immediate conclusions regarding the quality, power, and performance of an audio power amplifier by the size of its power transformer. This is a lot like judging how fast a car can go based on the size of the engine. In reality, size is not a very good determining factor in judging the true VA rating of a transformer, since a variety of manufacturing techniques can cause a substantial difference in size and weight between two power transformers of the same performance.

There are dissenting opinions regarding the required size of power transformers relative to given amplifier power ratings. It may seem odd that something so seemingly basic could be argued about, but there are valid rationalizations involved. Music and speech information inherently contains a high *peak-to-average power ratio.* In simple language, this means that an amplifier reproducing music at maximum volume (without excessive distortion) will not output its maximum power into a speaker load. This is because music or speech information contains a wide range of instantaneous amplitude levels, corresponding to the dynamic range of the program material. For example, if you sinewave tested a hypothetical 100-W RMS audio power amplifier at maximum output, you would be delivering 100-W RMS to a speaker load. In contrast, if you performed a maximum output test on the same amplifier using recorded music as a test signal, the lulls, voids, and low-amplitude peaks of the musical passage may cause only 20 to 40 W (on the average) to be delivered to the speaker load.

Manufacturers often incorporate transformers in commercial amplifiers rated for about 70% (sometimes much less) of the power required for continuous sinewave testing. This is not as bad as it seems, especially in cases of multichannel power amplifiers sharing a common power supply. The only time a multichannel power amplifier will be continuously sinewave tested with all channels driven simultaneously is during evaluation. In all other situations, the amplifier will be processing program material or sinewave tested on only one channel at a time (i.e., after a repair). Utilizing an underrated power transformer saves money and reduces the overall weight of commercial amplifiers, with both qualities being advantageous from marketing and convenience perspectives.

The only real-world problem that I am aware of pertaining to *practical levels* of transformer underrating is the possibility of overheating the transformer in situations where the amplifier is excessively overdriven. If transformer underrating is going to be practiced, the transformer should be equipped with an internal thermal cutout. Thermal cutouts come in two varieties. One type utilizes a thermal fuse buried in the outer winding layers as a means of meeting various safety standards. Safety standards are met, but the transformer is effectively destroyed the first time the fuse blows. For obvious reasons, I certainly do not recommend using this type of transformer. The other type of thermal cutout utilizes a bimetallic switch for this purpose (often referred to as *thermostatically protected*), which will provide the protective shutdown action, but will also reset after the transformer cools. Consequently, the transformer can be used again in the normal fashion.

Another method of protecting an underrated power transformer is the installation and proper selection of *secondary fuses.* I like to incorporate secondary fuses regardless of whether the transformer is underrated or not. There are problems that can occur internally with bridge rectifiers that can result in transformer damage before the problem is detected. Since transformers are a relatively expensive investment, the installation of a couple of secondary fuses seems a small price to pay for insurance against transformer damage.

Personally, I do not normally underrate transformers in the power amplifiers I construct. I have two reasons for this overindulgent attitude. First, I simply like the assurance of knowing that the power transformer will not overheat under any anticipated conditions. I am willing to sacrifice a few extra dollars and aggravate my lower back condition for this luxury. Second, the *regulation factor* of the transformer is improved, and this provides slightly better performance from the amplifier due to the lower *power supply droop* effects.

Power transformers do have a regulation factor, but this should not be confused with any type of feedback loop or active regulation circuitry. The regulation factor of a transformer is simply the ratio of the secondary voltage loaded (to maximum specifications) versus the unloaded secondary voltage. For example, if a transformer provided a secondary voltage of 30 V unloaded and 27 V fully loaded, the regulation factor would be 10% (27 V divided by 30 V = 0.9 or 90%. 100% − 90% = 10% regulation factor). Some transformer manufacturers appear to be somewhat reluctant about providing this specification, but it becomes important when analyzing the maximum drop in DC levels of a power supply when fully loaded (referred to as *power supply droop*). Obviously, an underrated transformer is going to cause a more severe droop in power supply DC levels at maximum power levels than a fully rated transformer.

Depending on your personal applications and location, it may be prudent to obtain power transformers with versatile primaries so that completed amplifiers can be easily adapted to both European and U.S. mains supplies. In commercial applications, it is also important to remember that federal emissions regulations vary widely throughout Europe, the United Kingdom, and the United States. It is possible to be in violation of certain emissions regulations by using unshielded E+I transformers in conjunction with some enclosure designs.

Transformer EMI Problems

For optimum hum reduction performance, the EMI (electromagnetic interference) radiation of large power supply transformers must be dealt with. EMI from transformers produces hum through induction of unwanted currents at the power line frequency. Basic electrical mathematics informs us that the induction coefficient is directly proportional to the inverse square of the distance, so the most effective means of reducing unwanted induction is *increasing the distance from the EMI source.* In the physical layout of a complete amplifier, keep the power transformer as far away as possible from sensitive circuitry (i.e., high-gain, low-current, low-voltage circuitry) and signal wiring, especially signal conductors terminating into high impedances. The typical way of accomplishing optimum performance in this respect is to mount the power transformer in the rear of the amplifier enclosure, close to the incoming AC line, reservoir capacitors, power switch, line fuse, rectifier, and all other power supply components. Sensitive electronic circuitry is typically mounted close to the front of the enclosure, putting as much distance as possible between it and the power transformer. In addition, input signal wires, and all other signal wiring, should be kept as far away from the power transformer as possible.

Toroidal transformer designs are the best at producing a minimum of EMI. The second best choice is a shielded E+I laminated transformer designed for use in electronic power supplies or power amplifiers. This type of transformer is usually encased in two metallic sides, with a copper *belly*band (or *hum strap,* as it is sometimes called) looped around the outside of the windings. An unshielded E+I laminated transformer is the worst choice in this regard, but can be effectively used in some situations.

Power transformers do not radiate EMI in an even pattern. It is quite possible to significantly reduce EMI problems by simply turning a power transformer in a different direction. The normal procedure for installing toroidal transformers is to leave enough primary and secondary wire length to rotate the transformer for minimum hum after it has been installed. The mounting techniques for E+I laminated transformers make this technique a little more difficult.

In some esoteric amplifier designs, the entire power supply is mounted in a different enclosure, which is physically placed in a remote location from the amplifier enclosure. The rectified, filtered, and fused rail supply power is provided to the amplifier via a heavy-duty cable. This method, of course, is the ultimate technique for eliminating EMI problems, but it is an extreme measure that cannot actually be justified on a performance basis. By utilizing a good quality transformer and following the few common sense construction techniques, it should be rather easy to push hum levels resulting from EMI down below the noise floor.

Rectification

In modern power supply designs, it is rare to come across a bridge rectifier made from discrete diodes. The modular types of diode bridges are small, inexpensive, and provide excellent performance in every respect, so there isn't any good reason not to use them. In most medium- to high-power amplifier designs, the bridge rectifier must be heatsinked. This is typically accomplished by mounting the diode bridge module to some unused area of the amplifier enclosure.

To reduce RF emissions from the hard and rapid turnoff of rectifier diodes, it is recommended that snubbing capacitors be soldered across all four connection posts of the rectifier module, as illustrated in Fig. 7-1. Snubbing capacitors are typically 0.01- to 0.1-μF ceramic disc types, and they should be soldered as close to the rectifier bridge terminals as possible for the optimum effect.

RF emissions external to the amplifier enclosure are significantly reduced with the use of an "X capacitor," illustrated on the primary side of the power transformer in Fig. 7-1; 0.1 μF is a typical value. Care should be exercised that the voltage rating of the X capacitor is sufficient for the maximum peak voltage level of the AC mains supply. Considering the typical AC variations above and below nominal RMS

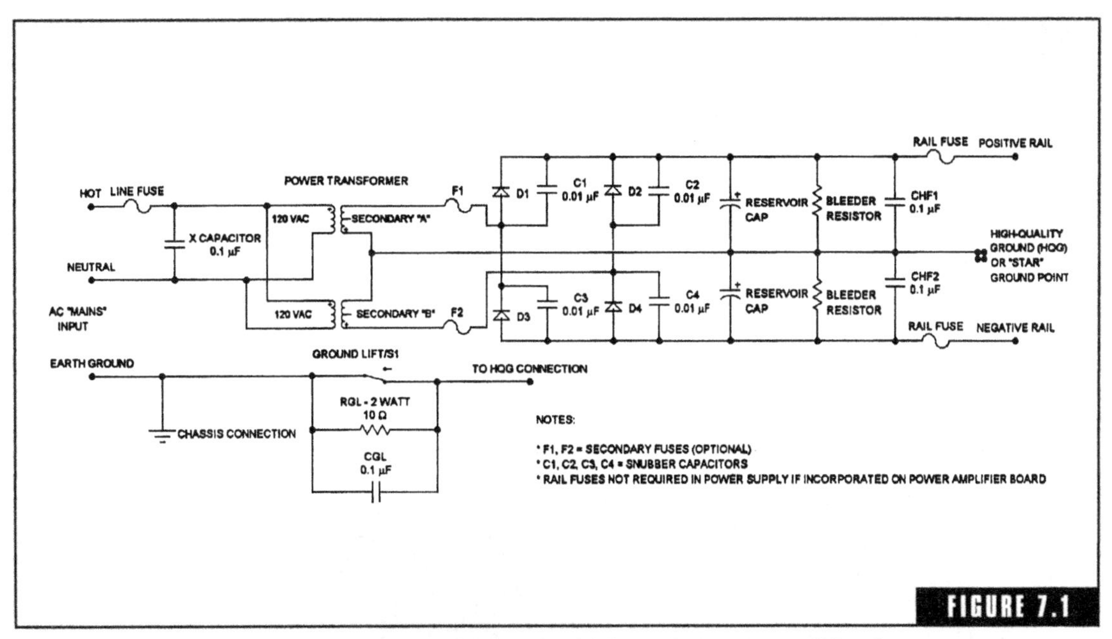

An illustration of a typical power supply as used with a high-quality power amplification system.

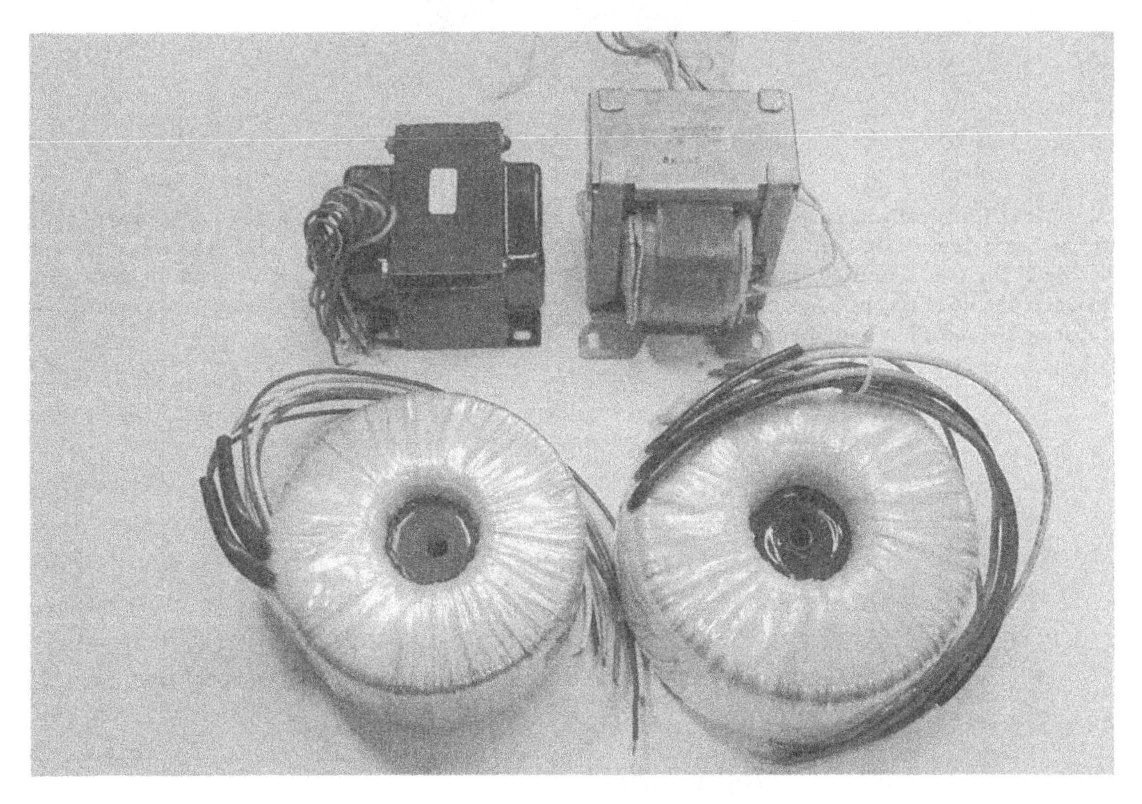

A variety of power transformers commonly used in power supplies for audio power amplifiers. Note the copper "belly-band" incorporated into the top right-hand transformer. The bottom two transformers are toroidal types; the larger of the two is rated for 1-KVA.

An enclosure originally housing a SMPS for a computer system, but converted to a dual-polarity power supply for an audio power amplifier. I mounted the power transformer, bridge rectifier, and reservoir capacitors inside of the metal housing. I cut a piece of circuit board material to cover the original fan holes, and mounted the line fuse and power switch to the circuit board material. The metal enclosure does a great job of reducing EMI radiation from the power transformer.

values, a good rule of thumb is to use a capacitor with a voltage rating of about 2½ to 3 times the nominal RMS value.

The total circuit resistance from the power transformer secondary to the reservoir capacitors should be as close to zero as possible, with the use of heavy-gauge connection wire and heavy-duty fuse holders recommended. The initial powerup surge currents that must be handled by the bridge rectifier will be quite substantial. Therefore, it is a good practice to generously overrate bridge rectifier modules for optimum reliability.

Power Supply Fusing

As illustrated in Fig. 7-1, the line fuse and secondary fuses will be in line with the huge surge currents resulting from the initial charging of

A variety of reservoir capacitors commonly found in audio power amplifier power supplies.

the reservoir capacitors. Therefore, these fuses should always be of the *slow-blow* variety.

Secondary fuses are not required for safety purposes, but they are certainly a good idea. If one of the diodes internal to the bridge rectifier happened to short, it is possible for the transformer secondary to overheat and be destroyed before the AC line fuse blows. Fuses and fuse holders are relatively cheap, so it is prudent to protect the most expensive component in the amplifier for this small investment.

During the initial testing phase of a newly constructed amplifier, it is wise to examine the fuses carefully during powerup. If you notice any movement or warping of the fuse element at this time, the fuse is underrated and should be replaced with the next higher value.

Reservoir Capacitors

The big controversy surrounding reservoir capacitors is the required capacity value for optimum amplifier performance. Many esoteric designs have gone to outrageous extremes in this regard, which boils down to a waste of money and available enclosure space. There is certainly a reasonable leeway in determining the total reservoir capacity you desire to use, but there are also a few myths floating around that have led to confusion.

There have been misunderstandings evolving from automotive sound systems involving the use of *stiffening capacitors* that have infiltrated into the other realms of professional audio. Some audiophiles have been misled to believe that large values of reservoir capacitors incorporated into AC line–powered audio amplifiers will provide a "firmer, stiffer bass response." This is only true if the original power supply reservoir capacitors were of inadequate capacity, which is rarely the case with line–powered equipment. Stiffening capacitors can help low-frequency performance in automotive power amplifiers, but this is due to the limited capabilities of most DC power converters.

The reduction of ripple content is desirable to practical limits, but excessive power supply capacity is not the prudent method of reducing ripple-induced hum problems in high-quality power amplifiers. As long as the ripple is reduced to reasonable levels, the amplifier's PSRR shouldn't have any trouble in effectively eliminating it. A typical value of ripple inherent to a quality power supply under normal loading is around 2 V peak-to-peak.

My rule of thumb for determining the size of the reservoir capacitors is to allow 1000 μF of rail capacity for every 10-W RMS of amplifier output power. For example, in the case of a 100-W RMS amplifier, this comes out to about 10,000 μF of capacitance per rail supply. In truth, this method is a little overindulgent on my part, since 6800 μF per rail will provide excellent performance from a 100-W RMS amplifier. A well-known commercial manufacturer of high-quality power amplifiers utilizes 20,000 μF per rail for their 225-W RMS power amplifiers. Another commercial manufacturer uses 13,000 μF per rail for their 300-W RMS design, but I consider this to be somewhat inadequate. Nevertheless, as can be seen, there is quite a bit of tolerance in this area. If you can come within plus or minus 30% of my aforementioned rule of thumb, you shouldn't have any concerns about significant degradation of performance due to inadequate reservoir capacitors. Of course, this assumes the power amplifier design is good and it inherently contains a good PSRR factor.

Regarding the allocation of large reservoir capacitors, the best methodology for the typical audiophile or hobbyist is to search through a variety of surplus catalogs and find the best deals on capacitors with adequate capacity and voltage ratings. Remember that capacitors can be connected in various parallel and series arrangements to provide the necessary ratings.

It is wise to overrate capacitor voltage by at least 20% to provide a reasonable safety margin. For example, most manufacturers utilize 63-V capacitors for 50-VDC rails, or 75-V capacitors for 60-VDC rails, etc.

Typical Power Supply Designs for Audio Power Amplifiers

Figure 7-1 is an illustration of a typical power supply design for a high-quality audio power amplifier. Note that *bleeder resistors* are installed in parallel with the reservoir capacitors. Their function is to safely discharge the reservoir capacitors when the operational power to the amplifier is turned off. Bleeder resistors are especially important in circumstances of amplifier failure, wherein one or both of the rail fuses blow. In these cases, the reservoir capacitors could remain charged at dangerous voltages for long periods of time. Depending on the power supply output voltages, the typical value for bleeder resistors are between 2.2 and 10 kohm, with power ratings between 2 and 5 W each.

CHF1 and CHF2 are high-frequency decoupling capacitors. These capacitors are a last defense at eliminating any RF components that could be applied to the amplifier circuitry. RF components can originate with the bridge rectifier or be inductively coupled into the power supply through the line cord or other external RF source. CHF capacitors are not always incorporated into power supply designs, but when they are, the typical values range from about 0.01 to 0.33 μF. Voltage ratings should be in accordance to the DC output voltage of the power supply.

Figure 7-1 also illustrates an optional *ground lift* circuit, incorporated into most high-quality professional audio power amplifiers. Ground lift circuits are typically not installed in audio power amplifiers intended for domestic hi-fi applications. Ground lift circuits are useful in professional circumstances for reducing earth grounding problems that may result from multiple connections of audio equipment that happen to be connected to multiple earth ground points.

You may note that Fig. 7-1 doesn't illustrate any type of AC power switch. For smaller power supplies that are typically used in conjunction with domestic power amplifiers, a simple single-pole, single-throw (SPST) power switch can be placed in series with the AC hot line. In larger audio power amplifier systems, a circuit breaker is sometimes substituted, serving the dual function of a power "on-off" switch and an AC line fuse. Large power supplies that don't incorporate any type of *soft-start* circuit can rapidly destroy ordinary switches, due to the significant arcing associated with the high in-rush currents. Figure 7-2 illustrates a method used by some manufacturers to overcome this problem. A power triac is placed in series with the primary of the power transformer, with the power switch controlling the application of gate current. With the power switch closed, the triac looks like a short-circuit, for all practical purposes, but the power switch contacts are only exposed to the very small gate currents required to fire the triac. When the power switch is opened, the triac will lose its holding current at the end of the next half-cycle, taking on the characteristic of an open circuit. This AC power control method is used with several very popular 400-W and 800-W professional audio power amplifiers.

Power Supply Calculations

I assume the reader is familiar with the standard Ohm's law equations, so I will not go into detail regarding the calculation procedures for determining the required voltage and current outputs for specific amplifier power levels. This section will assume the reader already knows (or can easily calculate) the output specifications needed for the power supply to be constructed.

It has been my experience that there are subtle design variations from one type (or manufacturer) of power transformers to another, making exact predictions of power supply output voltages a little difficult. To further complicate the issue, there are also slight variations in losses of bridge rectifier modules. A good rule of thumb is to multiply the AC secondary voltage of the power transformer by 1.4, and subtract about 1 V from the answer to compensate for the bridge rectifier. For example, a power transformer with two 30-V AC secondary windings (i.e., a 60-V.C.T. transformer) would provide about 41 V DC at the output of each rail of the power supply [$(30 \times 1.4) - 1 = 41$ V]. If you constructed such a power supply, the *unloaded* rail voltages would

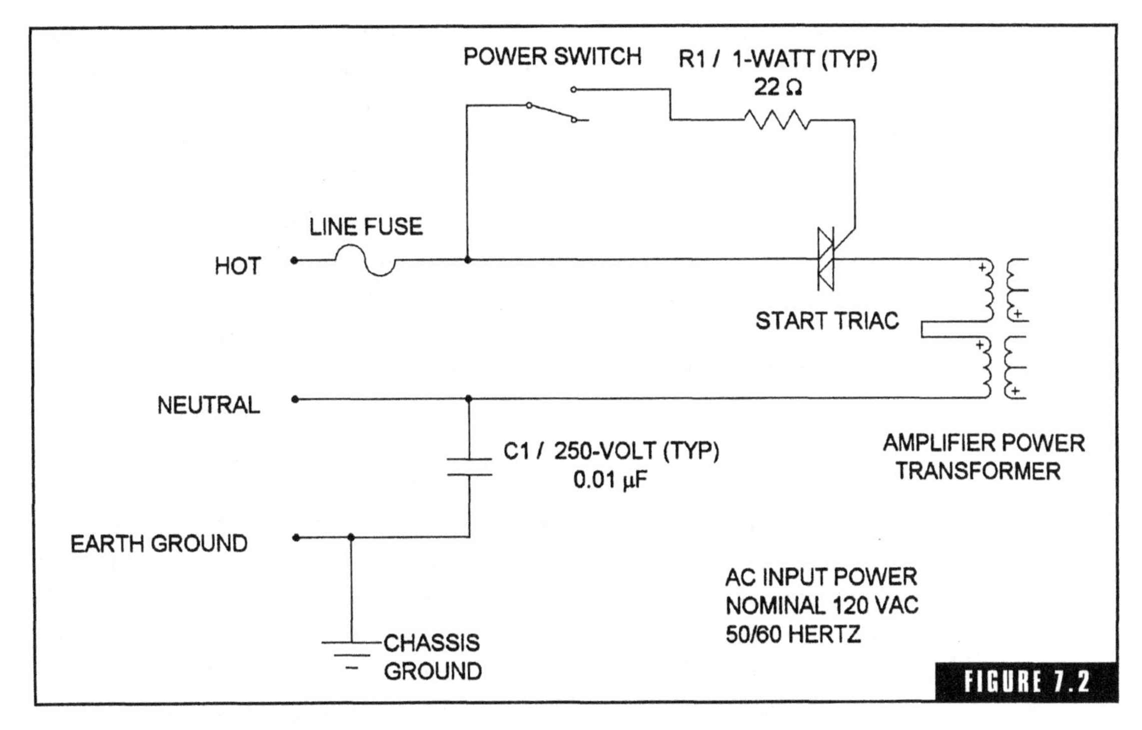

A high-power AC mains switch.

probably come out to about 46 V DC because the specified secondary voltages of power transformers are based on their maximum loaded conditions. Some high-quality power transformers have regulation factors of up to 4%, while some transformers provide regulation factors as poor as 15%, so power supply droop can vary significantly from one power supply design to another.

The area that seems to generate the greatest confusion among do-it-yourselfers regarding power supply design (i.e., for power amplifiers) is the required output current. It may seem easy enough to calculate the RMS current requirements for a 100-W RMS power amplifier, but what kind of speaker loading do you anticipate? Also, what level of efficiency is the power amplifier capable of providing? And to throw in an additional complication, the efficiency of an audio power amplifier will vary according to the speaker load.

Generally speaking, the current rating of the rail fuses incorporated into a power amplifier will be a fair estimate of the maximum required RMS output current rating of the power transformer's secondaries. For example, if you were intending on constructing a power supply for a monaural subwoofer power amplifier, and the power amplifier incorporated 5-A rail fuses, the secondary current rating of the power transformer should be about 5 A. If this transformer happened to be an 80-V center-tapped model (sometimes specified as a 40 + 40-V transformer), the VA rating of the transformer would be 400 VA (i.e., 5 A × 80 V = 400 VA). If you wanted to underrate the power transformer for this hypothetical application, you might multiply the 400-VA rating by 70%, resulting in 280 VA. If you underrate a power transformer, however, remember to install slow-blow secondary fuses according to the secondary rating of the power transformer, or make sure the transformer is thermally protected.

If you plan on constructing audio power amplifiers as an ongoing hobby, you will want to obtain a suitable bench power supply that you can use for testing and evaluation purposes. The easiest and least expensive way to go about this is to construct the bench power supply yourself. Figure 7-3 illustrates one of the bench power supplies that I use very frequently. It is a simple design, and I believe the illustration is mostly self-explanatory. The transformer specified is available from Plitron Manufacturing Co. (part number 107042201) and the rest of the components are commonly available. I mounted the entire power supply on a single piece of aluminum plate—it doesn't have to look fancy for bench applications. The Fig. 7-3 bench power supply plugs into a *variac* (an adjustable autotransformer), which provides adjustment of the dual DC outputs from zero to approximately 85 V DC (variacs typically provide adjustment up to about 120% of the nominal AC line voltage). If you don't have a variac, you should consider purchasing one if you are intending on doing significant work with audio power amplifiers.

Figure 7-4 illustrates another bench power supply I use frequently for testing audio power amplifiers. This design provides the advantage of four current-limited outputs, each having a maximum current limit of 1.5 A. Current limiting comes in very handy if you are bench testing a newly constructed power amplifier for the first time, or when you are initially testing a power amplifier after a repair. If there is something wrong with the amplifier under test, the current-limiting circuits will usually protect the amplifier from destruction as a result of applying operational power. When testing class B power amplifiers, 1.5 A is

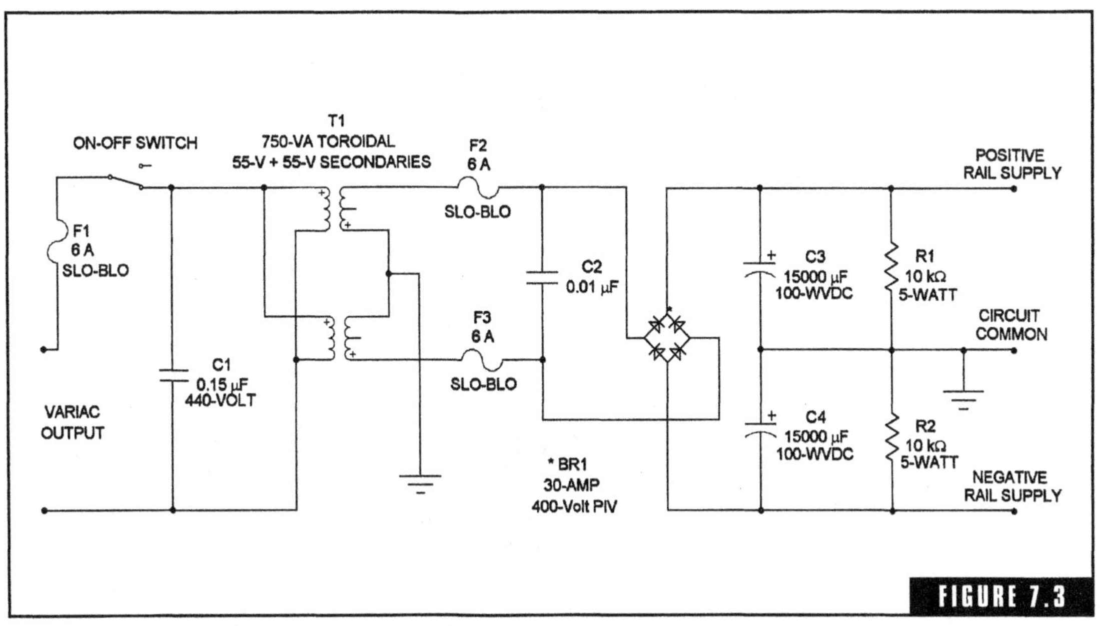

A dual-polarity bench power supply used for testing high-power amplifier designs.

typically more than enough to test for basic operation, and to ensure that there isn't a catastrophic fault with the amplifier. If you have the need to bench test a class A design, you can parallel the two positive outputs and the two negative outputs for a maximum current limit of 3 A. After you have tested the fundamental operation of the amplifier and are satisfied that it doesn't appear to have any severe problems, you can then load test the amplifier by connecting its power supply rails to the two "unlimited" DC outputs.

The bench power supply of Fig. 7-4 is intended to be used with a variac so that the dual-output voltage levels are infinitely adjustable up to about 52 V DC. I believe the illustration is mostly self-explanatory in regard to construction. The current-limiting transistors (Q5 through Q8) should be adequately heatsinked, but they will not dissipate any significant power unless a current-limit condition occurs. The series-pass transistors (Q1, Q2, Q3, and Q4) are intended to provide current-limiting action on an intermittent basis only; otherwise they will overheat (R_{THJC} of the series-pass transistors is such that even an ideal heatsink cannot keep the junction temperature from rising too high under all possible voltage and current conditions). Since a cur-

The Fig. 7-3 bench power supply.

rent-limit action indicates a fault in the amplifier under test, common sense dictates that the power supply will be turned off or disconnected from the amplifier very soon if a current-limit action occurs. I have found this method to be completely adequate for my needs. However, if you want something more elaborate, it is easy and inexpensive to incorporate four optical isolator ICs to monitor the collector-to-emitter voltage drop of the series pass transistors. The outputs of the optoisolators could then be OR-ed to a control relay that would shut down the power supply if a current-limit condition occurred (you would want a few seconds delayed response in this shut-down circuit to facilitate the tolerance of normal surge currents). Another option is to incorporate paralleled series-pass transistors. I leave these options to the imagination and ingenuity of my readers.

Power Supplies for Signal-Processing Applications

Compete audio power amplifier systems often include some type of dual-polarity regulated power supply to provide operational power for

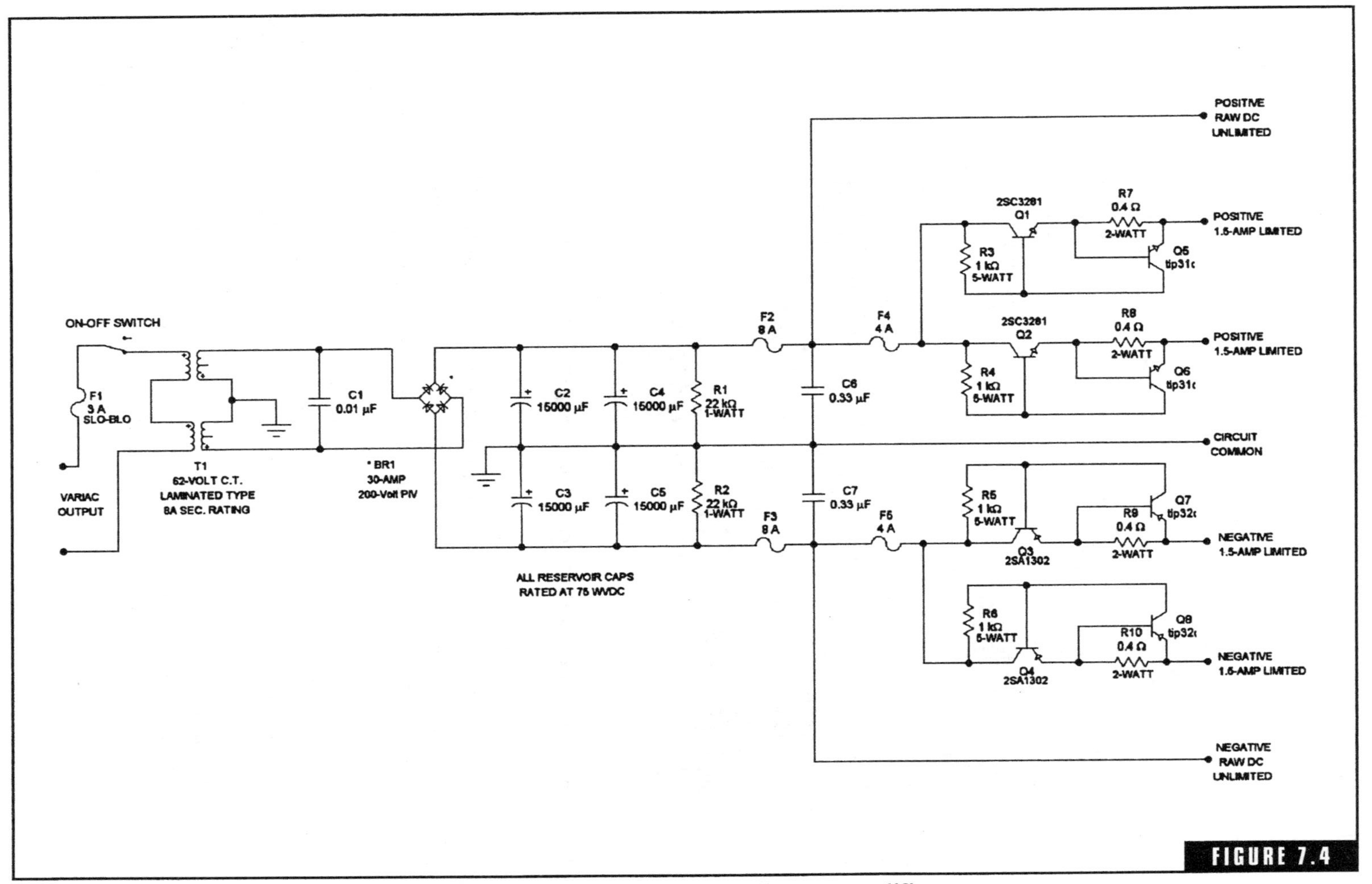

FIGURE 7.4

A versatile current-limited bench power supply for testing and setup of most audio power amplifiers.

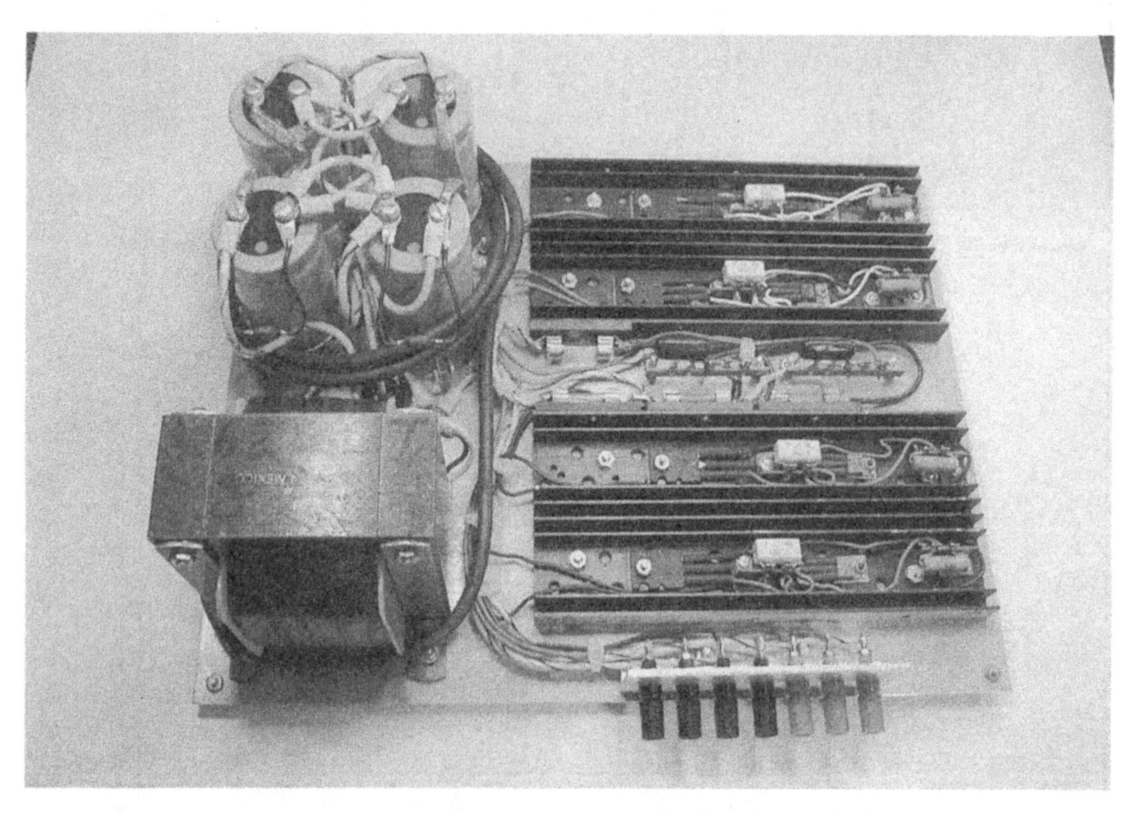

The Fig. 7-4 bench power supply. Note the additional cicuitry required for the current-limiting functions.

various types of internal signal processing circuitry. Most often, such signal-processing circuits are preamplifiers, differential line receivers, and a variety of auxiliary protection or monitoring circuits. It is often more convenient and less expensive to obtain low-voltage, dual-polarity regulated supplies by incorporating some type of *rail regulator circuit.* A rail regulator circuit will receive its operational power from the unregulated, dual-polarity DC power supply rails used to provide operational power to an associated audio power amplifier. Such a technique is usually less expensive because a separate power transformer and rectifier network is not required. It is important to remember that any type of rail regulator circuit should always have a separate, dedicated ground wire running to the HQG (high-quality ground) connection.

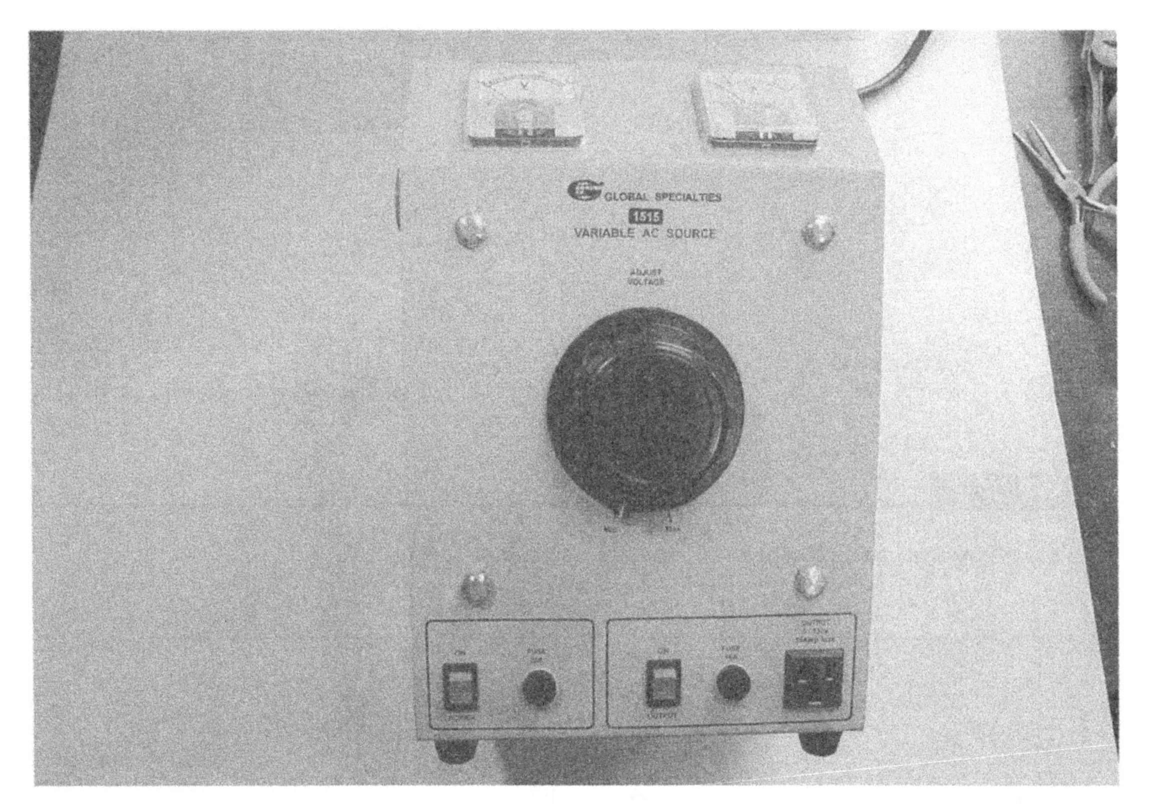

A common type of VARIAC used with bench power supplies, such as the Figs. 7-3 and 7-4 designs. The voltmeter and ammeter indicators come in very handy during test procedures.

A rail regulator circuit can be as simple as the one illustrated in Fig. 7-5. In this circuit, the majority of the rail potential is dropped across the series current-limiting resistors, R1 and R2, providing a zener regulated dual-polarity output. Figure 7-6 illustrates several improvements on the previous illustration, providing rail isolation filtering (R1, C1, R2, and C2), and improved high-frequency noise reduction provided by the tantalum capacitors (C3 and C4). The rail regulator types illustrated in Figs. 7-5 and 7-6 are commonly used in musical instrument amplifiers to provide operational power for a variety of preamplifier and tone control circuits (most incorporating multiple operational amplifiers).

In the case of some types of high-performance audio signal-processing equipment, it is desirable to provide better regulation than can be practically achieved with simple zener-regulated supplies. Figure 7-7

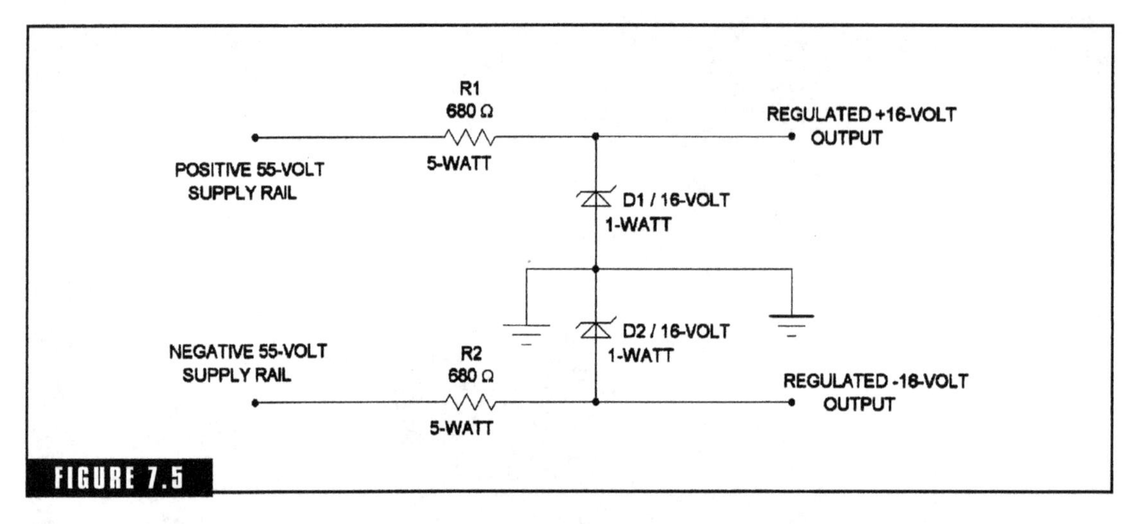

FIGURE 7.5

Simple "rail-powered" dual-polarity zener-regulated low-voltage power supply.

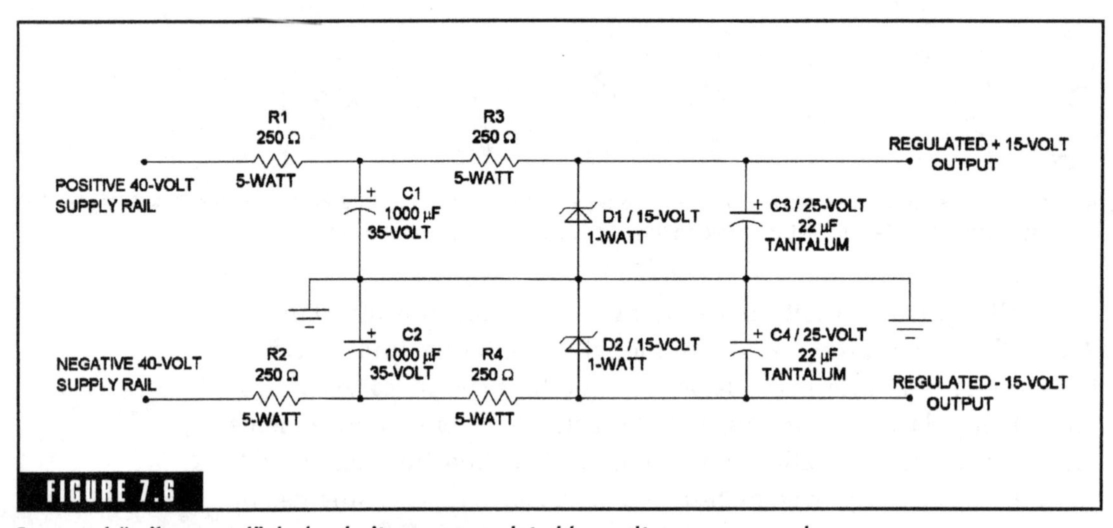

FIGURE 7.6

Improved "rail-powered" dual-polarity zener-regulated low-voltage power supply.

illustrates a high-performance rail regulator circuit, incorporating a *preregulator* circuit prior to the integrated circuit (IC) regulators. Such a preregulator circuit aids in isolating rail variations from the inputs of the IC regulators, as well as providing a reduced voltage level that will not exceed the IC regulators' maximum input/output voltage differential (usually about 40 V with most IC regulators). Referring to the positive regulator section of Fig. 7-7, D1 protects against the possibility of any polarity reversal to the regulator section that could result in component damage. R1 and R2 create a voltage divider, limiting the current to D3 and providing a desirable low-impedance pathway to circuit common. C1 and D3 provide a filtered voltage reference that is applied to the base of Q1, forcing the emitter of Q1 to be stabilized at approximately 23.3 V DC. C2 acts as a final filter stage. R5, D5, and D6 protect the positive half of the preregulator from overcurrent faults that could arise if the 7815 series IC regulator fails. The negative preregulator portion of the Fig. 7-7 design functions in an identical, but complementary, fashion. The maximum output current capability of each supply is approximately 150 milliamps.

Figure 7-8 illustrates a complete dual-polarity regulated power supply that is suitable for most signal-processing applications. This design can be used to operate signal-processing equipment requiring a little

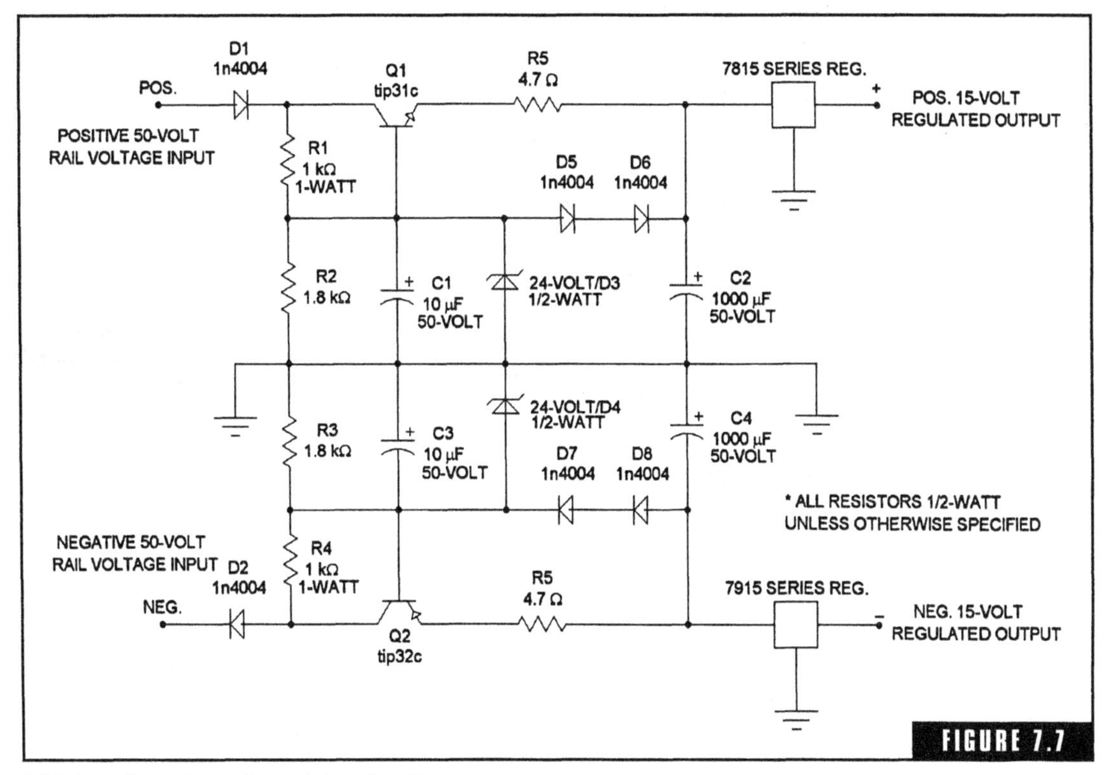

FIGURE 7.7

A high-performance rail regulator circuit.

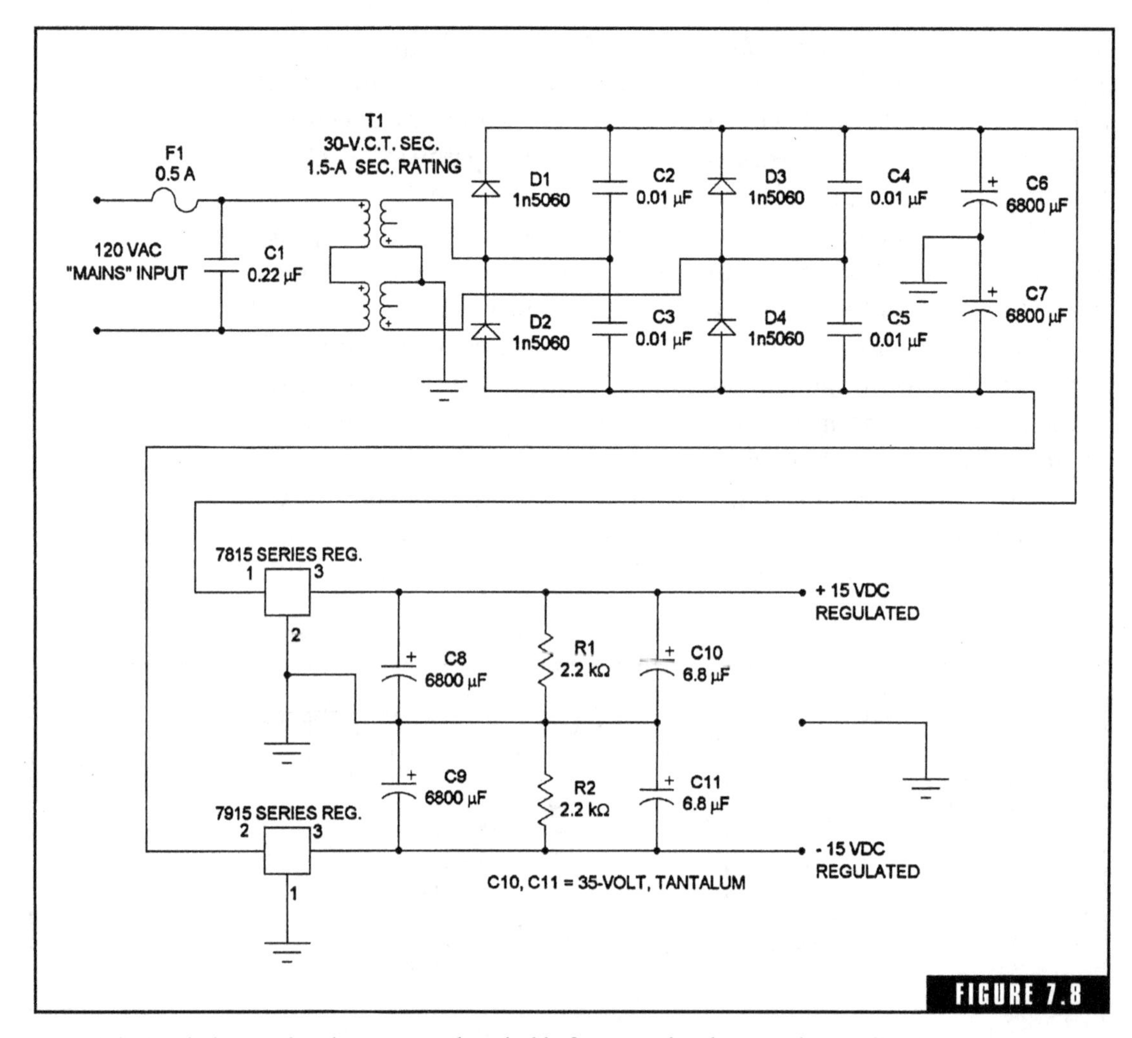

A good dual-polarity regulated power supply suitable for most signal-processing applications.

higher output current capability than normal (such as a sophisticated graphic equalizer that may incorporate up to 50 op-amp ICs), or it can be used as an excellent bench power supply for testing signal-processing equipment. The maximum current output capability is limited by the IC regulators, which is approximately 1.5 A.

The dual-output regulated supply of Fig. 7-8 is very conventional in design. The unregulated portion of the power supply consists of a discrete bridge rectifier (D1, D2, D3, and D4), snubbing capacitors (C2, C3, C4, and C5), and filter capacitors (C6 and C7). AC line filtering is provided by C1. An additional filter section is provided for the regulated

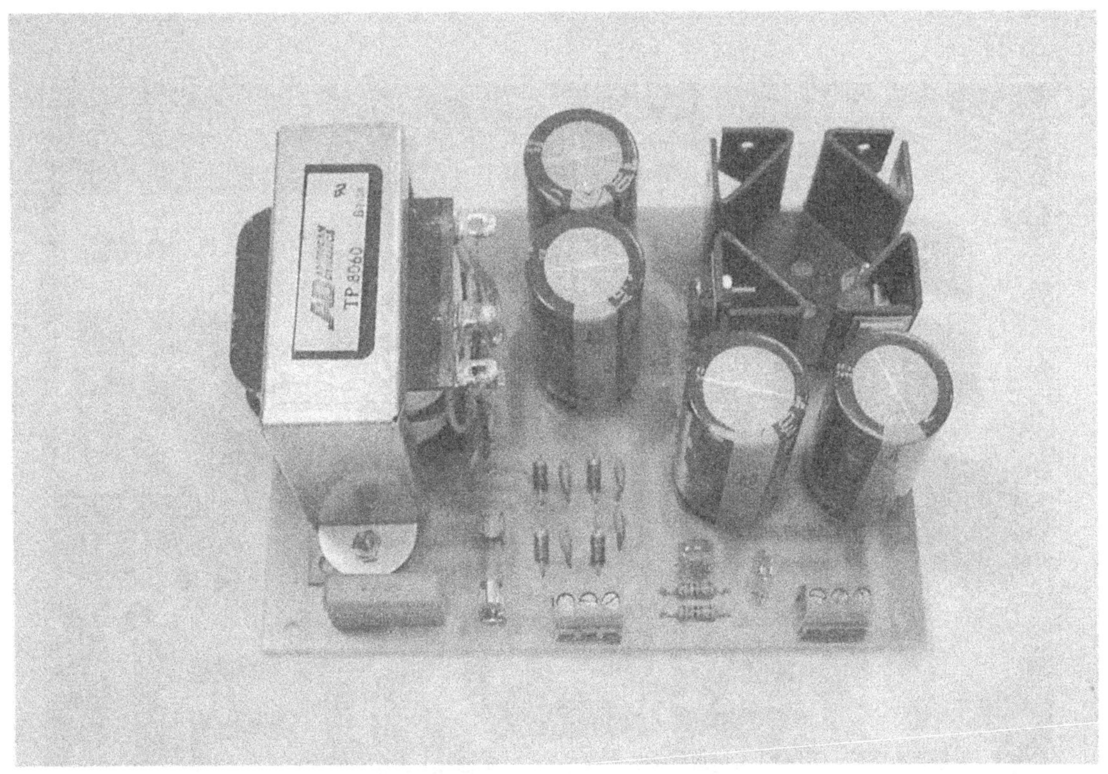

The completed Fig. 7-8 dual-polarity regulated power supply. Note the common TO-3 type heatsink used as a common heatsink for the two regulator ICs.

outputs of the IC regulators, consisting of C8, C9, C10, and C11. R1 and R2 are bleeder resistors.

Since the Fig. 7-8 power supply is a very versatile design, I have included the PC board layout and artwork illustrations (Figs. 7-9, 7-10, and 7-11). Referring to these illustrations, the pads for AC line inputs and the regulated outputs are spaced to accept a standard three-position terminal block with 0.197- to 0.2-in spacing. The heatsink mounting holes are spaced to accept a typical-style TO-3 transistor heatsink that is used to provide cooling and physical mounting for the two regulator ICs. I believe the remainder of the design is rather self-explanatory.

The following projects in this chapter have a dual purpose. They are cookbook projects that are suitable for a variety of applications. However, my primary motivation for including them is to illustrate a specific technique or methodology.

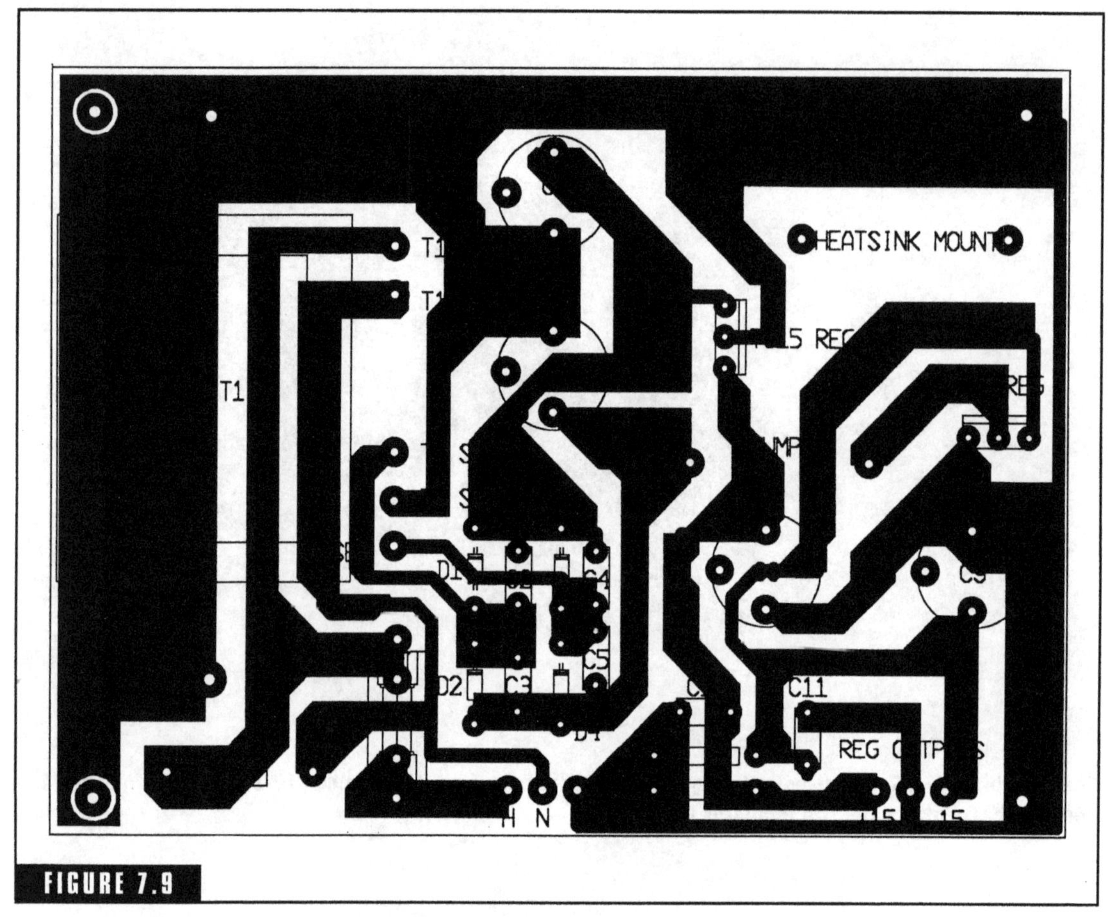

FIGURE 7.9

Top view silkscreen of the Fig. 7-8 power supply.

Figure 7-12 illustrates a technique for constructing a very low-noise rectifier/filter circuit for precision power supply applications. Diodes D1 through D4 make up a conventional discrete diode bridge rectifier. The incorporation of damping resistors (R1 through R4) and snubbing capacitors (C3 through C6) provide excellent RF and noise suppression. As general rule, the capacitance value of C3 through C6 should be increased as the current demands increase, up to a maximum of about 0.1 μF. Also, the resistance values of R1 through R4 can be decreased as current demands increase, down to minimum practical values of about 1 ohm. The remainder of Fig. 7-12 illustrates conventional filtering techniques.

Figure 7-13 illustrates how a high-performance dual-polarity regulated power supply can be constructed using a power transformer

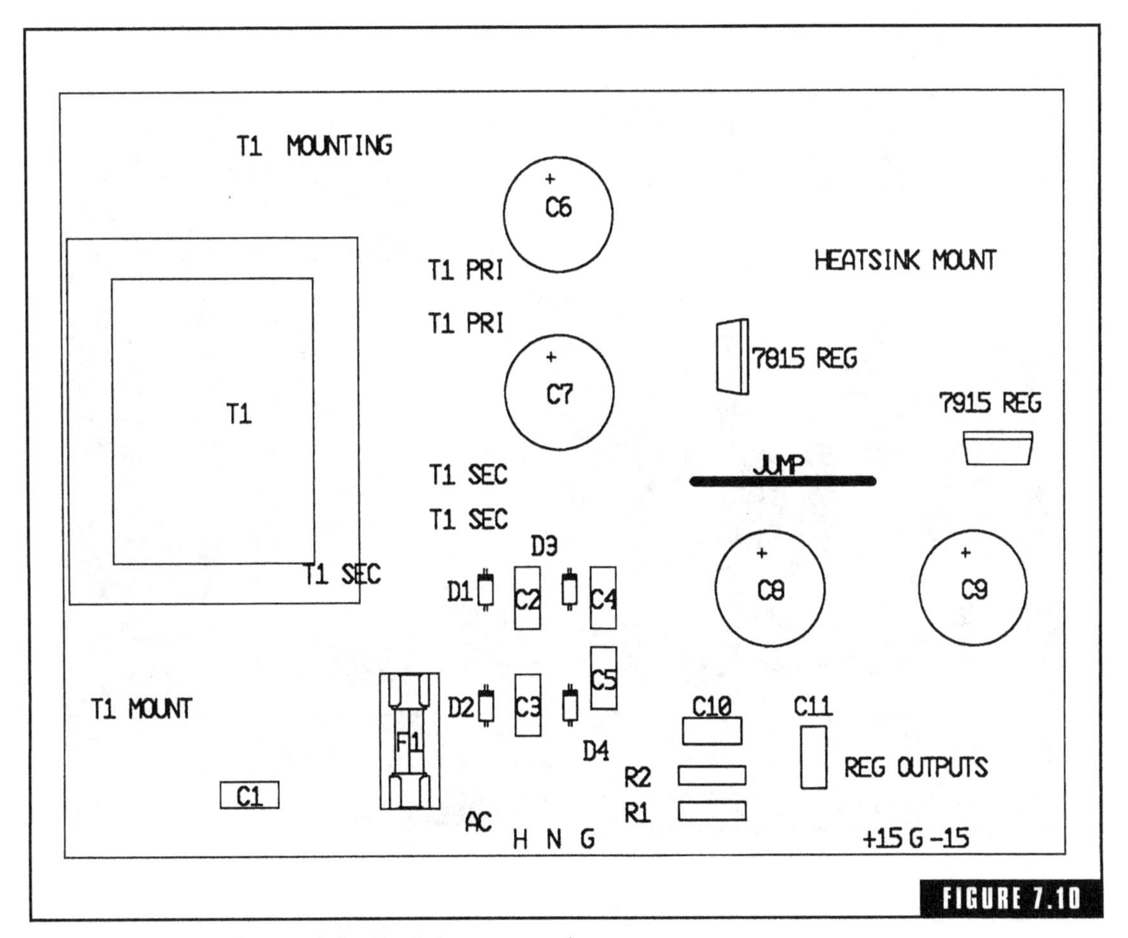

Top view component layout of the Fig. 7-8 power supply.

without a center-tapped secondary. The principles involved are a voltage doubler network (D1, D2, C2, and C3) with an input damping circuit (R1, R2, and C1), feeding a dual IC regulator circuit. The regulator circuits are of conventional design, utilizing the common LM317/LM337 adjustable regulator ICs. The regulated output voltages can be adjusted by varying the ratio of resistors R3 and R7 for the positive regulator, and R6 and R8 for the negative regulator. It is a good practice to incorporate reverse-biased diodes across the input/output terminals of IC regulators for applications wherein inductive loading might induce some high-level reverse-polarity transients (i.e., inductive kickback spikes). This is the protection function of diodes D3 and D4.

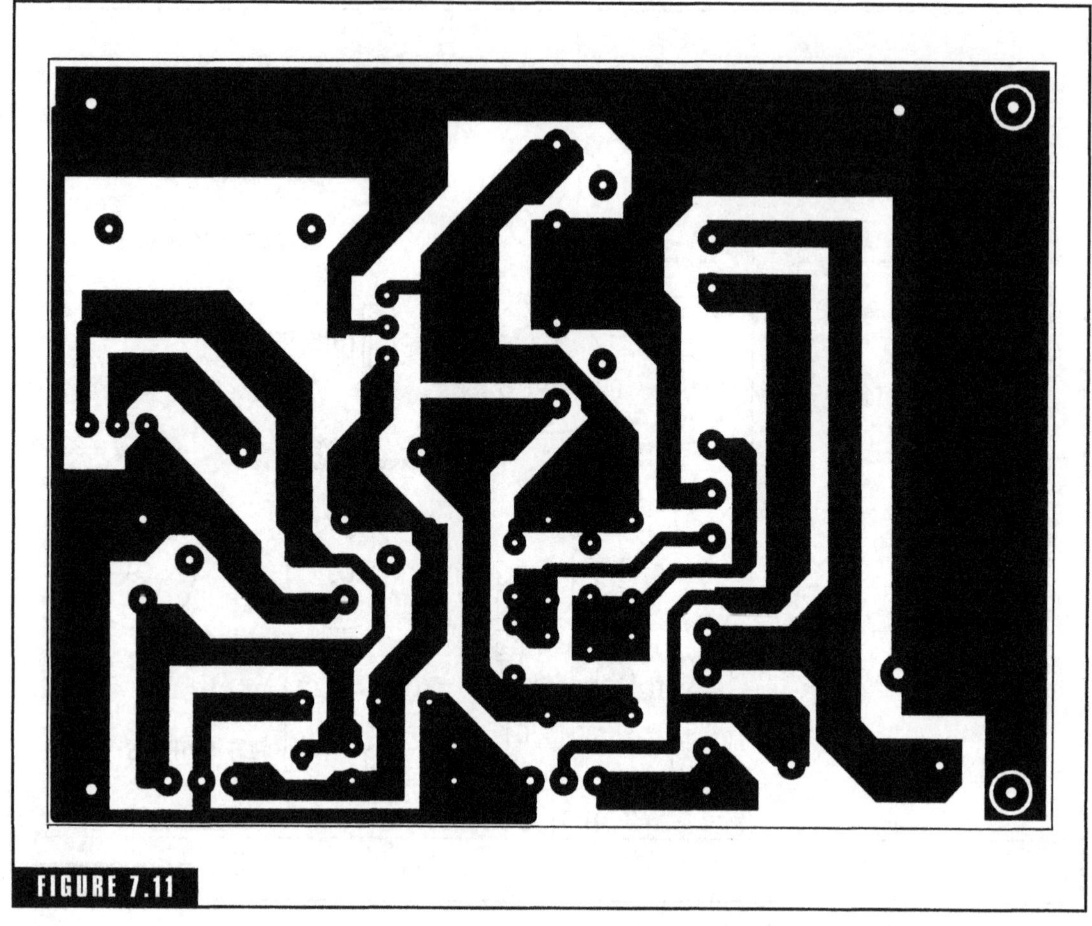

FIGURE 7.11

Bottom view reflected artwork of the Fig. 7-8 power supply.

The Fig. 7-13 power supply is intended for low-current applications, providing about 50 mA per supply. If higher output currents are required, the resistance values of R1 and R2 can be decreased accordingly, up to maximum practical output currents of about 200 mA per supply. Also, for high-performance operation under higher output current conditions, the capacitance value of C2 and C3 should be increased significantly.

Figure 7-14 illustrates a superior method of achieving high-performance operation from the common LM317/LM337 regulator ICs. The incorporation of transistors Q1 and Q2 improves the quality of the ref-

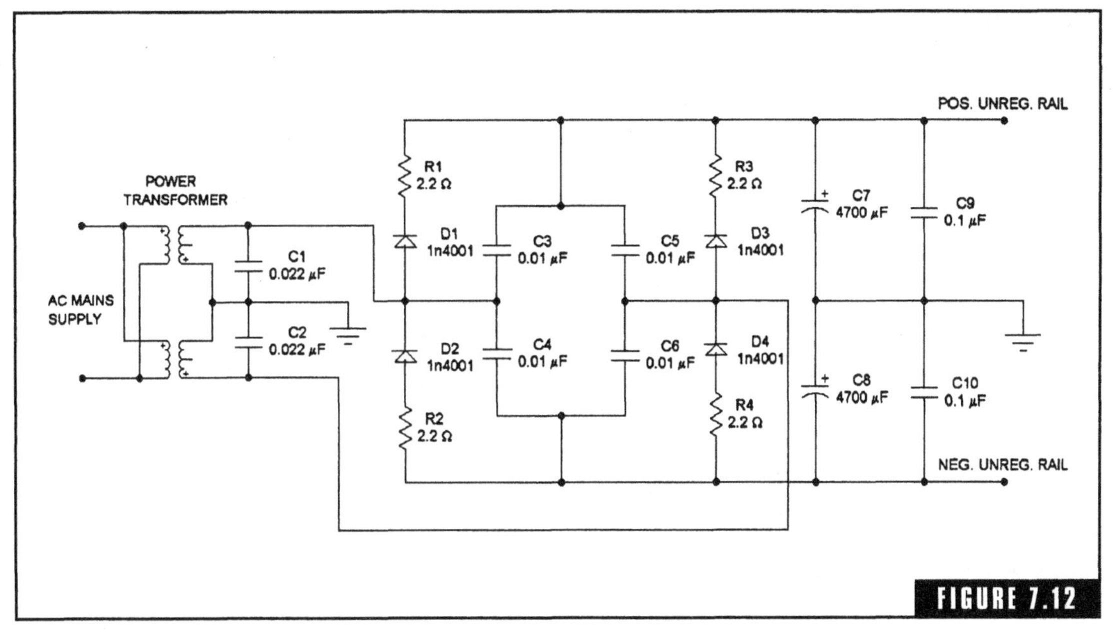

An example of a low-noise bridge rectifier for precision power supply applications.

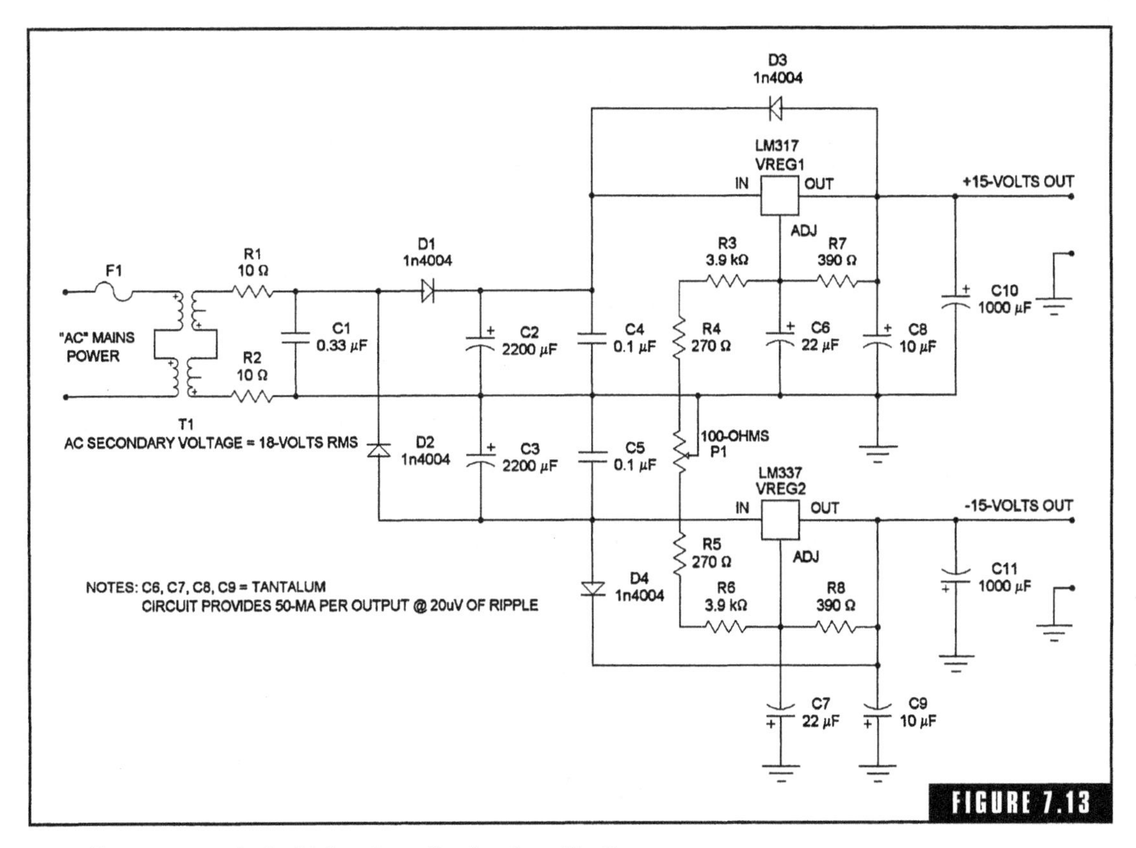

Versatile power supply for high-gain audio signal applications.

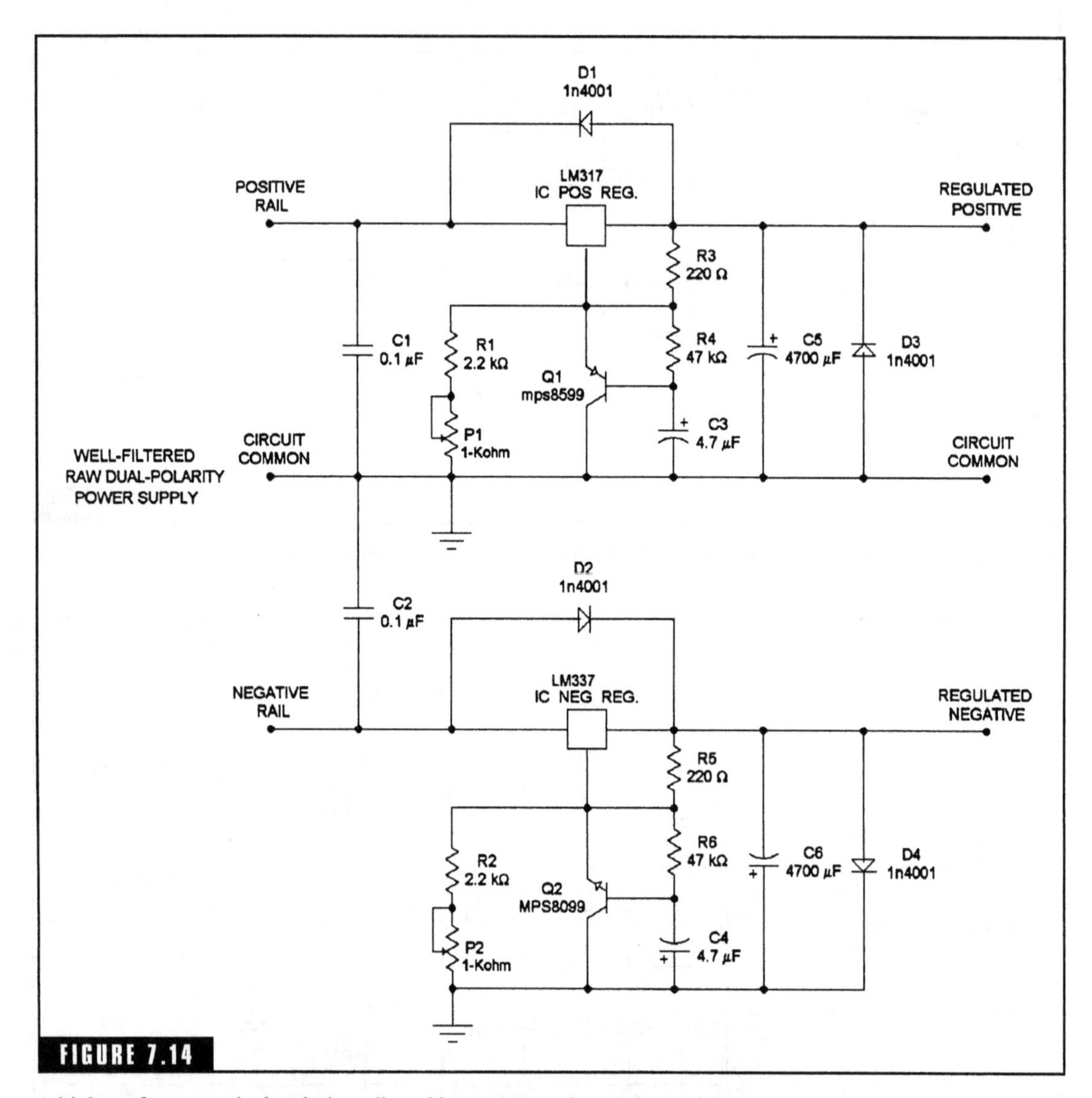

FIGURE 7.14

A high-performance dual-polarity adjustable power supply regulator circuit.

erence terminals of the IC regulators, by providing beta-enhanced voltage stabilization. The end result is improved adjustment precision and long-term stability. D1 and D2 provide the same inductive protection action as described for the previous design (i.e., Fig. 7-13). Diodes D3 and D4 provide additional reverse-transient protection.

CHAPTER 8

Protection Circuits

Protection is the afterthought of virtually every type of electronic circuit (except for protection circuits, of course). In situations involving signal-processing circuitry, the protective measures incorporated are very conventional throughout the broad realm of electrical engineering. Many of these types of conventional methodologies have already been touched upon in this book, such as zener diode overvoltage protection, the utilization of IC regulators with self-contained overcurrent and overtemperature shutdown, reverse-oriented diodes for inductive transient protection, and transistor/resistor current-sensing networks to suppress driving signals if overload conditions are detected. Coupling capacitors are a type of protection device in their function of separating destructive DC levels from desirable AC signals. Traditionally, input-output transformers have been used in the same way. A transformer-type balun provides the function of isolation protection. Resistors can also be protective devices, in that we typically design signal-processing equipment so that unanticipated short-circuit conditions will be current-limited by a properly chosen series resistance. Optical isolators are effectively used to sense various types of voltage and current conditions also.

A typical signal-processing piece of audio equipment (such as preamplifiers, equalizers, filters, effects circuits, baluns, etc.) will incorporate a well-protected power supply, with the protection usually provided by IC regulators, a few well-placed diodes, and possibly a surge suppression device on the AC power line. Consequently, if an overload or short-circuit condition should develop within the signal-processing circuitry, the power supply automatically drops to very low output values, thus minimizing various types of collateral damage. Overvoltage devices (usually diodes or transistors), overcurrent limiters (usually resistors), and/or isolation devices (either transformers or capacitors) are installed on the signal input and output lines of signal-processing equipment, thus protecting the internal circuitry from external sources of damage, as well as protecting external devices from any internal faults.

All of the aforementioned protection devices and methods are illustrated throughout this book, and are well-documented in a host of other electronic reference sources. The major protection problems relevant to high-performance audio are associated with the high-power links of the audio chain; namely, audio power amplifiers and speaker systems. The problems associated with protection are not limited to achieving effective measures of protection, but also to utilizing effective measures that are transparent to the audio signals.

DC and Transient Speaker Protection

Modern audio speaker systems are much more durable and reliable than their earlier predecessors. However, they are still quite vulnerable to destruction resulting from significant DC current levels and extreme "thumps" resulting from power-on/power-off transients (colloquially referred to as *turn-on thumps*). Of course, speaker systems are also susceptible to damage from the occasional audiophile who wants to discover how well a 400-W RMS power amplifier can drive a 50-W speaker. Usually, the audiophile will discover that such an amplifier will drive a speaker cone very well, oftentimes into the next room. The speaker protection methods discussed in this chapter will assume that the audiophile or hobbyist has the good sense to match the power output rating of the amplifier to the power-handling capability of the speaker system, so this particular issue will not be discussed further.

The types of speaker fuses installed in most commercial speaker systems are next to worthless in actually protecting the speaker drivers. Oftentimes, such fusing is only installed to meet various safety codes. Fusing, in general, cannot provide reliable protection to speaker systems, because any fuse of sufficient rating to handle the high AC currents involved will also allow significant DC currents to pass without blowing. Even fairly low DC current levels can gradually cook the voice coil of a loudspeaker into charred ruins. Also, when a speaker system is rapidly and excessively overdriven, speaker fuses will blow at about the same time that the speaker cone "pops" into speaker heaven, so the level of protection afforded by speaker fuses in this regard is not too favorable either. (By the way, is there a resurrection hope for speakers? Yes, there is! In fact, many hobbyists have discovered a very profitable business in repairing or replacing speaker cones and rubber surrounds.)

Figure 8-1 illustrates a very crude and low-performance type of DC speaker protection. Its desirable attributes are simplicity, cost, and an operational power supply is not required, but it doesn't actually provide adequate protection and it will almost certainly interject some levels of distortion into the audio signal. The principle of operation is very simple. If a DC level of sufficient amplitude should exist at the speaker terminals, the zener diode action will cause a DC level buildup across the nonpolarized capacitor (C1 and C2). If this DC level is high enough to energize the 12-V relay (CR1), the contacts of the relay will disconnect the amplifier output from the speaker system. The problems with this design are severalfold. For one, at a variety of signal-level output conditions, the protection circuitry will present an uneven loading effect, and interject some levels of distortion into the audio signal. Another problem is the requirement of about 10 V DC before the protection circuit will even activate. Such DC levels are much too high for many speaker systems. A third problem is the tendency of these circuits to false activate at very low infrabass frequencies.

I included the Fig. 8-1 protection circuit in this chapter for a couple of reasons. One, it could have an application in a higher-power automotive sound system, and possibly in a few rare cases where some protection is desired but there is no means of providing operational power to an external protection network. Also, this is the type of DC speaker protection that is found in the most popular electronic "kits" available from electronics distributors. If you were considering the purchase of one of these kits, you might want to save your money.

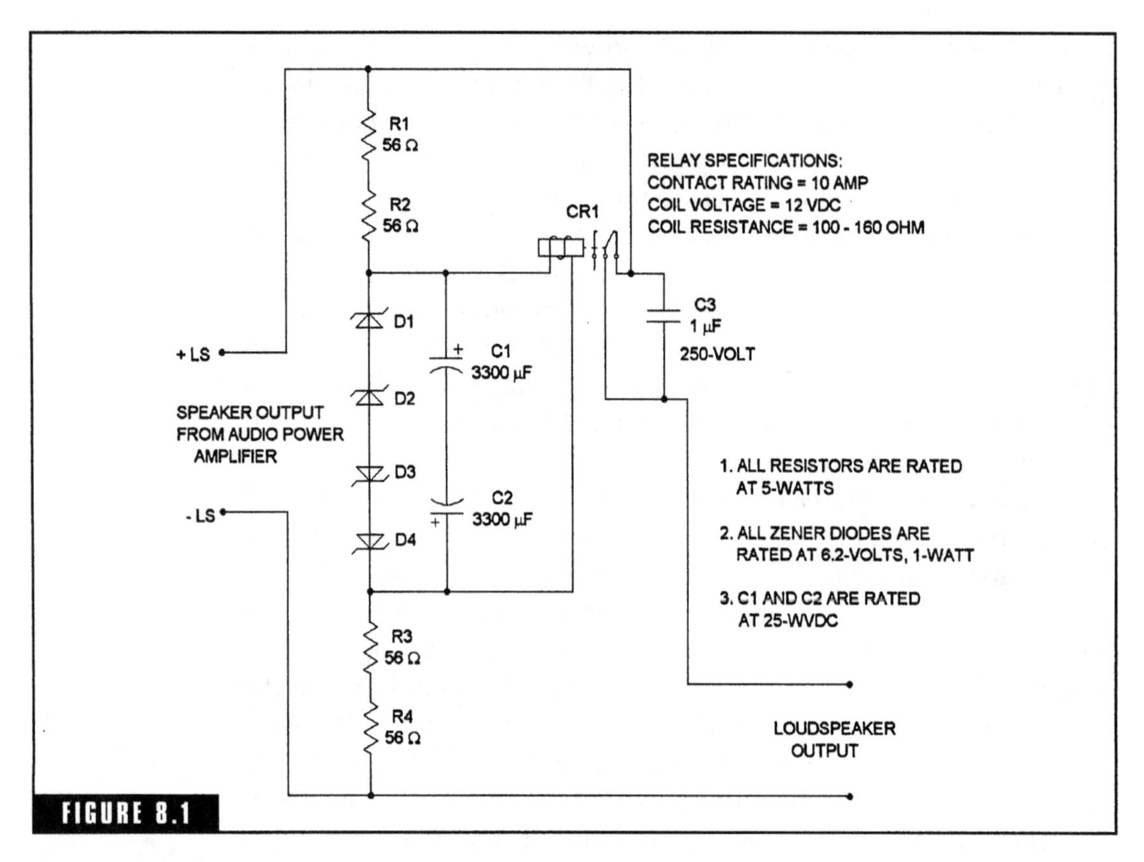

FIGURE 8.1

Simple, "self-powered" DC speaker protection circuit.

Figure 8-2 illustrates a type of DC speaker protection circuit used in many professional audio power amplifiers. It is commonly called a *crowbar* circuit. (Such crowbar circuits are also commonly used in DC power supplies to provide overvoltage protection.) Figure 8-2 illustrates the output section of a typical audio power amplifier, consisting of the Zobel network (RZ and CZ) and the output coil with damping resistor (L1 and R damping). If a significant DC level occurs on the output rail, it will charge capacitor C1 according to the time constant of the R2/C1 network. If the DC level across C1 reaches about 30 V, the diac will avalanche, providing the necessary gate current to fire the triac, and this will result in the output rail being short-circuited to ground. Normal AC audio signals could not develop a significant voltage across C1 due to the long RC time constant.

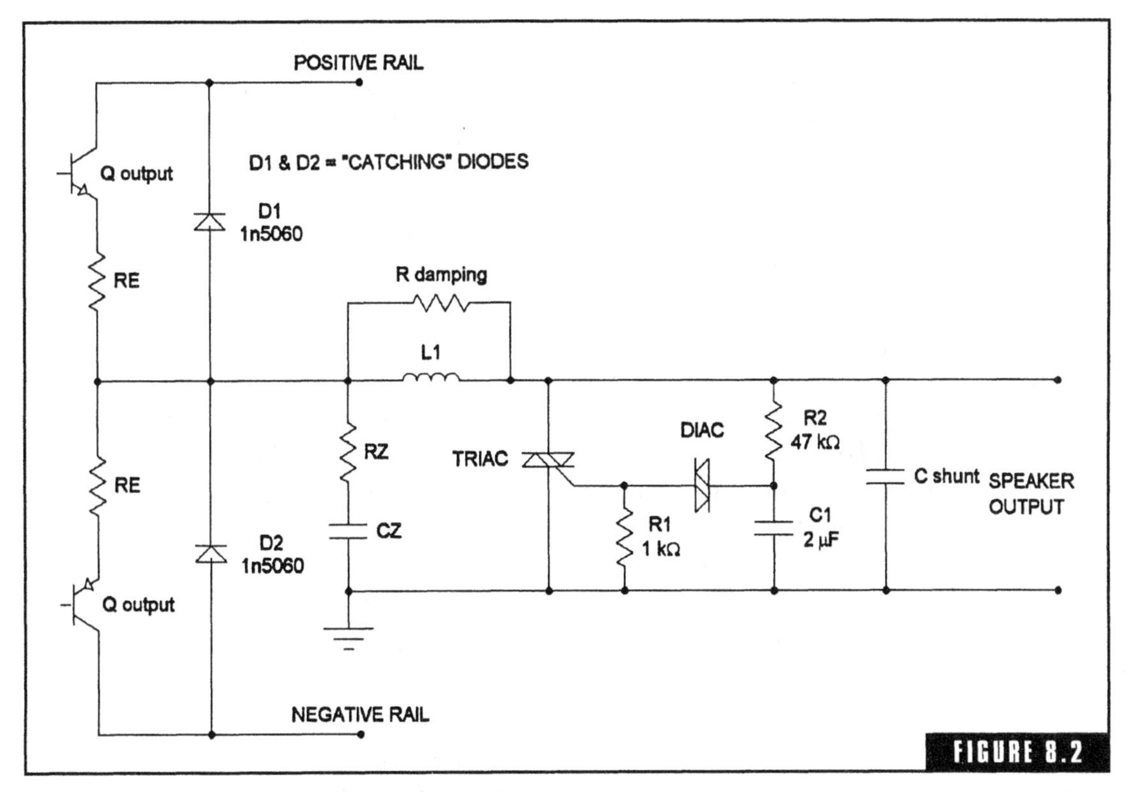

Illustration of incorporating "catching" diodes for OPS protection and the implementation of a "crowbar" circuit for DC speaker protection.

There are a couple of major disadvantages associated with crowbar-type DC protection circuits. The DC level at the output of the power amplifier has to be fairly high to build up enough DC potential across C1 to fire the diac (most diacs avalanche at between 28 and 32 V). Again, such DC voltage levels are much too high for many speaker systems. It should be remembered that undesirable DC voltages typically result due to a failure of one or more of the output devices, and output devices often fail "hard" (i.e., they develop an internal short circuit). Therefore, if these conditions arise and the triac fires, it will essentially short-circuit the power supply rail to ground, and, depending on the value of any series resistances involved, such action could destroy the triac and possibly the bridge rectifier of the power supply before the rail fuse had time to blow. In any event, such a drastic shorting action will result in enormous short-term current flow, since the reservoir capacitors in the power supply will dump all of their stored energy through the triac for the relatively long time period required to blow the rail fuse. I've experienced disintegrated components and blown-up PC boards from many high-powered audio amplifiers resulting from crowbar action.

Another problem with the Fig. 8-2 circuit is one that is seldom thought of until it is too late. It is very common (almost normal) for the triac to be destroyed when the crowbar circuit is activated (due to the previously detailed high-current conditions). Unfortunately, the triac will probably fail by developing an open-circuit between its MT1 and MT2 terminals. Since the reason behind the crowbar activation is often a failed output device, the defective amplifier will be sent to the repair shop. *Very few repair personnel will check the functional status of the triac—they will simply take a resistance measurement across it, and if it looks open, they'll pronounce it good.* So the output device is replaced, the amplifier checks good on the bench, it is given back to the owner, and the next time an output device fails (which is common in high-power audio amplifiers using bipolar outputs), the amplifier could easily destroy the speaker system it is driving. For the most part, crowbar systems are not a good method of providing DC speaker protection.

Figure 8-3 illustrates an improved method of DC speaker protection. Resistors R1 and R2 are connected to the right and left channel speaker output rails. If a DC condition arises on the output rail of either channel, it will charge the nonpolarized capacitor, consisting of C1 and C2, according to the time constant of the R1 or R2/$C_{NONPOLARIZED}$ network. This DC level will be *steered* by the diode bridge (D1, D2, D3, and D4) to impose a forward-bias polarity to the base-emitter junction of Q1, causing it to saturate. This saturation action turns off Q2, resulting in a high collector voltage which saturates Q3 and energizes the control relay (CR1). Upon being energized, CR1 disconnects the normally closed contacts connecting the amplifier to the speaker system, and will remain in this state until the DC fault is removed. Another option is to use CR1 to shut down the operational power to the amplifier with an associated latching circuit of some type (not illustrated in Fig. 8-3).

The Fig. 8-3 protection circuit provides excellent speaker protection against any DC levels that might exist at the output of a power amplifier, needing to sense only about 2 V for protection activation. The time constant established by the input sensing network is much

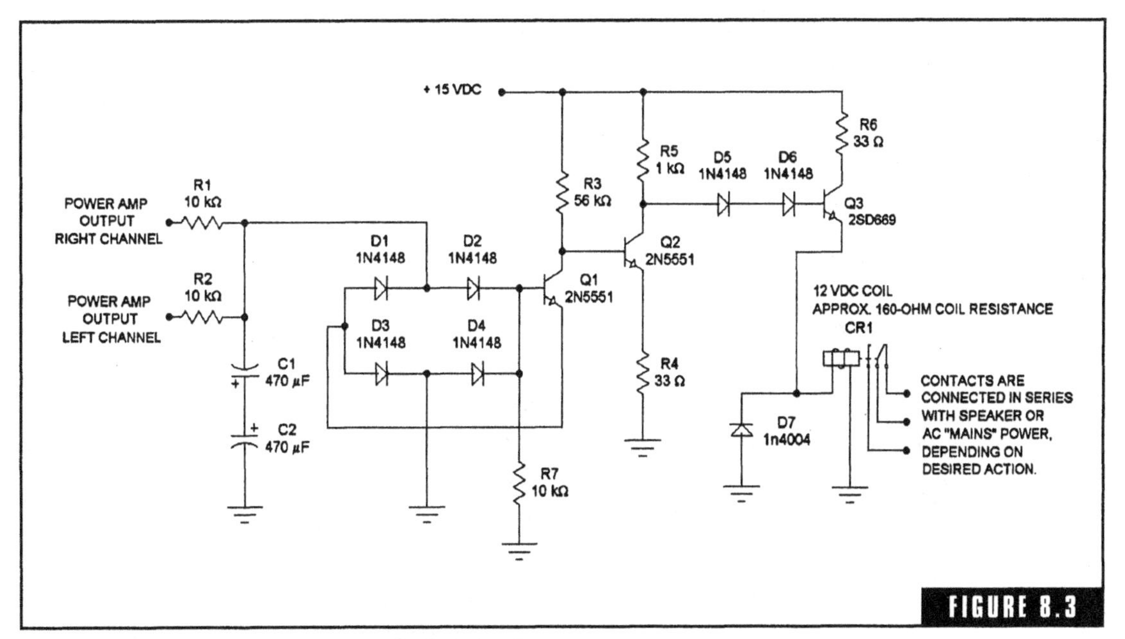

FIGURE 8.3

A versatile automatic DC speaker protection circuit.

too long to false-activate with normal audio signals. If false activation did occur during extremely low frequency infrabass passages, the capacitance values of C1 and C2 could be increased. However, I've found the illustrated values to be more than adequate. Some DC protection circuits similar to the Fig. 8-3 circuit incorporate dual-polarity-input optical isolators as a substitute for the diode bridge and Q1.

Figure 8-4 illustrates a type of DC protection circuit providing the dual function of sensing instability problems as well as DC output levels. Under adverse conditions, it is possible for a high-power audio amplifier to degenerate into self-sustaining oscillations. If this happens, the frequency will be very high, ranging from about 200 kHz in bipolar amplifiers up to the megahertz region for some MOSFET amplifiers. The Fig. 8-4 protection circuit monitors for excessive DC levels as well as AC levels that would occur across L1 in the event of high-frequency oscillation. In addition, it provides a time-delay turn on to the speaker system to avoid irritating and potentially destructive power-on/power-off transients.

Referring to Fig. 8-4, RZ, CZ, L1, and R_{DAMP} make up the Zobel network and output coil/damping resistor of any typical audio power amplifier. Resistors R1 and R2 sense excessive AC or DC levels across the series network of L1 and RS. The sensed voltage level will be steered or rectified by the diode bridge BR1. This rectified voltage will be applied to the RC network of R3 and C1, and also to the bleeder resistor R4. Thus, the only way that C1 can charge to a sufficient level to turn on Q1 is if the sensed voltage condition is sustained for a substantial period of time. Such a condition would only occur if the amplifier was outputting a DC level or sustained high-frequency oscillations.

Note that R6 and R7 form a voltage divider from the negative power supply rail (of the power amplifier), which tries to apply a negative base bias to Q2. However, this negative bias is canceled out by a positive voltage developed across C2, from the rectification action of D2, which is connected to one of the power transformer secondaries. The summing action at the base of Q2 is such that the positive potential applied by R5 is higher than the negative potential applied by the R6/R7 voltage divider. Thus, the base of Q2 is held at a positive potential and remains cut off under normal circumstances.

If a sustained error condition does occur, causing C1 to charge to sufficient levels to saturate Q1, Q1 shorts the positive potential at the base of Q2 to the output rail of the amplifier, through the D1 blocking diode. This results in a negative potential appearing at the base of Q2, which saturates it and cuts off Q3, deenergizing the control relay, CR1. When CR1 deenergizes, it disconnects the speaker from the output of the power amplifier. Note also that the *failsafe* condition of CR1 is such that the speaker is disconnected from the amplifier when the relay is deenergized, which is a desirable attribute.

The turn-on delay provided by the Fig. 8-4 circuit is achieved by the simple charging action of C3 from the negative supply rail. When power is first applied to the amplifier, the speaker is disconnected from the power amplifier due to the failsafe status of CR1. The base voltage of Q3 is very low until C3 has time to charge through R8. After about 1 s, the voltage at the base of Q3 will increase to a sufficient level to saturate Q3, thus energizing CR1 and connecting the speaker to the output of the audio power amplifier. R9 is incorporated to limit C3's discharge current when Q2 is saturated during an error condition. D4 protects the circuit against inductive transients from the coil of CR1, and D3 is installed to improve the circuit's rejection of false triggering (i.e., it decreases the sensitivity of Q3's base voltage). With all things considered, this is a pretty nifty little protection circuit.

The biggest disadvantage to the Fig. 8-4 protection circuit is that it requires a resistor (RS) to be placed in series with the speaker load.

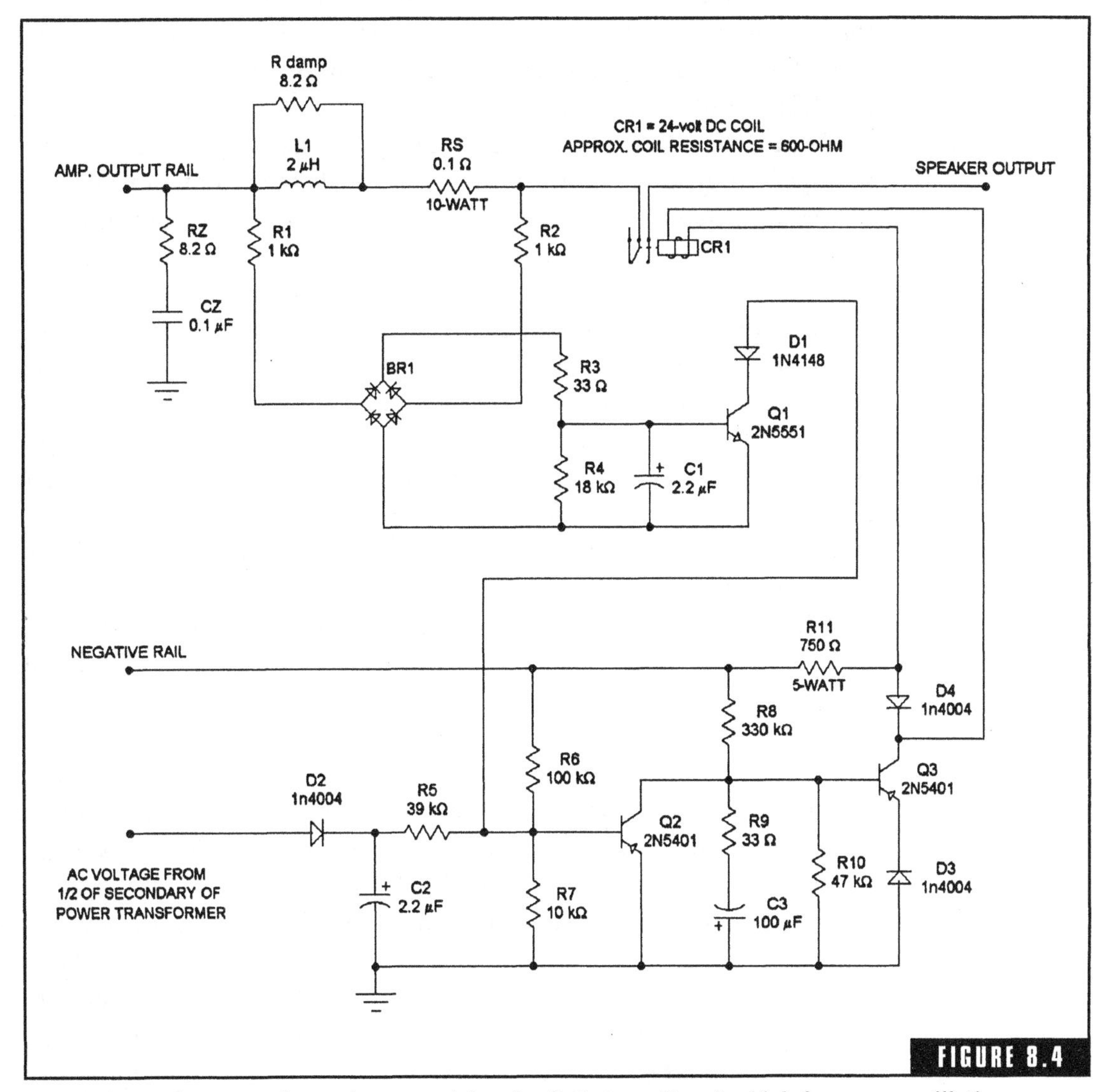

FIGURE 8.4

Automatic speaker protection and turn-on delay circuit that monitors for high-frequency oscillations.

Any resistance in series with a speaker system will reduce the damping factor of the power amplifier. However, the very small 0.1-ohm resistance required for the Fig. 8-4 circuit is typically insignificant relative to most speaker systems.

Figure 8-5 illustrates one of my favorite DC speaker protection systems. In addition to DC protection, it provides on-delay speaker connection, which avoids power-on/power-off transients, and it also includes a unique status display. Another nice feature of this design is that it operates entirely from its own power supply, which is more desirable than having the protection circuitry operate from the power amplifier's rail supplies.

Beginning at the left-hand side of Fig. 8-5, the AC power inputs can be supplied by tapping off a portion of the AC secondary power of the power amplifier transformer. For example, if the power amplifier rail supplies are being powered by a toroidal transformer with two 35-V secondaries (equating to approximately 48-V rails), one of the AC power inputs of the protection circuit would be connected to center-tap connection point of the transformer, and the other input would be connected to one secondary connection. This would provide a half-wave rectified voltage across C1 of about 48 V, which would be regulated down to 24 V DC by the simple zener regulator circuit of R1 and ZD1. Depending on the secondary output voltage of the power transformer, you may have to adjust the resistance value and power rating of R1. Another option is to provide AC power from a small dedicated transformer, with a secondary voltage rating of 24 V and a current rating of 250 mA. If this option is chosen, R1 should be changed to a 220-ohm, 2-W value.

The two R_{IN} resistors connect to the speaker outputs of the right and left channel of a typical stereo power amplifier. Only one resistor is needed for monaural systems, and more input resistors can be added for multichannel power amplifiers. As in the previous DC speaker protection systems that have been detailed, any DC levels appearing at the speaker outputs will charge the nonpolarized capacitor (C5 and C6) according to the time constant of the R_{IN} resistors and the capacitance value of the nonpolarized capacitor. This DC level will be steered by the diode bridge (BR1) and applied to the base of Q5, saturating Q5 if the DC input voltage exceeds about 2 V. If Q5 saturates, the input to the Darlington pair (Q3 and Q4) is pulled low, cutting off transistor Q4,

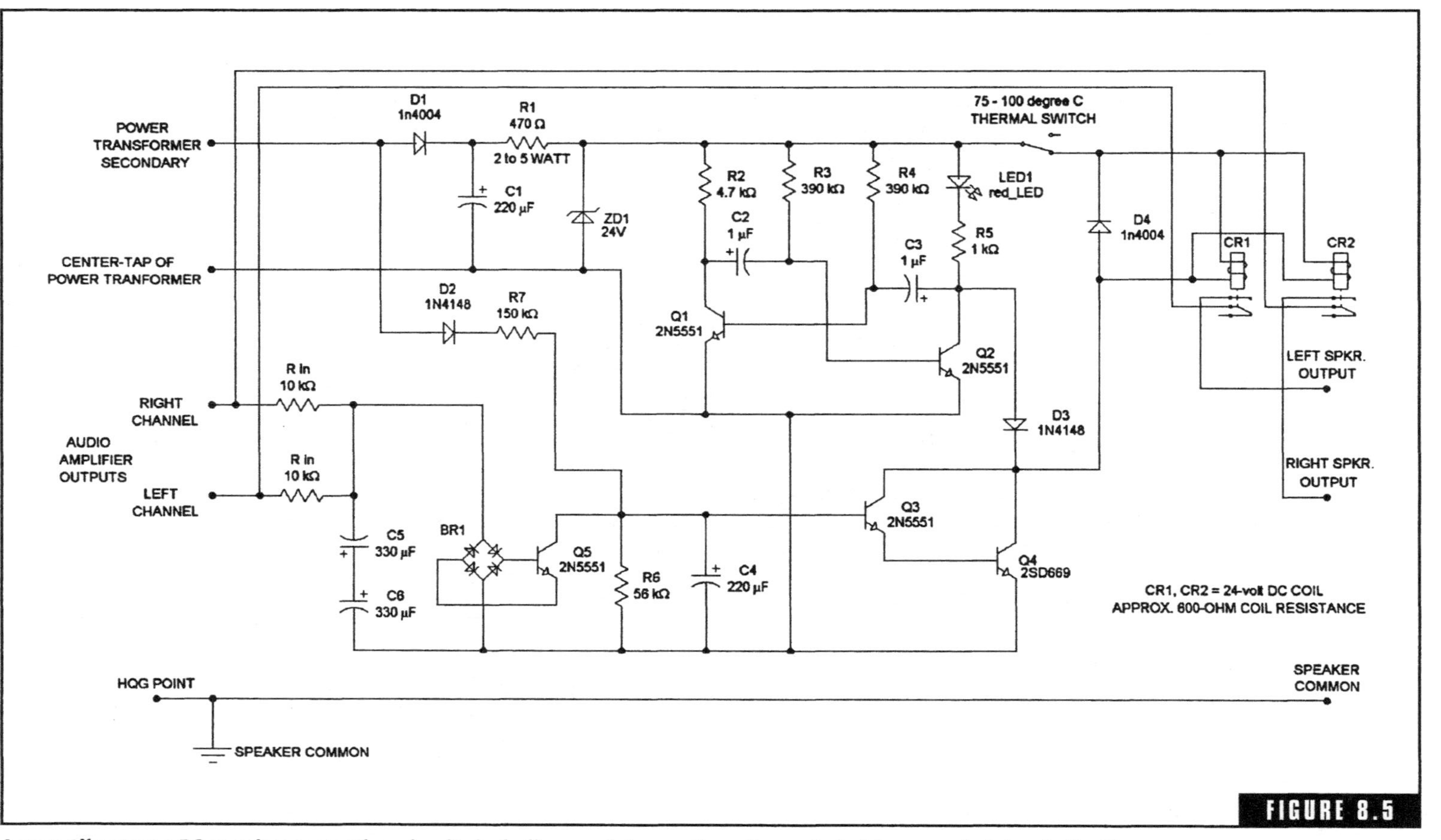

FIGURE 8.5

A versatile stereo DC speaker protection circuit, including on-delay muting, thermal shutdown, and status indication.

and deenergizing the control relays, CR1 and CR2. When the relays are deenergized, the speaker systems are disconnected from the power amplifiers. A single double-pole, double-throw (DPDT) relay can be substituted for the two single-pole, double-throw (SPDT) relays illustrated in Fig. 8-5.

The on-delay time period is a function of the charge rate of C4. When circuit power is first applied, D2 rectifies a portion of the AC operational power, charging C4 according to the time constant of R7 and C4. After 1 to 2 seconds, C4 will charge to a high enough potential to saturate the Darlington pair (Q3 and Q4), with Q4 energizing the control relays through the circuit's 24-V power supply, which connects the power amplifiers to the speaker systems. R6 is a bleeder resistor to discharge C4 rapidly if operational power is turned off, so that the protection circuit cannot immediately energize the control relays if the control circuit happens to be turned off and back on quickly. The thermal switch is normally mounted to the power amplifier's main heatsink. If the heatsink overheats, the thermal switch will open, disconnecting operational power from the control relays, causing them to deenergize and open the connection between the power amplifier and the speaker systems. Since the power amplifier will no longer be applying power to the speaker systems under these conditions, the main heatsink should rapidly cool down. If you want to thermally protect more than one power amplifier, you can place several thermal switches in series in the same circuit path.

The status indication circuit of Fig. 8-5 consists of Q1, R2, C2, R3, R4, C3, LED1, R5, Q2, and D3. These components make up a conventional astable multivibrator that will flash LED1 at a rate of about 2 Hz. When operational power is first applied to the protection circuit, the astable multivibrator will immediately activate and begin flashing LED1. This is an indication that the on-delay time period is in effect. After several seconds, when Q4 saturates, energizing the control relays, the collector of Q4 will go low, inhibiting the oscillations of the astable multivibrator, causing LED1 to light continuously. This is an indication that the speaker systems are connected and the status is normal. If a thermal overload occurs, opening the thermal switch, the control relays will lose their operational power and de-energize. Without the smoothing effect of the coil inductances, the collector load of Q4 is able to rapidly follow the oscillations of the multivibrator circuit, but the collector of Q2 sees a substantially lower impedance than when Q4 is cutoff (i.e., during the on-delay period). The end result is a very rapid blinking of LED1, indicative of a thermal overload. If a DC voltage appears at one of the amplifier outputs, the DC protection function of the circuit will activate, as previously described, causing Q4 to go into cutoff. This action causes the multivibrator circuit to begin blinking at its normal on-delay rate, but it will continue blinking for as long as the DC fault condition exists. Thus, the one LED can provide a four-function status indication of the protection circuit. In summary, they are:

1. Slow-blinking, stopping after a few seconds—normal power on-delay
2. Continuously lit—normal operation, speakers connected
3. Fast-blinking—thermal overload, speakers disconnected
4. Slow-blinking, does not stop after a few seconds—DC fault

When incorporating the protection circuit of Fig. 8-5, LED1 is typically mounted to the front panel of the associated power amplifier(s), becoming the "power" indicator as well as status indicator.

Before ending this section, I'd like to discuss a few general concerns of auxiliary protection circuits. It is extremely important to remember to provide dedicated ground lines back to the HQG point for all auxiliary circuits (if required). In other words, never attach the circuit ground of auxiliary circuits to any kind of signal ground or sensitive ground area. A large portion of the hum problems that I hear about from audiophiles and hobbyists relates to erroneous grounding connections of auxiliary circuitry.

A controversy that keeps surfacing from time to time is debate over the degrading influence of relay contacts in line with the audio output signal. There have been some weird claims made in this regard, with so-called discoveries of "microdiodes" inherent to relay contacts, and a whole raft of sonic problems supposedly created from contact corrosion. In truth, relays are simple electromechanical devices, performing a simple function. It is possible for relay contacts to become pitted, or corroded, and provide poor conductivity. Extreme cases of such problems will drastically affect sonic quality.

However, due to the simple physics involved with relay contact operation, if a significant conduction problem develops with the contacts, they will degrade rapidly, and the levels of distortion will be very obvious. Audiophiles seem to worry that subtle levels of relay contact distortion will build up progressively in time, but after more than 22 years of working with high-power audio amplifiers, I have not personally experienced even one such case. In short, relays function well while they are functioning, and when they fail, they fail in a "hard" obvious manner. I have no reservations about incorporating muting relays and protection relays in my personal audio systems.

One of the lessons I have learned over the years is that there doesn't always have to be a good reason for doing something. When my father resigned from the Army, his ambition was to buy a new car with *holes in the side.* So he got a job and saved his money until he could afford to buy a brand new 1954 Buick, with big holes in the sides. Then he was happy! There wasn't any practical reason at all for the holes. Likewise, there are many audiophiles who do not want any relay contacts in series with the audio signal, for whatever reason they harbor. There isn't anything wrong with wanting to have things *your way,* especially when it pertains to your own personal hobby. So if you fall into this category, just remember that the control relays incorporated into the protection circuits of Figs. 8-3, 8-4, and 8-5 can be used in a variety of ways to shut down the operation of the power amplifier in the case of a fault. I leave these options to your ingenuity and imagination.

Soft-Start Circuits

Soft start is a general name given to any kind of circuitry that limits current surges when the operational power is first applied to an electronic or electrical system. Soft-start circuits are most commonly used to limit the huge surge currents associated with starting up large electric motors. They are also commonly used in audio power amplifiers to limit the surge currents associated with the initial charging of the reservoir capacitors in the power supply. Depending on your applications and the quantities and types of audio power amplifiers you employ in your system, you may want to install a soft-start circuit of some type. In the case of most domestic hi-fi systems, soft-start circuitry is usually not necessary, but it is a luxury. It can get irritating to have to put up with the lights flickering every time you turn on your audio system. Surprisingly, not all large professional amplifiers incorporate soft-start systems, and this is a real pain to sound engineers. Typically, after you power up about three such professional amplifiers that are operating from the same AC distribution panel, and then turn on the fourth, the other three go off due to AC line fluctuations. So you have to start powering up all over again. As a general rule, all professional audio power amplifiers should utilize some type of soft-start system.

Figure 8-6 illustrates a common type of thyristor soft-start system. Essentially, this is nothing more than a high-power light dimmer with an automatic ramp-up feature. When the power switch is initially closed, C2 begins to charge through the resistance of R1 according to the polarity established by the bridge rectifier of D1, D2, D3, and D4. C2 charges between the voltage levels of ZD1 and ZD2, which help to make the soft-start action begin immediately and advance smoothly. As C2 charges, the AC voltage drop across the entire bridge rectifier rises (i.e., because the impedance of the bridge rectifier is increasing), causing the AC voltage level across P1 to rise. The rising voltage across P1 varies the phase/amplitude characteristics of the AC voltage developed across C1 (as in the case of any common light dimmer), varying the avalanche pulses through the diac, and eventually causing the triac to conduct for virtually the entire AC cycle. P1 is adjusted so that the end of the ramp sequence results in the full line voltage across the primary of the power transformer. The capacitance value of C2 controls the duration time of the ramp sequence, while R2 acts as a bleeder resistor to discharge C2 rapidly when the AC power is turned off.

The purpose of the Fig. 8-6 soft-start circuit is to increase the line voltage across the power transformer primary in a gradual manner, allowing the reservoir capacitors to charge slowly, and thus eliminate any significant AC line surges.

Thyristor soft-start circuits are the most common types of soft-start systems incorporated into commercial power amplifiers (if the power amplifier happens to utilize a soft-start system). They are inexpensive, very reliable, and relatively small in size. However, they are not without their faults. Thyristors (i.e., triacs) are extraordinarily noisy during the ramp-up sequence, impressing high levels of line noise on the outgoing AC line as well as internally to the amplifier components. This situation is usually not a problem if the power amplifier contains a

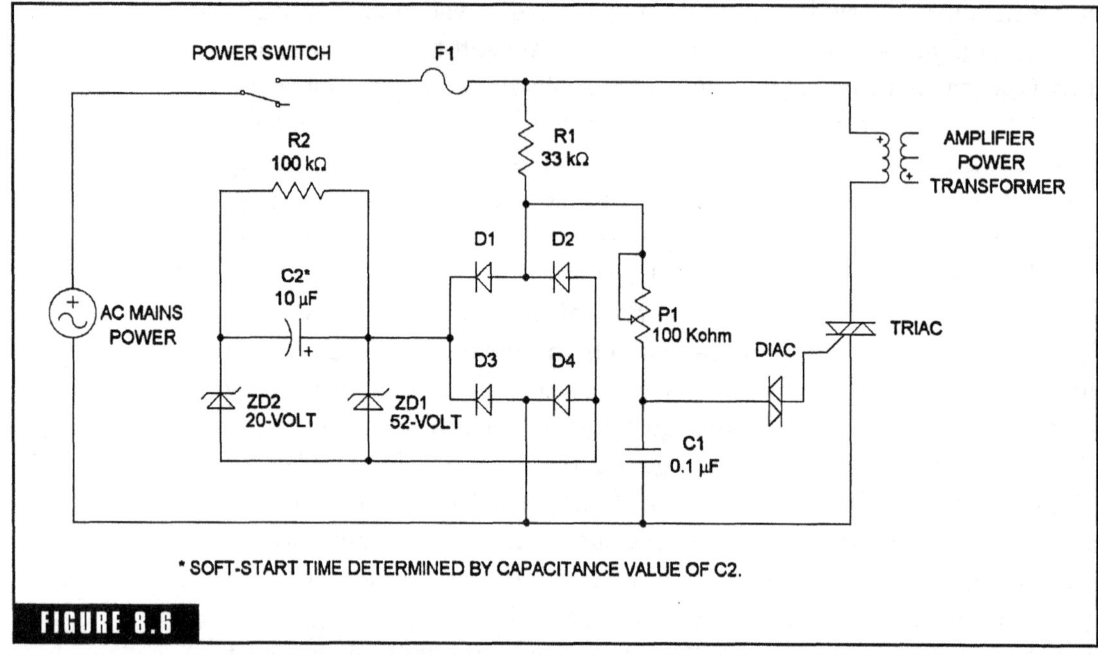

FIGURE 8.6

A simple thyristor-based soft-start circuit.

muting relay with on-delay activation. But if the power amplifier is directly coupled to the speaker during the ramp sequence, the speaker noise can be disturbing and potentially destructive to sensitive tweeter systems. In addition, the noise feeding back into the AC line can affect other equipment that happens to be plugged into the same AC distribution system.

Figure 8-7 illustrates another method of providing soft-start action to a typical audio power amplifier. Such systems are fairly common in esoteric audiophile amplifiers. When the power switch (S1) is initially closed, C3 begins to charge from the rectified DC current provided by the bridge rectifier of D1, D2, D3, and D4. Rather than using a high resistance value to slow down the charge rate of C3, this design uses the capacitive reactance of C2 to impede the charge current flow. It takes about 1 second for C3 to charge to a voltage level sufficient to energize CR1. During this 1-second period, the AC mains power is applied to the primary of power transformer T2 through current limiting resistors R3, R4, R5, and R6, which provide a parallel equivalent resistance of about 45 ohms. This resistance is sufficient to limit the

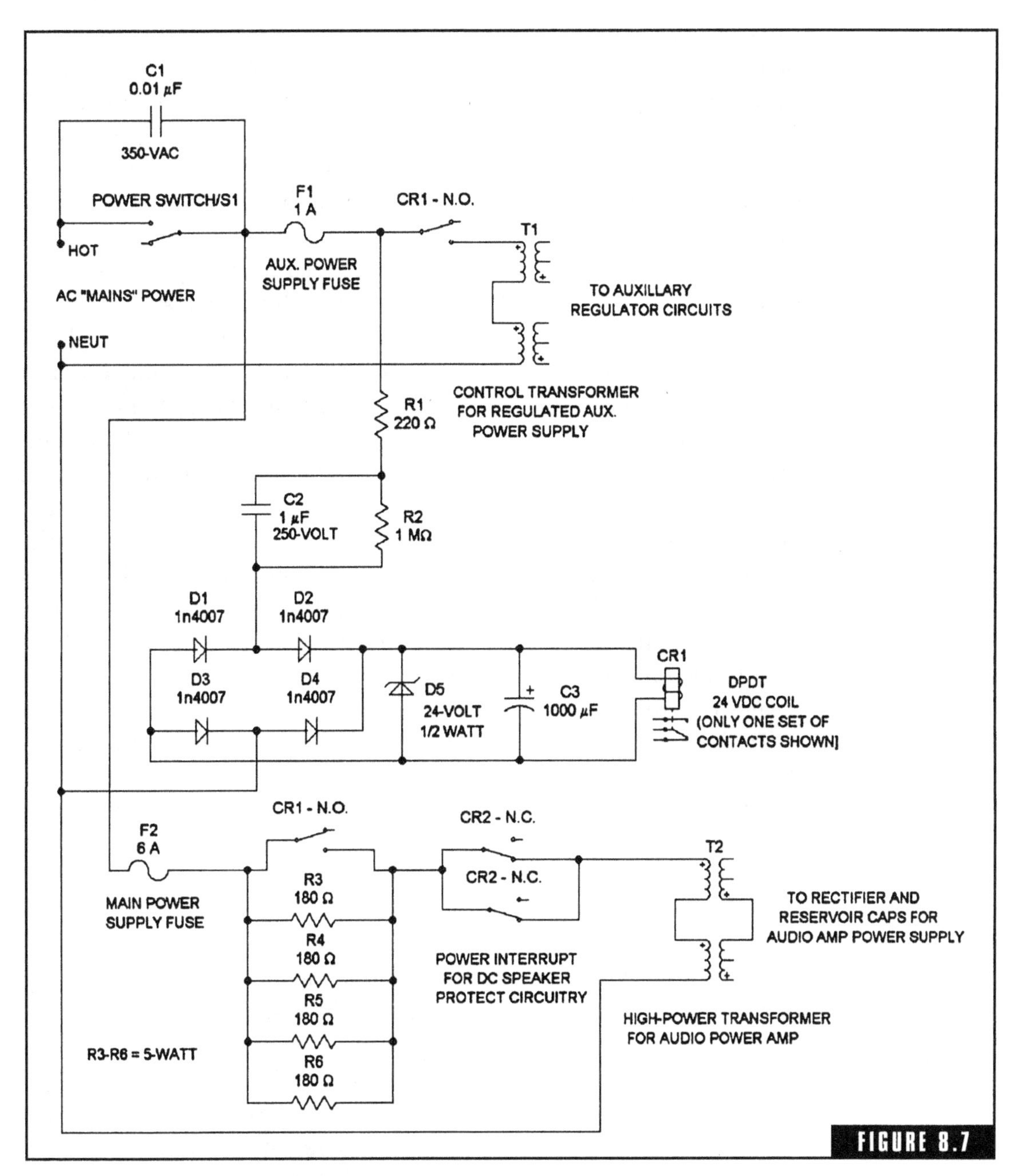

A relay controlled soft-start system.

surge current to under 3 A. After the 1-second delay is completed and CR1 energizes, CR1's contacts short out the current-limiting resistor bank, allowing unimpeded current flow to the primary of power transformer T2. By this time, the power supply's reservoir capacitors are charged to a sufficient level to minimize line surges, and there is virtually no electrical noise generated from this type of resistive current limiting.

There are some other refinements included in the Fig. 8-7 circuitry that should be mentioned. C1 is placed across the power switch (S1) to eliminate annoying "pops" that could occur when S1 is turned off. Note that one set of CR1's contacts is used to apply AC mains power to the auxiliary transformer T1. In this particular design, the auxiliary transformer is used to provide operational power to a solid-state balun input circuit. Thus, by not providing operational power to the balun circuit until after the soft-start sequence is over, all input signals to the audio power amplifier are inhibited (or gated) during this time. Such an action is desirable, because the power amplifier in this case is directly connected to the speaker system (without any muting relay action), and its circuitry is stabilizing during the soft-start routine, which would cause any simultaneous audio input signals to be severely distorted.

In the Fig. 8-8 illustration, I have included all of the individual circuits in a typical auxiliary control and protection system. Such a system is intended for a directly coupled speaker system (i.e., without a muting or protection relay between the power amplifier and the speaker system), and it is designed to be constructed on a single PC board assembly. The groupings of three dots indicate the edge connections of the PC board.

The soft-start design is identical to the previously described Fig. 8-7 circuit. Upon closing the on-off switch, the resistor bank (R3, R4, R5, and R6) will be in series with the primary of the high-power transformer (i.e., the audio power amplifier transformer) until the soft-start sequence is completed, after which time the resistor bank will be short-circuited by CR1's relay contacts. When CR1 energizes, it will also provide AC operational power to the auxiliary transformer, which provides operational power to a solid-state balun circuit, allowing the audio input signal to start being processed by the power amplifier. If a thermal overload should occur, the normally closed contacts of the thermal switch will

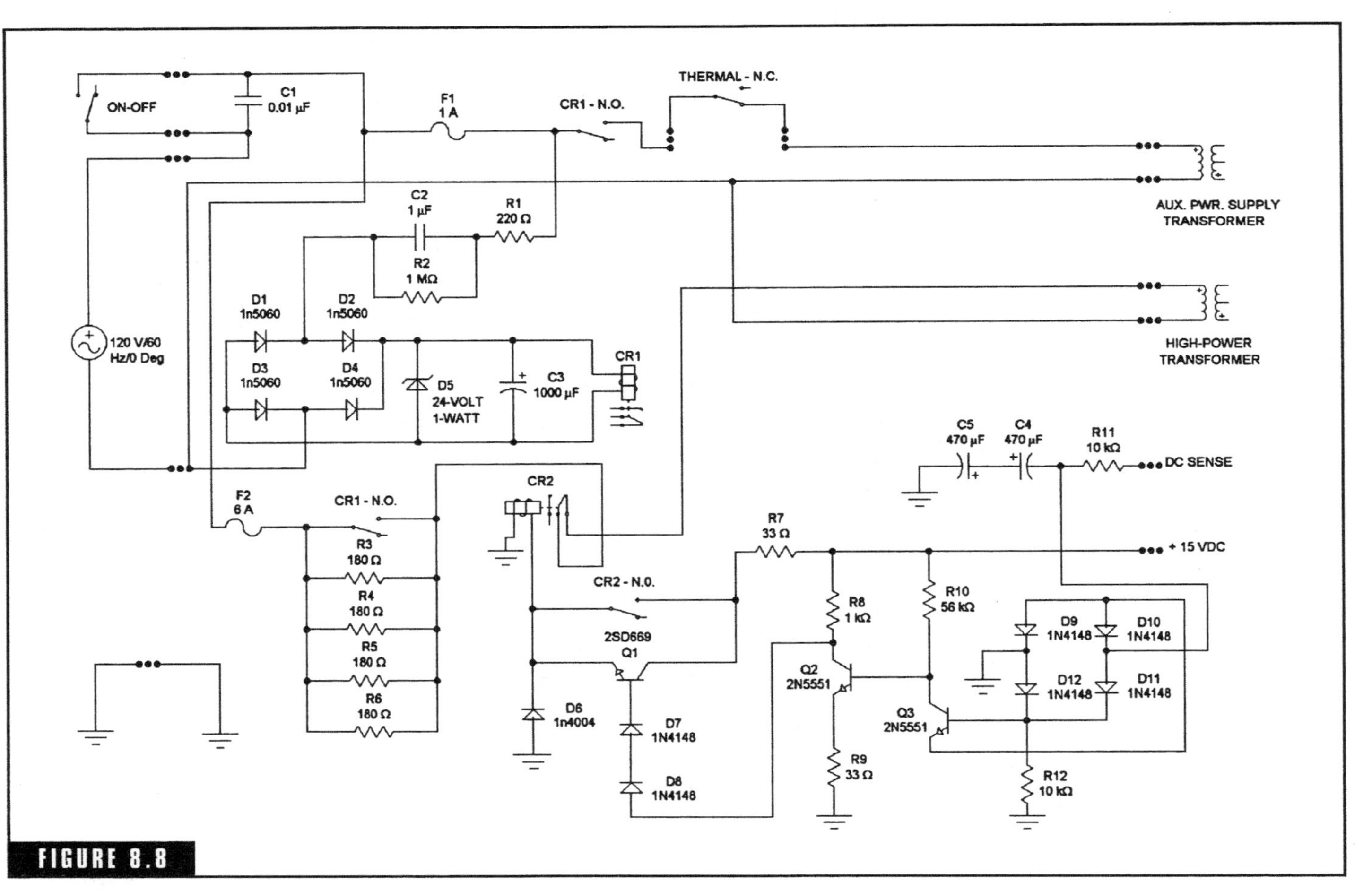

FIGURE 8.8

A complete auxiliary control system.

open, eliminating the applied power to the auxiliary transformer, disabling the balun circuit, and inhibiting the audio input signal. Without any audio signal to process, the heatsink(s) of the power amplifier should begin to cool.

The DC protection circuitry of Fig. 8-8 is identical to the Fig. 8-3 illustration, causing an interruption of the AC mains power to the high-power transformer if a significant DC offset is detected at the speaker output [more sense resistors can be connected in parallel with R11 if multiple amplifier outputs need to be monitored]. The +15-V operational power is obtained from the auxiliary power supply. If a significant DC level is detected at the amplifier output, the DC protection circuit will energize CR2. Once CR2 is energized, it will open the AC mains circuit to the primary of the high-power transformer, shutting down the audio power amplifier. Note also that the normally open contacts of CR2 will latch it into a continuously energized state. Consequently, the amplifier system must be manually switched off and back on again before AC power can be reapplied to the primary of the high-power transformer. The latching function of CR2 is to keep the AC mains power from being automatically reapplied to the primary of the high-power transformer without the benefit of the soft-start action (the DC fault at the output of the power amplifier will disappear when the amplifier's operational power is removed). An SCR can be substituted for the CR2 latching contacts if desired.

Figure 8-9 illustrates the fundamental architecture of the auxiliary control system illustrated in Fig. 8-8. Note how the external connections would typically apply. I didn't include any type of PC board artwork or layout because the ratings and types of relays will vary quite radically depending on the size of the audio power amplifier. I also illustrated typical fuse ratings, which would likewise vary depending on the specifications of the power amplifier. The important aspects of Figs. 8-8 and 8-9 are the concepts and methodologies of designing a good-quality auxiliary control system. Obviously, there are many optional ways of incorporating similar circuitry into a variety of control and protection schemes for audio power amplifier systems.

One point to keep in mind is that the control complexity of the Fig. 8-8 auxiliary control system is partially a result of wanting to eliminate the need for a muting/DC protection relay at the output of the audio power amplifier. If you don't have any objections to placing a control

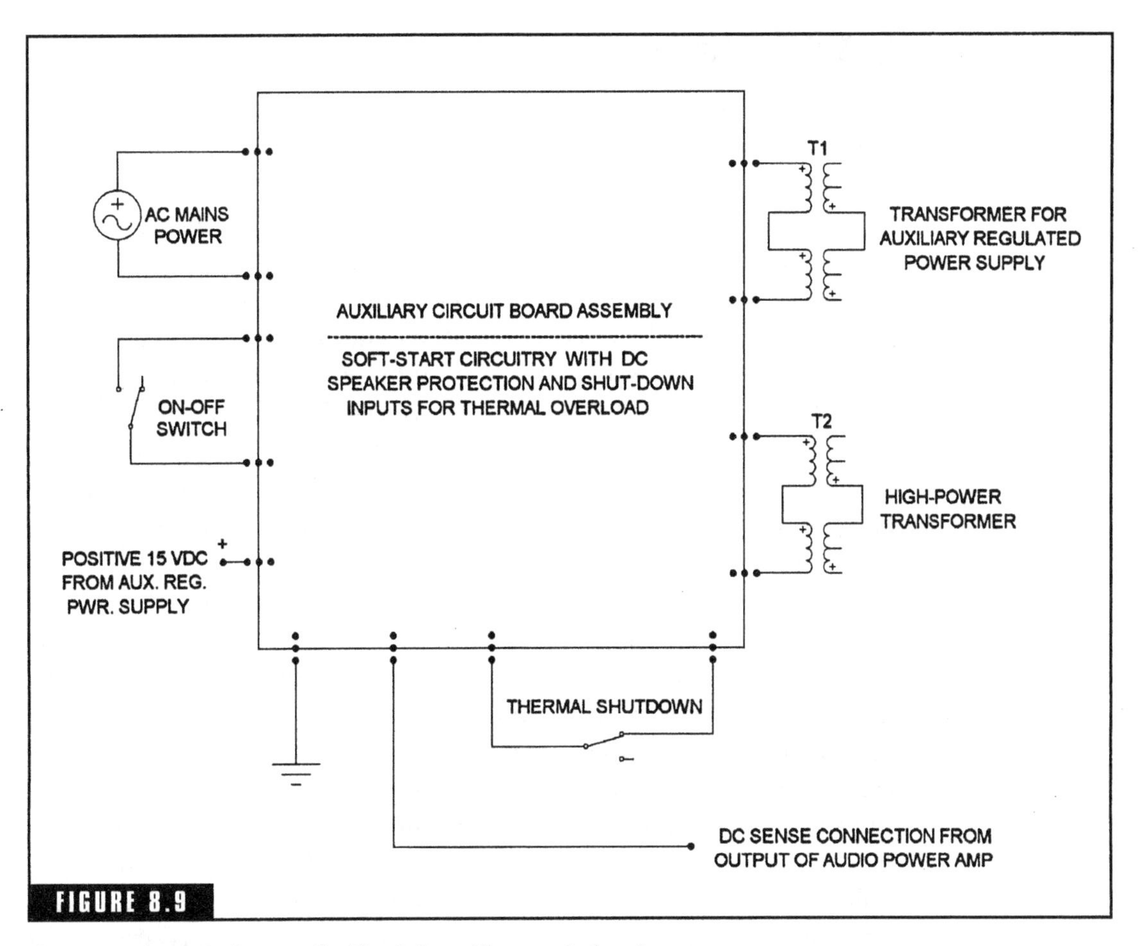

FIGURE 8.9

The external connections to the Fig. 8-8 auxiliary control system.

relay in the power amplifier's output signal path, then there isn't any need to tie the control functions of the soft-start system and the DC speaker protection system together—they simply become two independent control/protection actions that are not associated with each other.

Another modern option that would reduce the size and complexity of the auxiliary control system is to utilize a versatile microcontroller (such as a Basic Stamp or PIC Microcontroller) to perform all of the logical sequences. Such a system has many more options than would typically be found in a discrete auxiliary control system.

Clip Detection Circuits

Clip detection circuits are not normally considered protection circuits, but they are commonly used as a visual aid to manually adjust volume levels so that sensitive tweeter systems are not inadvertently destroyed. They also provide the user with a general idea of transient distortion levels.

Figure 8-10 illustrates a very good clip indicator design that is high-performance as well as easy to construct and implement. It can also be used in conjunction with almost any audio power amplifier design. The series circuit of D1 and R5 causes the voltage drop across R5 to be 5.1 V less than the power amplifier rail voltage. The voltage across R5 is reduced by the voltage divider of R7 and R8, with the voltage across R8 being applied to the inverting input of A1 as a reference voltage. Another voltage divider, identical to the R7/R8 divider, consisting of resistors R1 and R2, is connected from the emitter of the NPN output transistor to ground. The voltage drop across R2 is applied to the noninverting input of A1. Note that the reference voltage applied to the inverting input of A1 is a constant reference, based on a portion of the rail voltage. However, the voltage applied to the noninverting input is dynamic, and will represent a portion of the variable emitter voltage of the output device. If, during a positive output signal peak, the emitter voltage of the NPN output device rises to a voltage level closer than 5.1 V to the rail voltage, the voltage drop across R2 will exceed the voltage drop across R8. When this happens, the output of A1 will go high (i.e., positive), forward-biasing D3, and lighting D5 through the current-limiting resistor R11. In other words, A1 is acting as a simple comparator circuit, going high when the emitter voltage of the NPN output device exceeds a level that is within 5.1 V of the rail voltage. It is assumed, in this case, that clipping action will start when the peak output signal level reaches a voltage level that is 5.1 V less than the rail voltage. Depending on the audio power amplifier design, the actual clipping level could be varied by simply changing the two zener diodes (D1 and D2) to a more appropriate zener voltage. For example, if a hypothetical audio power amplifier happened to start clipping at about 3.9 V from the rail voltage, then 3.9-V zeners should be used for D1 and D2. The lower half of the Fig. 8-10 circuit functions in an identical but complementary fashion, with the output of A2 going high (i.e., positive) whenever a negative peak exceeds the point that is 5.1 V less than the negative rail voltage, which also causes D5 to light.

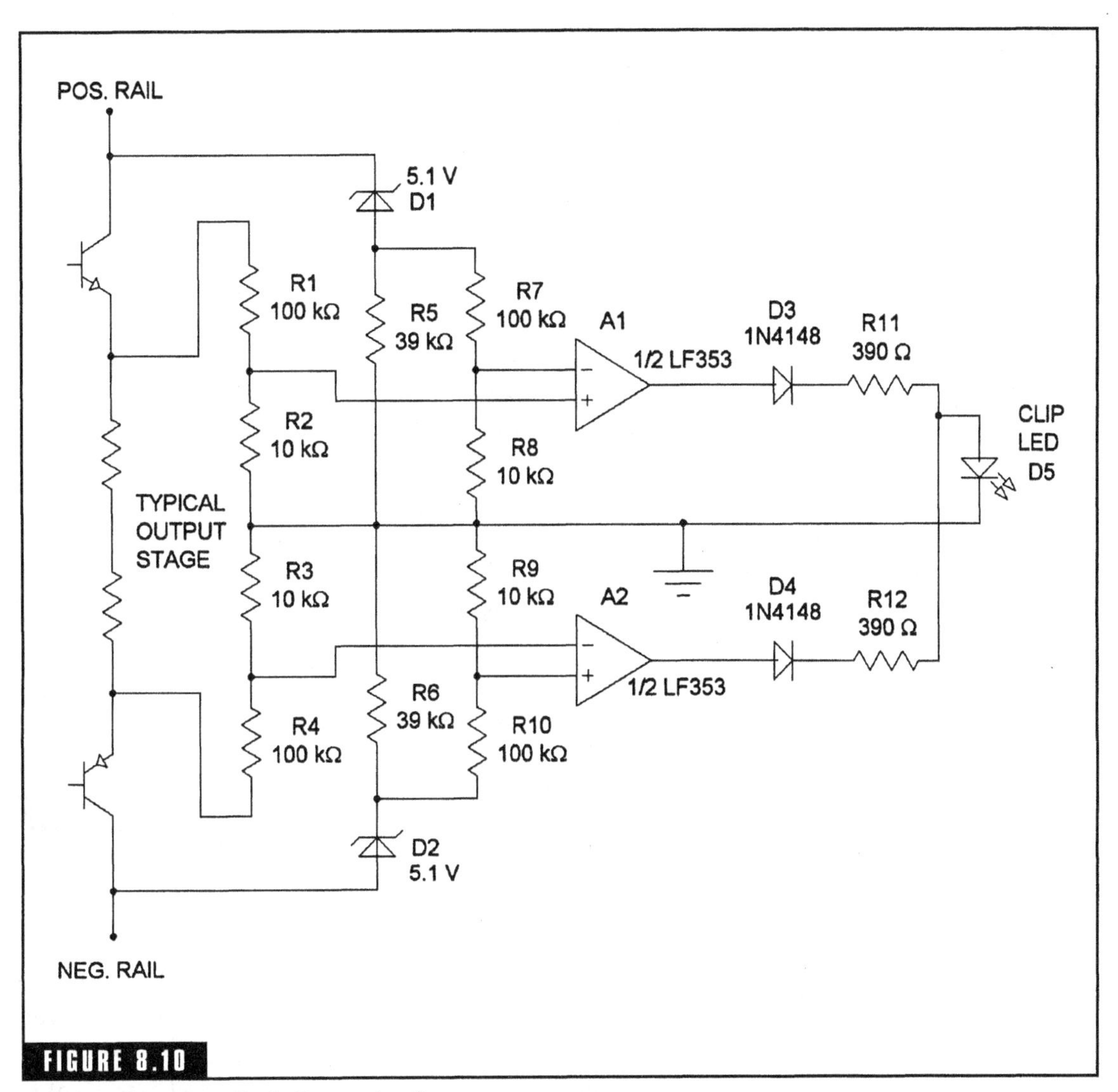

FIGURE 8.10

A high-performance clip indicator circuit.

The Fig. 8-10 clip indicator circuit is superior to common types of clip indicators that measure input signal levels, because it will indicate actual clipping circumstances under adverse conditions, such as excessive loading on the output stage or low-voltage rail conditions. The intensity variance of D5 also provides a very good "feel" for the frequency and severity of the clipping conditions.

Figure 8-11 illustrates an improvement to the previous Fig. 8-10 clip indication circuit. This circuit is identical to the Fig. 8-10 circuit, with the inclusion of two additional voltage dividers, two additional

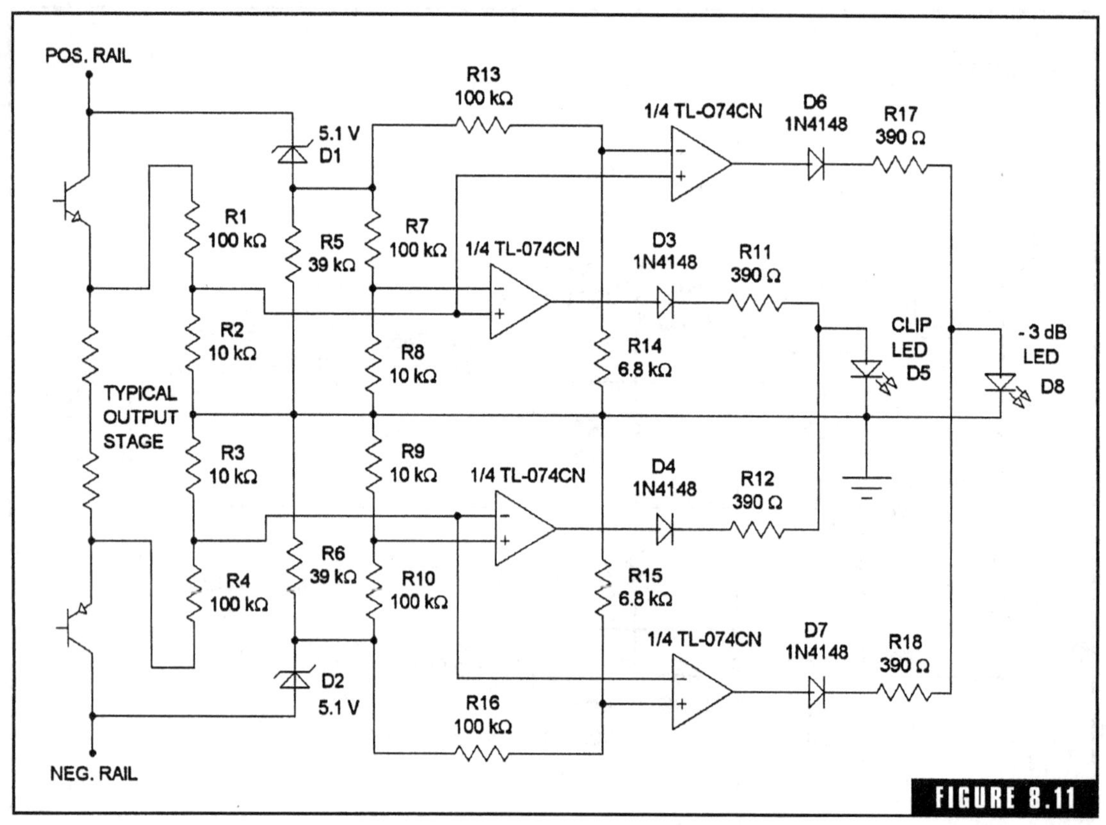

FIGURE 8.11

A clip indicator circuit with a −3-dB indicator.

op-amps, and an additional LED lighting circuit. All of the biasing and input/output conditions are identical for the additional set of op-amps, with the exception that R14 is roughly 70% of the value of R8, and R15 is roughly 70% of the value of R10. This modification in the resistance value of the added voltage dividers will cause D8 to light at an amplifier output voltage level that is approximately 70% of clip level, or approximately −3 dB.

The ability to view the frequency and intensity level of half-power peaks adds an additional dimension to the clip indication, in that it allows the user to estimate how much of the program material is "approaching" clip levels. A nice touch is to use green LEDs for the −3-dB indicators, and red LEDs for the clip indicators. Of course, if you wanted to get really fancy, you could also add additional voltage dividers and op-amps for additional level indications.

CHAPTER

9

Miscellaneous Audio Circuits

This chapter includes a number of circuit designs that didn't seem to fit into any other major category.

Display Circuits

In the early days of hi-fi audio systems, audio signal amplitude indicators were primarily D'Arsonval-type meter movements calibrated to provide a rough estimate of the signal levels in decibels. Such meter movements were colloquially called *VU meters,* which is short for *volume-unit meter.* Electromechanical D'Arsonval movements were comparatively fragile and prone to defective operation due to corrosion and various types of contamination. With the advent of cheap, readily available digital devices, VU meters began to be replaced by *bar/graph indicators* and *digital panel meters* (*DPMs*). Digital panel meters tend to be difficult to read in conjunction with rapidly changing dynamic variables (such as musical program material), so bar/graph-type displays became the favorite replacement for the more obsolete electromechanical level indicators. The LED indicators for some bar/graph displays are arranged in a circular arc to imitate the action of the needle movement in older VU meter displays.

Bar/graph displays can be very useful and accurate for analyzing the dynamics of musical programs relative to applications requiring the monitoring or adjustment of *signal voltage levels.* Such applications include signal processing and recording functions. However, almost every serious audiophile or electronics hobbyist already knows that bar/graph displays are close to useless for the application they are most commonly used for, which is the attempt to indicate output power levels of audio power amplifiers. This is because bar/graph displays simply indicate *voltage levels* at the output of the power amplifier, but *power output* is dependent upon the impedance of the speaker load. The impedance of a typical speaker system can vary, depending on frequency, by as much as 600% throughout the audio bandwidth, so a simple voltage measurement to indicate power output levels will virtually always be very inaccurate.

Even though the inaccuracy of bar/graph displays in indicating power levels is a well-documented subject, many audiophiles and hobbyists still like to incorporate them into audio power amplifier systems, because they do provide a *relative* indication of output levels (i.e., based on voltage dynamics). Even though the literal power indications are probably inaccurate, bar/graph displays incorporated into stereo or multichannel audio systems can aid in adjusting balance levels. Also, they add a type of visual effect to the aural experience. As one audiophile once told me, "I like 'em cause those green, yellow, and red LEDs are darn pretty!" I agree.

Figure 9-1 illustrates a simple bar/graph LED output level display designed to receive its operational power directly from the audio output signal. Thus, no type of operational power supply is required. This circuit functions surprisingly well, drawing only about 30-mA from the audio output signal at full-scale display. Unfortunately, the output intensity of the LEDs is relatively weak, and it does present an uneven loading effect upon the audio output signal, and will consequently inject some distortion effects. Although I wouldn't recommend this design for a high-quality domestic audiophile system, it does have a practical application in high-power automotive sound systems.

Referring to Fig. 9-1, diodes D1 and D2, together with capacitors C1 and C2, make up a voltage-doubler circuit; rectifying and doubling the signal voltage applied to the speaker system, which it uses for operational power. The operational power is applied to the anode of LED1, as well as the voltage reference source of R2, Q1, R3, D3, and D4. Q1 helps to regulate the current flow through the voltage reference source

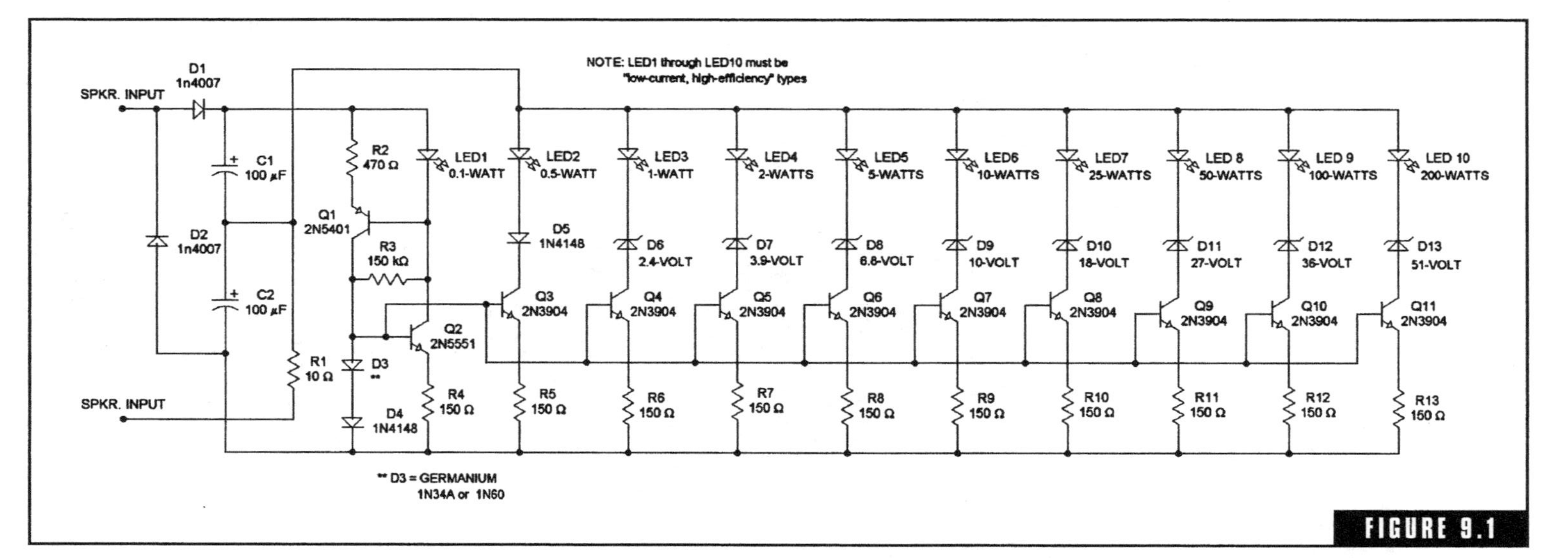

FIGURE 9.1

"Speaker-powered" (no power supply needed) audio power level indicator. Power indications are approximate, and based on a typical 8-ohm speaker system. Power indications for 4-ohm loads are approximately double.

of D3 and D4, with the voltage reference being applied to all of the base inputs of the driving transistors. Since a stable reference voltage is applied to the base of each driving transistor, the collector current of each driving transistor is held to a safe maximum so that the LEDs indicating lower power levels won't be destroyed when the operational power rises to amplitude levels sufficient to illuminate the LEDs indicating higher power levels. The voltage level applied to the anode of LED1 is the doubled voltage amplitude, which makes LED1 the most sensitive to lower power levels. The remaining nine LEDs receive half of the operational voltage applied to their anodes, supplied from the connection point of C1 and C2. As the operational voltage level increases (from increasing output signal voltage levels of the associated power amplifier), LEDs 2 through 10 will light in ascending order as zener diodes D6 through D13 reach their associated zener avalanche voltage ratings. In the case of LED2, it will light as soon as the 0.7-V forward threshold voltage of D5 is exceeded.

Note that the indicated power levels are not truly logarithmic, but they are about the best obtainable from commonly available components and the restrictions that apply to having to tap off some of the power amplifier's output power to operate from. The LEDs chosen for this project must be the *low-current, high-efficiency* types, requiring approximately 1 to 2 mA for nominal luminescence. For optimum operation, D3 should be a germanium-type diode, such as a 1N34A or 1N60.

Figure 9-2 illustrates a very versatile bar/graph display based on the popular LM3916 bar/dot display driver IC. As illustrated, the LM3916 is logarithmic in its output response, but if a *linear output response* is desired, the LM3914 IC can be substituted. It is also very easy to cascade two or more LM3916 chips for bar/graph displays of 20 to 50 LEDs.

The Fig. 9-2 display circuit is designed to operate from +12- to +15-V DC power supplies (note that a dual-polarity power supply is not required). The output current sink for lighting each LED will be limited to approximately 5.5 mA as illustrated, but higher LED currents can be obtained by reducing the value of R6. For example, changing the value of R6 to a 1-kohm resistor will provide about 10 mA per LED of sink current. The "bar/dot" switch between pins 3 and 9 can provide either "bar" or "dot" operation; as illustrated, the circuit is in "bar" mode. The circuit's operational power is approximately halved

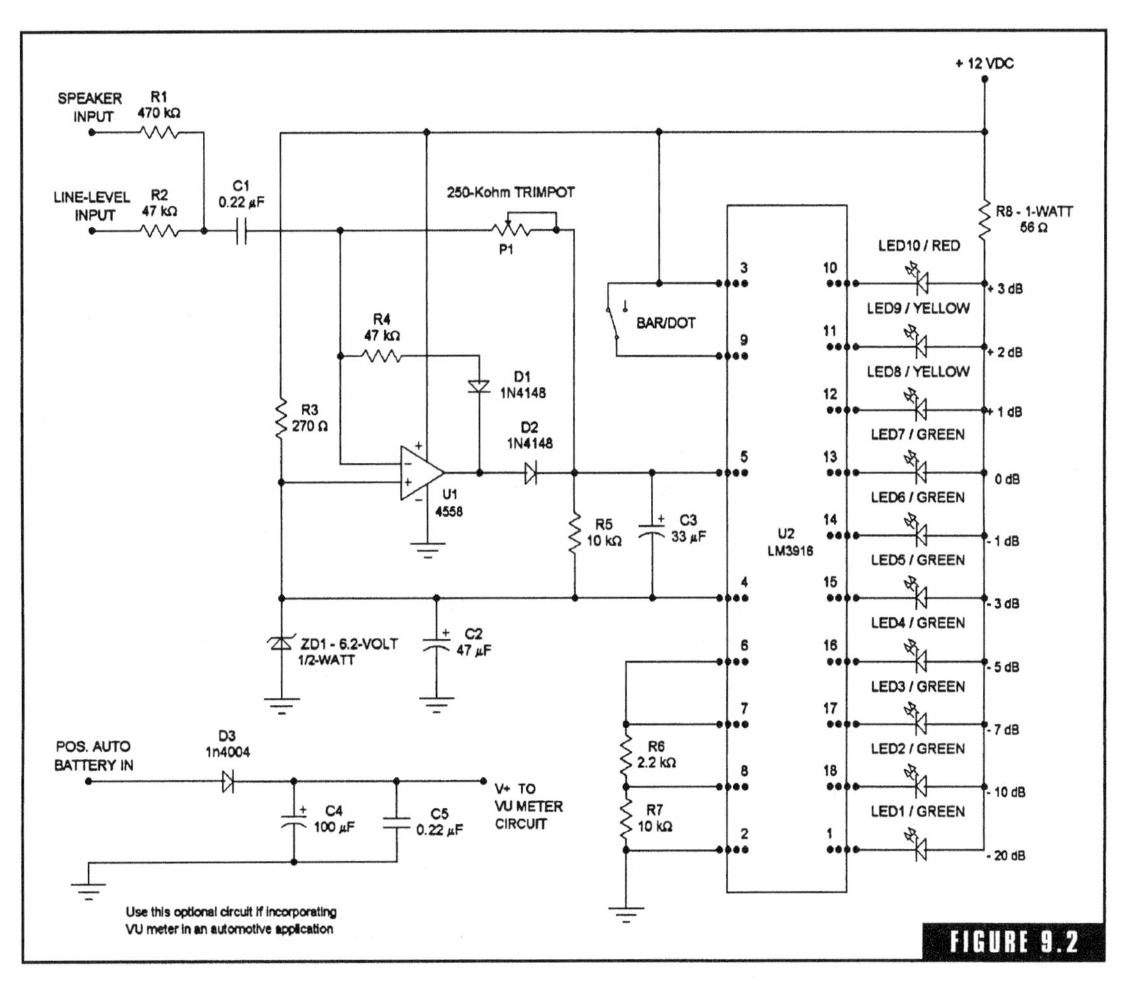

A bar/graph display based on the LM3916 IC.

by the action of ZD1, R3, and C2. This provides a half-voltage reference for the op-amp (U1) so that it doesn't require a dual-polarity power supply.

Note that U1 contains a rectification circuit (D1 and D2) that provides a positive-polarity output signal equivalent to the AC input amplitudes. C3 filters the output pulses from U1 to provide a smoother action from the LED circuits. Level calibration is provided by P1. This circuit will function with either line-level inputs or speaker-level inputs, which is determined by the resistance value of the input dropping resistor (R1 or R2). Almost any common op-amp can be used in place of the 4558 op-amp shown in the illustration. A simple circuit for incorporating the Fig. 9-2 design into automotive sound applications is provided in the illustration insert. The Fig. 9-2 display circuit *should not* be used in high-powered automotive sound systems incorporating equipment being operated from an isolated ground.

The LM3916 can be easily interfaced with small PC board–type relays for a wide variety of control and display functions. (This holds true for the LM3914 devices as well.) Choose relays that don't require coil currents of any more than 10 mA for operation. Connect the relay coils between the power supply and the current-sink pins of the IC in exactly the same manner that LEDs are normally connected. Place a reverse-biased diode across each coil to prevent inductive kickback spikes (1N4148 diodes function well for this application). Also place a 47-μF capacitor across each coil, which eliminates "chatter" problems (aluminum electrolytics are fine for this application—watch the polarity!). Once accomplished, the LM3916 can be used to control high-voltage incandescent or fluorescent devices by means of the relay contacts. There are also many similar applications outside of the audio fields for this same type of control action.

Phase Control

Figure 9-3 illustrates an operational amplifier circuit designed to provide manual adjustment of the phase relationships of audio signals. U1, U3, and their associated resistors form two isolation networks to buffer any impedance variations to the phase control circuit. Phase control is provided by the RC network of C1 and P1, with C1 providing an adjustable phase lag from (approximately) 0 to 180 degrees. This circuit can be cascaded with another identical circuit to provide a full 360-degree phase control. In such a circuit, you would probably want to use a two-gang potentiometer for the P1 adjustment control.

The Fig. 9-3 phase control circuit is useful in subwoofer applications wherein you would like an infinitely adjustable phase control. This circuit could be modified to 360-degree operation, as stated in the previous paragraph, or the simpler circuit as illustrated could be incorporated into a subwoofer system containing an "inverting" switch. Thus, 180-degree phase control could be achieved during "noninverting" operation, with the remaining 180-degree phase con-

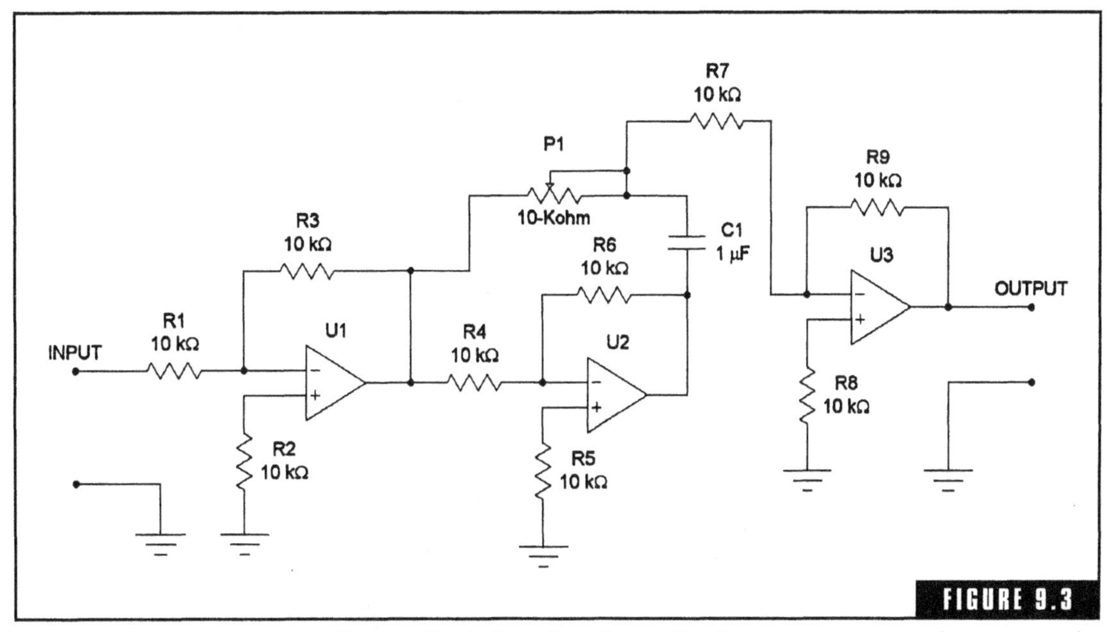

A simple 0- to 180-degree phase adjust circuit for subwoofer applications.

trol provided during "inverting" operation. High-performance operation from the Fig. 9-3 circuit will require the use of low-noise, high-performance operational amplifiers.

Delay Effects

The final two circuits provided in this chapter will be of great interest to audiophiles or electronic hobbyists who desire to construct musical instrument amplifiers. These circuits are normally considered "effects" circuits.

In the earlier days of musical instrument amplifier design, long-term delay of the musical signal was accomplished with mechanical springs. These springs were excited on one end with a magnetic transducer, similar to the motor structure in a typical loudspeaker, and the signal would be received on the other end of the springs, with another magnetic transducer acting as a type of microphone. The mechanical springs provided a finite time delay of the musical signal depending on their physical length, which was typically in the range of 40 to 80 ms. The delayed musical signal was mixed back into the original signal, and the effect was similar to the echo effects that can be heard in a large room or auditorium. This delay effect is commonly called *reverb*.

The mechanical spring system used for reverb effects had some very severe faults. One major fault was the limited frequency response, which was really due to the "resonant" effect of the springs themselves (in the classic reverb designs, the length of the springs was a major game player in the overall tonal effects). Another fault was the nonadjustable finite delay period, which was permanently set by the length of the springs incorporated into the specific system. A third problem was the sensitivity to physical shock and vibration—an inadvertent bump to the amplifier could result in a very loud "crashing" sound.

Some modern musical instrument amplifiers still incorporate spring reverb systems, probably due to the desire held by many musicians to recreate many of the classic sounds of the past. Also, many musicians are simply accustomed to using spring reverb systems, and they may not want to change their musical sounds. However, a higher quality of reverb effects, with more adjustment versatility, can be accomplished with solid-state circuitry.

Figure 9-4 illustrates a solid-state reverb system based on a technique called *bucket brigade*. The mental imagery is one of a long line of firefighters stretching from a source of water to a fire, passing buckets of water hand-to-hand in the effort to get the water to the location of the fire. U2, the "bucket brigade device" (abbreviated BBD), operates just like the name implies. It contains 1024 charge storage devices (or *bucket holders*) that will charge to an instantaneous level of the incoming audio signal, comparable to storing a small "piece" of the audio signal. This small signal portion will be coupled from one charge storage device to the next with each clock pulse, so the speed of the incoming clock pulses will determine how long it will take for the sequential signal portions to be coupled through all 1024 devices. All of the sequential signal portions are summed at the output of the BBD chip, recreating the original audio signal. However, the time required for the audio signal to make it through all 1024 devices can be quite substantial.

The MN3101 IC chip is the clock generator for the MN3007 BBD chip. The speed of the clock frequency, which corresponds to the length of the time delay, is controlled by the "speed" potentiometer. The summing network for the BBD chip (U2) consists of resistors R8, R9, R10,

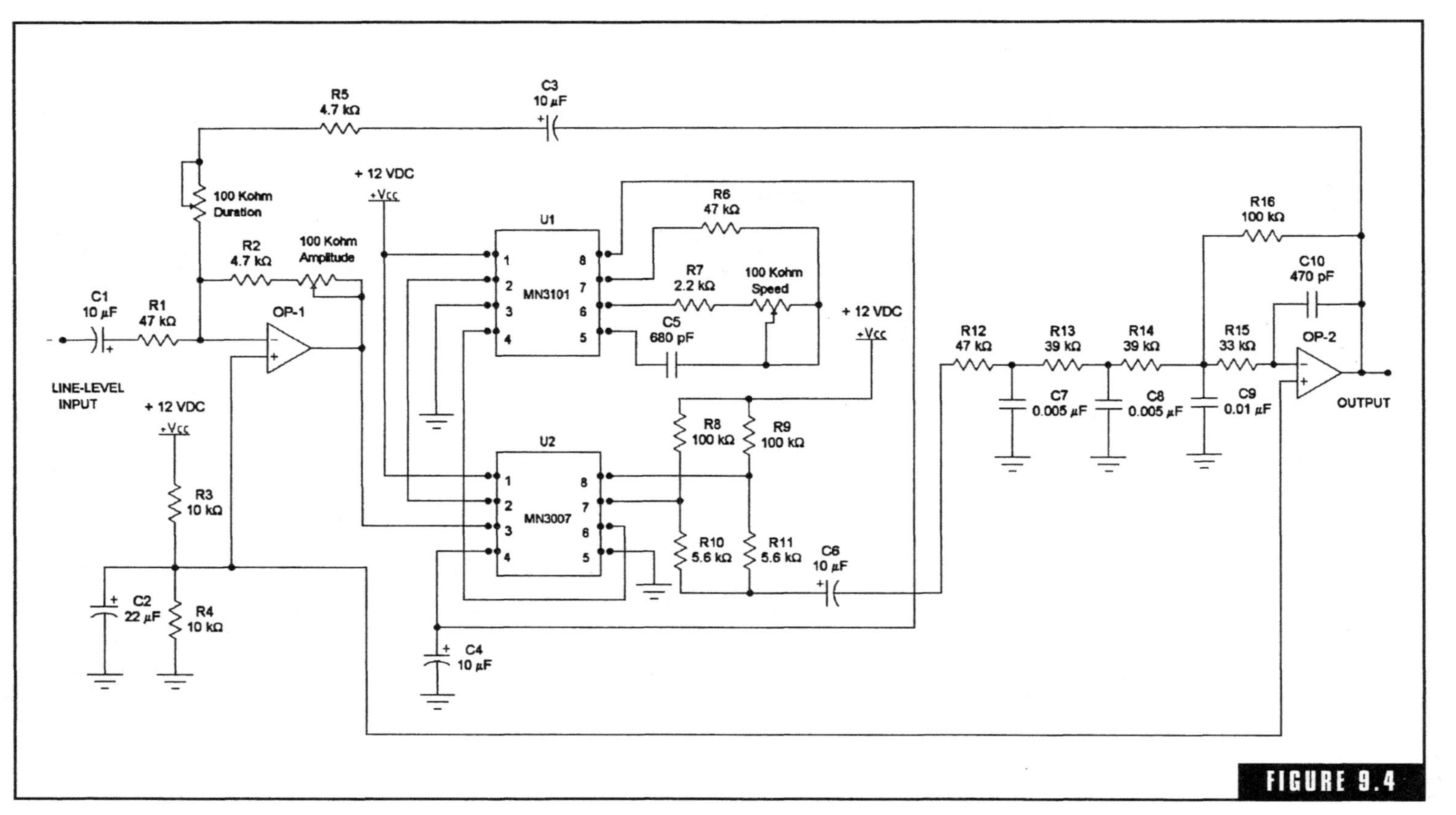

Monophonic adjustable analog delay unit (reverb).

and R11. The recreated audio signal is capacitor-coupled to the low-pass filter of OP-2 and its associated components. The recreated audio signal contains a significant content of high-frequency "chopping" components that would provide a distorted sound if not removed by the low-pass filter of OP-2. The output of OP-2 is the delayed audio signal, which has to be mixed back with the original audio signal in a later stage. This can be accomplished with a simple op-amp summing circuit, or the delayed signal can be applied to one channel of a stereo system while the original audio signal is processed through a different channel. The options for this type of circuit are many.

The "amplitude" potentiometer adjusts the delayed audio signal level, which will vary the reverb "depth" when mixed with the original stationary-level audio signal. R3, R4, and C2 make up a simple voltage-divider network to provide 50% of the power supply voltage to the noninverting pins of the op-amps. This allows the op-amps to be powered from the same single-polarity power supply as needed for the other IC chips. Thus, the negative power supply pin of each op-amp will go to circuit common, and the positive power supply pins will connect to the +12-V Vcc supply. (I didn't illustrate the op-amp power supply connections in Fig. 9-4 for the sake of clarity.)

Referring back to Fig. 9-4, note that the delayed audio signal at the output of OP-2 is also applied back to the input of OP-1 through capacitor C3 and the "duration" potentiometer. The setting of the duration potentiometer controls how much of the delayed signal is applied back to the input of the circuitry for *reprocessing*. This delayed feedback signal will always be at lower amplitude than the input signal, or else the circuit will go into a low-frequency oscillation. The purpose of the delayed signal feedback is to provide the "decay" effect associated with natural reverberation. For example, if you shouted into a large cave, you might hear your shout echoed back to you several times. Each time you heard the shout repeated, it would be at a lower volume level. This is the same effect accomplished by the setting of the duration potentiometer, which feeds back less that 100% of the delayed signal, causing an incremental decay of the delayed signal with each reprocessing step. When the speed potentiometer is set for short delay periods, the duration potentiometer will provide a sustaining effect to the reverberation sound.

The Fig. 9-4 audio delay, or reverb, unit is capable of an enormous variety of sound effects, from long-duration echo sounds to really "far-out" science fiction movie sounds. The less radical adjustment capabilities provide a superior, high-quality reverb unit for musical instruments, PA systems, or surround-sound simulations. As illustrated, the design is capable of audio signal delays up to a half second. If such long delays are not required and you desire improved high-frequency performance of the delayed audio signal, you can lower the capacitance values of the capacitors in the low-pass filter (i.e., C7, C8, and C9) according to your application needs.

The final design entry into this chapter is the "chorus" sound effect circuit of Fig. 9-5. If you have never experienced the sound effects produced from a chorus effects circuit, it is very difficult to describe with words. Chorus circuits create an elegant swirling and compounding of the musical harmonics inherent to a musical instrument. The effect is especially prominent for musical instruments that do not contain an overabundance of harmonic content to begin with, such as electric guitars and similar electric stringed instruments. The mechanics involved in creating the effect are relatively simple. Referring back to the previous reverb circuit of Fig. 9-4, a chorus effect would be created if you rapidly adjusted the "speed" potentiometer back and forth, changing the analog delay time at a rapid and periodic rate. When the analog delay time is varied up and down, while mixed with the original music signal, the delay is actually creating extreme phase shifts between the original signal and the delayed signal. This action causes the harmonics contained in the original signal to add and subtract rapidly as the phasing relationship seems to *slide* back and forth.

Referring back to Fig. 9-5, you will note that this circuit is very similar to the reverb circuit of the previous design, expect for the addition of the NE555 timer (U3) and the triangle-wave oscillator composed of U5, U6, and their associated components. The output of U6 is a triangle-wave frequency, with a peak-to-peak amplitude of about 10 V, and a frequency range (adjusted by the "rate" potentiometer) of approximately 2.4 to 0.35 Hz. The amplitude of the triangle wave applied to pin 5 of U3 is controlled by the "depth" potentiometer. U3 is configured as a voltage-to-frequency converter, so as the low-frequency triangle wave ramps up and down, the output frequency of U3 increases and decreases, following the control voltage of the triangle wave. The

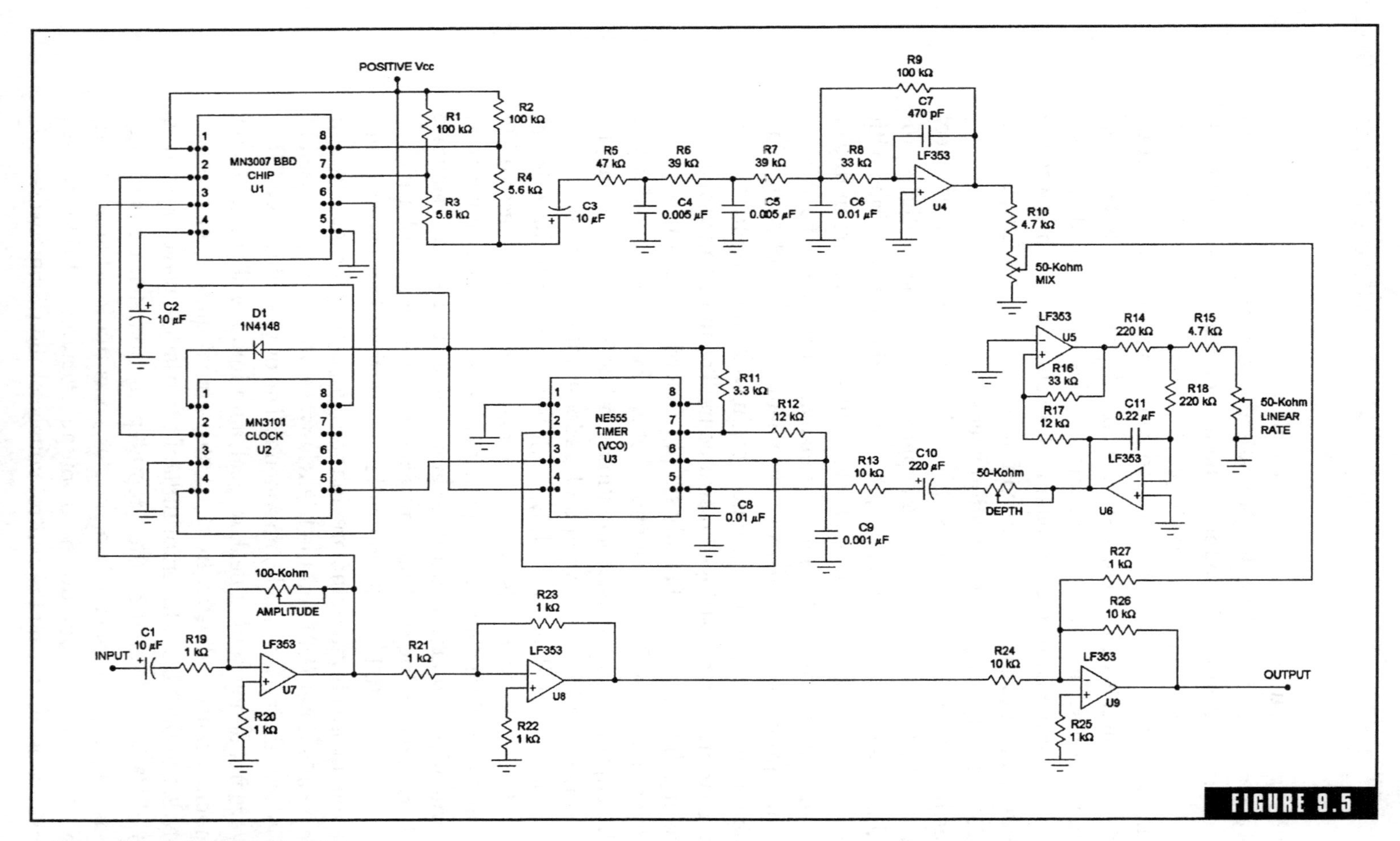

A "chorus" sound effects circuit.

output of U3 (pin 3) drives the input of the U2 BBD clock (pin 5) directly, so the BBD clock frequency is periodically varying up and down according to the output of U3. Therefore, the analog delay time provided by U1 is varying up and down in synchronization with the triangle wave output of U6. The remaining principles of the analog delay circuit were previously detailed in the description of Fig. 9-4.

U7 buffers and amplifies the input signal, with the overall gain provided by the "amplitude" potentiometer (this is typically a trim potentiometer, adjusted to provide the best overall response from the "mix" potentiometer). The output of U7 is applied to the input (pin 3) of the U1 BBD analog delay chip. The output of U1 is summed and applied to the low-pass filter of U4 and its associated components. The output of U4, which is now the varying-delay analog signal, is applied to the "mix" potentiometer, with the attenuated signal from the mix control being mixed with the original audio signal in the summing amplifier of U9. U8 is incorporated to keep any of the chorus signal at the input of U9 from feeding back into the input of the delay processing electronics.

When using the Fig. 9-5 chorus circuit, the "mix" control adjusts the amplitude level of the chorus effect, the "rate" control adjusts how fast the delay shifting will occur, and the "depth" control adjusts the time delay limits. The input to the chorus circuit should be a line-level audio signal input. Unlike the previous 9-4 reverb circuit, this design will require a typical dual-polarity power supply to provide operational power to the op-amps. For the sake of clarity, I have not illustrated the op-amp power supply connections.

CHAPTER 10

General Construction Information

I continually receive many questions from audiophiles and electronics hobbyists regarding a variety of construction concerns—some electrical and some mechanical. This chapter is dedicated to answering as many of those concerns as possible, taking into consideration the limited resources that may be available to the home electronics lab.

Hum, Noise, and Grounding Considerations

Over half of the problems that typically arise in "home-brew" audio projects can be traced back to grounding problems. The effects of poor grounding techniques will be hum, noise, hum pickup from external devices, and in severe cases, it can even make the project inoperable, or present a safety hazard. When most hobbyists encounter a hum problem, they tend to think of the problem as originating from a defective filter capacitor or decoupling capacitor. A defective capacitor could certainly cause such a problem, but more often than not, it ends up being a grounding problem. Even in the case of older systems that have been in operation for a while, various types of hum and noise problems develop from corrosion, broken wiring, and loose connections.

There are a number of misconceptions relating to grounding systems. One common misconception is that the ground wire has to be an enormous gauge, even for small-signal applications. I've seen many custom audio projects with #8-AWG solid-core wire and massive buss bars incorporated all through them, and they still might have a low-level hum problem. Increasing the gauge of ground wiring will usually reduce hum problems resulting from ground inadequacies, but this is typically the wrong approach. Another common misconception is that almost all of the internal wiring must be shielded. Again, shielding can help in some instances of hum or noise, but when improperly implemented, it can also do more harm than good.

Speaking from a physics point of view, hum problems arise because undesirable AC signals get mixed into the program signal. In most cases, the undesirable AC signal is a power supply component, so it will be at the frequency of the AC mains power or the frequency of the rectified AC power line. It is possible for other types of signals from adjacent circuitry to get mixed in with the program signal, but in such cases, the frequency is often so high that it is no longer called hum, and it is an easy problem to correct since the source of the problem is so easily located. Virtually all cases of hum problems within audio equipment can be traced back to two types of phenomena—the hum signal is either being *induced* or *injected.*

I would guess that at least 90% of the common hum problems encountered by hobbyists result from signal injection. The term *signal injection* refers to AC signals that are mixed with another signal by means of a *conductive path.* Inadequate ground wiring is the most common source of signal injection problems. To understand how signal injection problems arise in the ground wiring, and the techniques to eliminate them, refer to Fig. 10-1. This illustration shows a "bare-bones" type of audio power amplifier schematic, together with the basic power supply and interconnecting wiring. I am using an audio power amplifier system for this example because they are typically more prone to hum problems, due to the large power supply and relatively high levels of ripple on the power supply rails of the power amplifier.

Figure 10-1 is a very good example of the "wrong method" of wiring the grounding system. I have added five little AC symbols (i.e., G1 through G5) to illustrate how a hum signal can be injected into the common ground (or circuit common). Note that there are four decoupling capacitors (often called *bypass capacitors*) connected between the two power supply rails and ground. The two electrolytic decou-

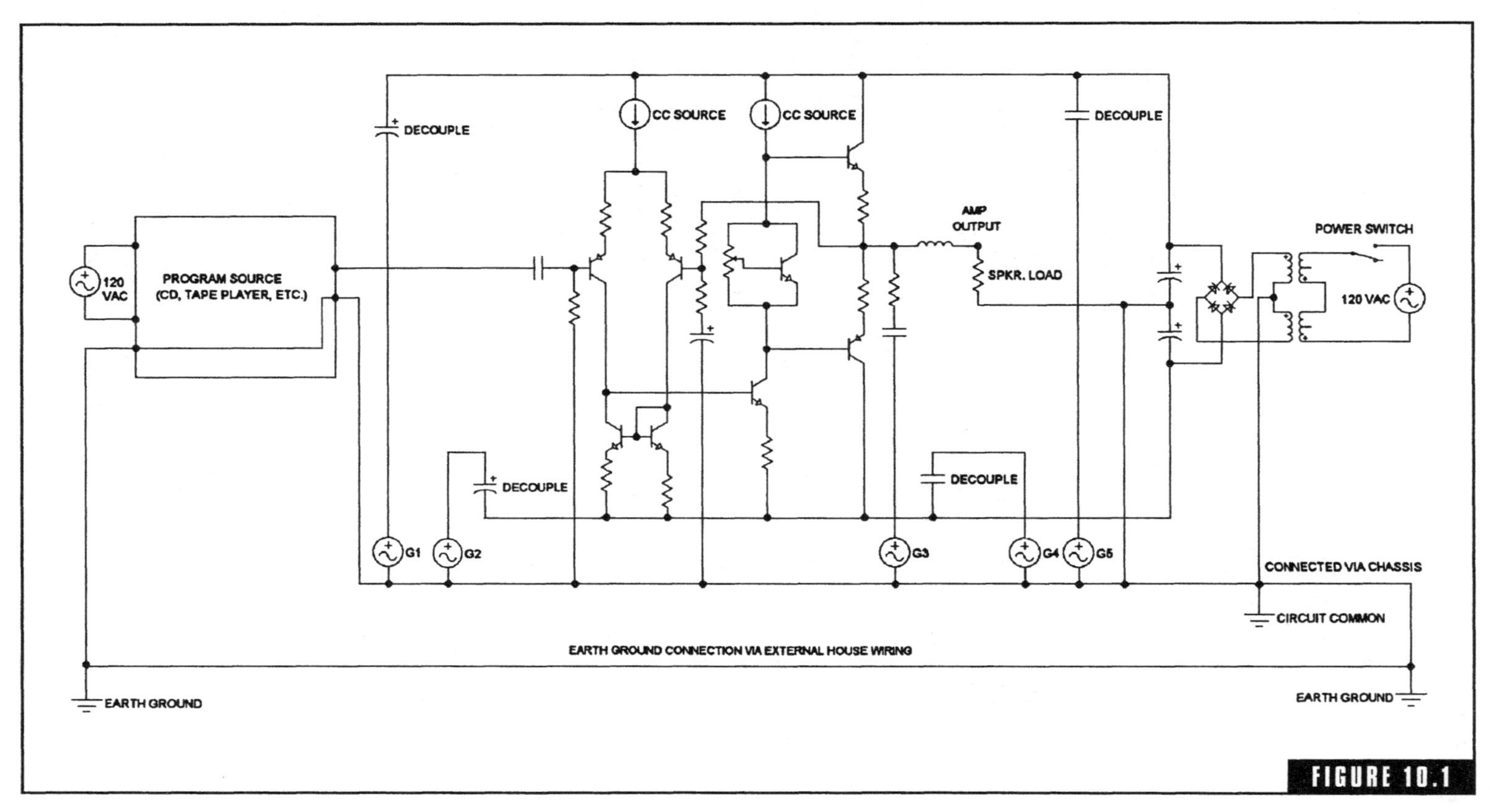

FIGURE 10.1

Illustration of how "hum" and "buzz" problems develop within a high-gain amplification system.

pling capacitors, close to the differential input stage, are usually fairly high in capacitance value, ranging from about 47 to 220 μF. These capacitors aid in stabilizing the power supply rails for the differential input stage, and they will look like a short circuit to AC ripple components. I have illustrated the AC ripple from these decoupling capacitors as sources G1 and G2. On the output rail of the power amplifier is the Zobel network, typically consisting of an 8-ohm power resistor in series with a 0.1-μF capacitor. Higher audio frequencies that can freely pass through the Zobel capacitor are shown as source G3. Close to the point where the power supply connects to the power amplifier's power supply rails, it is typical to incorporate two additional decoupling capacitors for the reduction of high-frequency signals that may be originating in the rectification process. These high-frequency AC components are shown as sources G4 and G5.

If you have considerable experience in the electronics fields, you have probably gotten very used to always considering a wire as representing zero resistance. In 99% of the cases, such "ideal" theory and calculation methods are totally appropriate. However, in real life, a wire represents a finite resistance. This resistance may be very low, but it still exists. Keeping this point in mind, reexamine Fig. 10-1 and mentally place a small resistor between each connection node on the common ground line as well as the earth ground line.

Begin by taking a look at the G1 signal injection source. It has two directions to flow. It can return to ground through the program source, or it can travel along the common ground for the amplifier and eventually return to ground at the power supply end. It should also be remembered that the quality of the earth ground in many structures is very poor. So, in reality, the G1 signal might be flowing in multiple directions depending on the wiring resistance, since it is essentially a current source looking at a variety of parallel paths to flow through. The G1 and G2 signals will almost certainly contaminate the input and feedback grounds for the differential amplifier input stage, which is the most voltage-sensitive stage in the amplifier.

The G3 high-frequency signal from the Zobel network can create some varieties of high-frequency intermodulation distortion. The G4 and G5 signals will pass high-frequency components of the rectification process that look like small spikes on an oscilloscope. These spikes will cause a "buzz" sound from the audio amplifier. When you consider that all of these AC signals are riding on the common ground line for the power amplifier, and add to that the possibility of generating some ground-loop problems via the external earth ground wiring, it is no wonder that the Fig. 10-1 audio power amplifier system would hum, or buzz, or both. Keep in mind that it only takes a few milliwatts to create very annoying levels of hum and buzz when operating into high-efficiency speaker systems. And remember that the *power gain* of a 200-W RMS audio power amplifier compared to a 2-mW output is 50 dB, which equals 100,000 times! It should now be obvious why increasing the gauge of the ground wiring will "improve" hum problems. Essentially, you are reducing the finite resistances of the ground wiring between the various connection nodes. However, it should be just as obvious that hum problems cannot be practically eliminated by this method, because the finite wiring resistance can only be reduced, not eliminated.

Figure 10-2 illustrates the "correct method" of wiring the grounding system, using the same example audio power amplifier as in the previous illustration. To begin, notice that the circuit common point, which is the junction of the two reservoir capacitors in the power supply, is connected to a *common connection point,* usually referred to as the *high-quality ground* (HQG) or *star ground* point. The ground wire connections to all of the decoupling capacitors, as well as the Zobel network, are connected to the circuit common ground line, which runs back to the HQG point. However, the signal input ground and the feedback ground (i.e., the two inputs to the differential input stage) are connected to their own dedicated ground wire, also running back to the HQG point. Notice how it is now impossible for any hum originating from a signal injection problem (via the ground wiring) to occur. Using this technique, signal injection problems originating from the grounding system can be virtually eliminated (and it's cheaper than buying heavy-gauge wire).

An optional technique that is often employed in high-gain audio circuitry is to install a low-value resistor between the signal-input ground and the HQG point. Such a resistor is illustrated in Fig. 10-2 as resistor RSG. Allow me to explain how this resistor functions based on a hypothetical situation. Suppose a poor electrical connection developed in the HQG point so that a relatively high resistance resulted between the HQG and circuit common. Assume this resistance to be 0.1 ohm (a

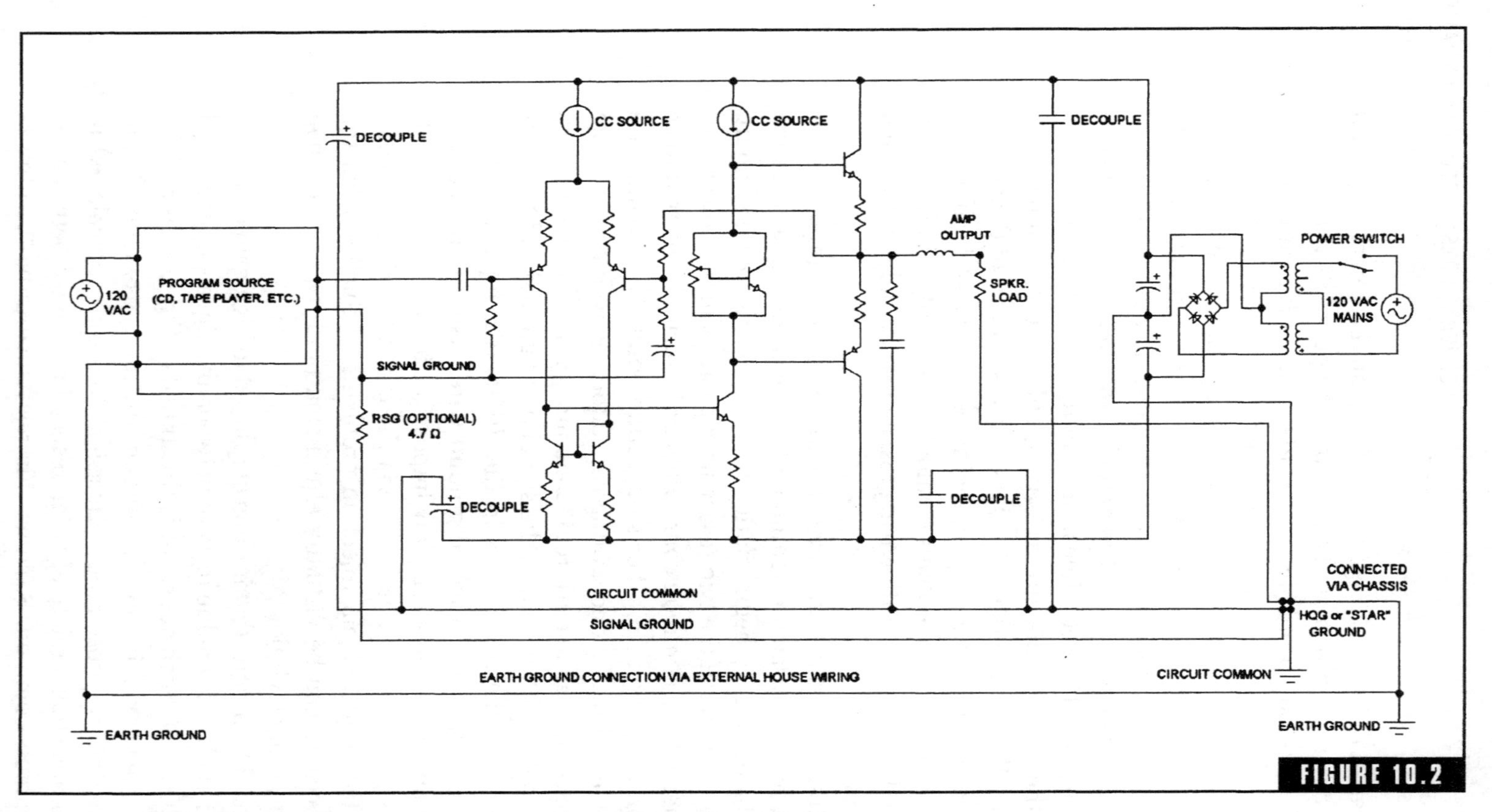

Illustration of the correct method of externally wiring an audio power amplifier.

tenth of an ohm would be a very high resistance for a ground connection). The AC injected voltages on the circuit common line (coming from the amplifier circuitry) would see this resistance, but at the same time, they would see the parallel resistance of going back through the signal ground and through the ground of the signal source. This second pathway might represent an equal resistance, causing half of the "hum" signal to feed back through the signal ground and create a hum problem in the amplifier. However, by installing a low-value resistance in the signal ground path, such as the 4.7-ohm RSG resistor illustrated, the signal ground would represent a resistance that is 47 times higher than the poor 0.1-ohm ground connection at the HQG point. Thus, virtually all of the "hum" signal would still return to ground through the HQG point. RSG resistors, as illustrated in Fig. 10-2, typically vary in value from 4.7 to around 15 ohms.

My experience indicates that if you utilize good materials and good grounding techniques, an RSG resistor isn't needed. In reality, the technique of using such a resistor is an old one, stemming from necessity when amplifier designs were not as well evolved. There are a few manufacturers who still incorporate them, however.

As can be seen from Fig. 10-2, we can eliminate signal injection hum problems with correct wiring techniques, and we can effectively isolate the signal ground from signal injection interference by installing an RSG resistor. However, the amplifier still has one area vulnerable to signal injection problems. If a ground loop develops on the external earth ground wiring, hum can be fed into the amplifier input via the program source. There are two common methods of eliminating this problem. One is to use a transformer-type balun, and simply configure it to accept a single-ended input. The other method is to install a *ground-lift circuit* in the audio power amplifier system. A typical ground-lift circuit is illustrated in Fig. 7-1.

The wire gauge I use for high-current grounding applications varies between #14-AWG and #18-AWG, depending on the size and power ratings of the audio design. Avoid using any type of *solderless crimp connectors* for ground wiring (in fact, I avoid crimp connectors for almost any application). Their connection resistance is not as low as other connection methods, and they can internally corrode in time, increasing the connection resistance even further. In addition, if you do not have a professional-quality crimping tool coupled with a good deal of experience in working with solderless connectors, the chances are good that some of your connections will not be of high integrity. I simply solder wires together and use insulating heatshrink tubing for "butt" splices. For other types of connections I use solder lugs, screw-down terminal blocks, or solder posts. Such wiring techniques take a "wee bit" more time, but the long-term integrity and reliability are worth the extra effort.

Hum or noise problems resulting from *induced* AC signals are caused by electromagnetic fields creating induced currents into the signal ground, signal wiring, or sensitive components on the PC board. The most prolific source of EMI fields is usually the power transformer, and this is almost certainly the case with audio power amplifiers. However, it should be remembered that even relatively small transformers can produce significant levels of EMI if placed too close to high-gain circuitry. Many preamplifier designs house the power supply transformer in a ferrous cage of some type. I have discovered that many switching power supply enclosures (i.e., the type that are most often used in computer systems) make excellent EMI shields for large toroidal or laminated transformers in power amplifier applications. It is true that toroidal power transformers produce less EMI than most laminated types, but toroidal EMI emissions are still very formidable. (EMI problems associated with power transformers have been discussed at length in Chap. 7.)

There are other sources of induced hum in audio circuits besides transformer EMI fields. Generally speaking, almost anything associated with the AC power lines can produce radiated fields strong enough to cause some problems. It is a good idea to house everything associated with the incoming AC power and the power supply in a rear corner of the project enclosure. In addition, mount PC boards in such a way that the most sensitive circuitry is as far away as possible from the power supply section, and be careful not to route any wiring through the power supply area. If you install a power on/off switch (switching the AC power line) in the front of a project enclosure, remember to keep the switch wires well away from sensitive wiring or circuitry. The same holds true for any type of AC-powered indicators.

General Wiring Considerations

When it comes to low-level signal-processing projects, the majority of hobbyists and audiophiles run into few problems regarding fundamental wiring techniques. Most of these projects only require a small dual-polarity power supply that can be remotely stuck in a corner of the enclosure, with almost any type of connection wire suitable for the low currents and low voltages involved. Typically, if the power supply rails and operational amplifiers are properly decoupled, and the input/output wiring is a good grade of shielded cable, hum and noise problems are practically nonexistent.

When wiring concerns do arise, they almost always relate to the audio power amplifier system, due to the high-current, high-voltage demands. A secondary source of confusion relating to general wiring considerations of audio power amplifiers is misunderstandings of how all of the various subcircuits work together. If you have a good "block diagram" concept of audio power amplifiers, it will help to resolve questions about how to connect it all together.

Figure 10-3 illustrates a complete block diagram of a dual-channel audio power amplifier. I believe the diagram is mostly self-explanatory, but there are a few points that should be touched on. You'll note that two individual power supplies are incorporated: one for each power amplifier. This is optional. The decision to utilize dual power supplies is typically not based on performance issues, but rather on the availability of certain types of power transformers and the advantage of being able to maintain the operation of one power amplifier even if the other one fails (this allows a professional entertainer to continue to use the amplifier, even though at reduced performance capacities). For most domestic applications, there really isn't any good reason to provide dedicated power supplies for each audio power amplifier channel. If properly wired and constructed, crosstalk will never be a problem with modern power amplifiers exhibiting high PSRR characteristics.

You'll note that the "low-voltage regulated power supplies" block (often referred to as the *auxiliary power supply*) will normally provide operational power to the clip-detection circuits, the protection circuits, and the input stage electronics (i.e., solid-state baluns and input buffer/bridging circuits).

As a final entry to this section, the following is a short list of general wiring tips.

- Low-level and line-level signal wires should be of a good-quality shielded variety. Take care to dress the shield connections

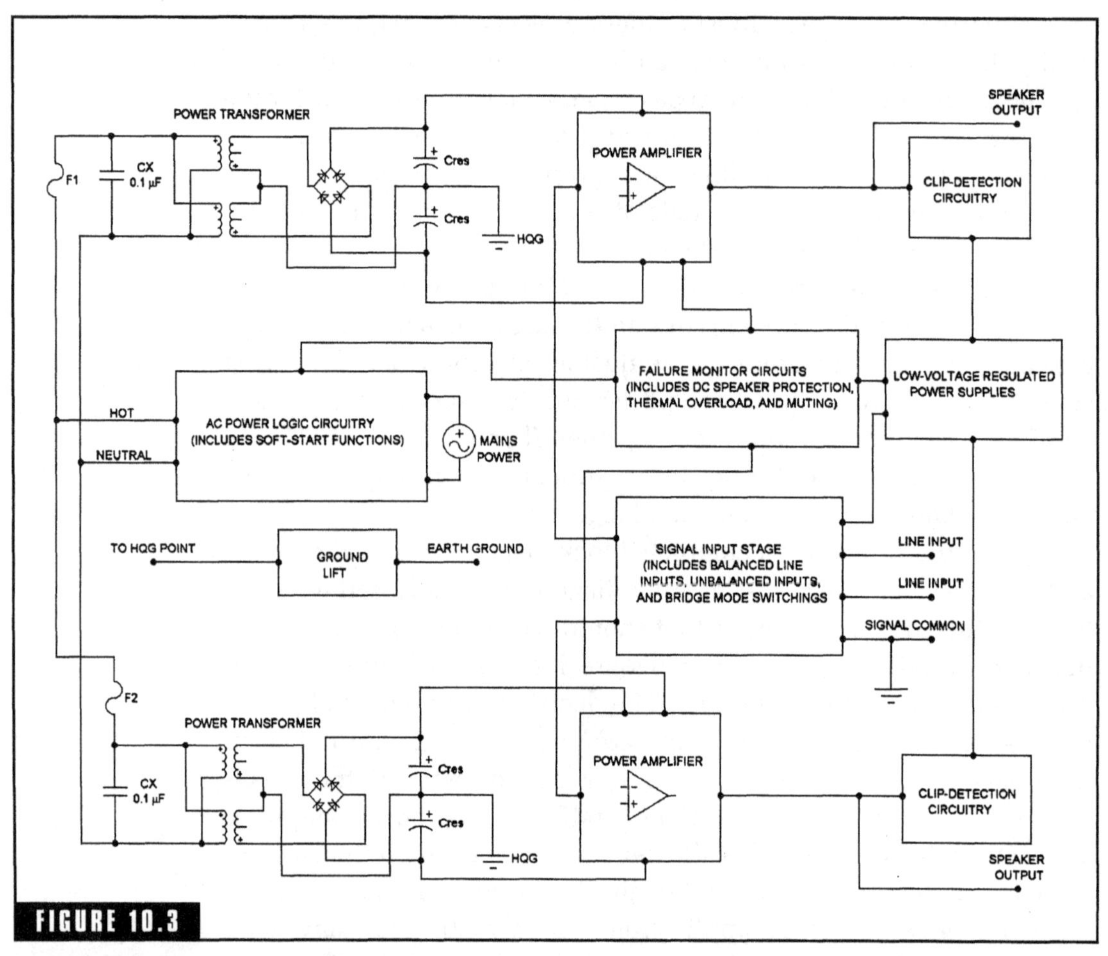

FIGURE 10.3

A block diagram of a typical professional stereo audio power amplifier.

properly so that there aren't any wire-hairs sticking out and the connection is mechanically solid.

- If you have to run the power supply rail connection wires in an audio power amplifier for a significant distance, tightly twist them together for as much of the wire run as possible. (Do not twist the rail power supply wires around the ground wire.) This helps to cancel out radiated fields produced by the ripple content.
- Regarding input/out RCA jacks, it usually won't make any difference if the outside ground connection is electrically connected to the chassis. Personally, I prefer to use the types that have the "hot signal" and "signal ground" connections isolated from the

mounting base. In the case of professional audio power amplifiers, a dedicated earth ground wire should be connected directly to the ground connection pin of signal input jacks, regardless of whether they are the RCA type or 1/4-in phone plugs.

- There are many good options for loudspeaker output connectors. From a performance perspective, the only real concern is that they are heavy-duty enough to adequately handle the peak current outputs. Quarter-inch phone plugs are not a good choice for this application. I like to use the heavy-duty "dual-banana" binding posts set to 0.72-in centers. Besides adapting to almost any type of speaker cable, the binding post spacing is standardized to many professional speaker cables.

Printed Circuit Board Construction Techniques

Making Printed Circuit Boards by Hand

Making PC boards is not as difficult as you may have been led to believe, or as your past experiences may have indicated. Circuit board fabrication is a learned technique, and like any technique, there are several "correct" ways of going about it and many "wrong" ways of accomplishing disaster. In this section, I'm going to detail several methods that can be successfully used by the home-based hobbyist or audiophile. With a little practice, either of these methods should provide excellent results.

To begin, you will need to acquire some basic tools and materials. If you are a complete novice relative to PC board fabrication, I recommend that you start by purchasing a "PC board kit" that will allow you to experiment at making a few simple PC board designs by hand. Typically, such kits will contain a bottle of *etchant* (an acid solution used to dissolve any exposed copper areas on the PC board), a *resist ink pen* (a marking pen used to draw a protective ink coating over any copper areas that you don't want dissolved by the etchant), several blank (i.e., unetched) pieces of PC board material, and some miscellaneous supplies that you may or may not need. In addition, you will need a few very small drill bits (#61 is a good size), an electric hand drill, some very fine emery paper, a small pin punch, a tack hammer, and a glass tray.

Choose a very simple PC board design as your first project. Most magazines or textbooks providing PC board artwork will provide a *bottom-view reflected image* of the artwork. This is the view that you will need to construct the PC board. Make a good copy of the reflected image illustration, and pay close attention to the *scale* of the illustration. If it is not a *full-scale* illustration (sometimes termed 1:1 scale, or 100% scale), you will need to adjust its size to full scale with the copy machine. Cut a piece of the PC board material to the same size as the illustration. Cut out the illustration from the copy and tape it securely on the "foil" side of the PC board material. Using a small pin punch and tack hammer, make a dimple in the PC board copper at each spot where a hole is to be drilled. When finished, you should be able to remove the copy from the PC board and find a dimple in the copper corresponding to every hole shown in the artwork diagram. Next, drill the component lead holes through the PC board at each dimple position. When finished, hold the PC board up to a light, with the diagram placed over the foil side, to make certain you haven't missed any holes and that all of the holes are drilled in the right positions. If everything looks good, lightly sand the entire surface of the copper foil with 600-grit emery paper to remove any burrs and surface corrosion.

Using the resist ink pen, draw a "solder pad" area around every hole. Make these pads very small for now; you can always go back and make selected ones larger if need be. Using a pencil instead of the resist ink pen, connect the pads as shown in the artwork diagram in the following manner. Being sure you have the board turned correctly to match the diagram, start at one end and connect the simplest points first. Using these first points as a reference, eventually proceed on to the more difficult connections. When finished, you'll have a diagram that looks like a "connect-the-dots" picture in a coloring book. When you are satisfied that all of the "pencil" connections are accurate with the artwork illustration, go over the pencil lines with the resist ink pen. Finally, go back and "color" in the wide foil areas (if applicable) and fill in the wider tracks with the resist ink pen. This technique is easier than it seems at first thought. If you happen to make a major mistake, just remove all of the ink with ink solvent or a steel wool pad and start over again—nothing is lost but a little time. You'll be surprised at how accomplished you will become at this after only a few experiences. Also, keep in mind that your PC board artwork need only correspond electrically with the illustrated artwork; it doesn't need to look exactly like it.

Place the PC board in a plastic or glass tray (not metal!) that corresponds to the size of the PC board (if your tray is too large, you'll waste a lot of etchant). Pour about an inch of etchant solution over the board. (Be careful with this etchant solution. It permanently stains everything it comes in contact with, including skin. Wear goggles to protect your eyes, and don't breathe the fumes.) After about 15 to 20 minutes, check the board using a pair of tongs to lift it out of the etchant solution. Continue checking it every few minutes until all of the unwanted copper has been etched away. When this is accomplished, wash the board under cold water, and remove the ink with solvent or steel wool. When finished, the PC board will be ready for component installation.

If you construct a PC board using the aforementioned procedure, you will discover that the copper artwork on the finished PC board is slightly "pitted" in areas where the resist ink did not adequately protect the copper from the etchant. If you want to improve the finished quality of your PC boards, use inexpensive fingernail polish instead of resist ink. Obtain a dried-out felt-tipped marking pen (one with an extrafine point), repeatedly dip the pen in the fingernail polish (like you would a quill pen), and use it to draw your pads and traces in the same way that you would use the resist ink pen (as described previously). After etching, the fingernail polish can be removed with ordinary fingernail polish remover, and the copper foil surface of the PC board will be totally free of any pits or corrosion from the etchant.

Many perfectionist-type audiophiles will consider the previously described technique to be rather "crude." However, from a functional and quality standpoint, I found this method to be far superior to the rub-on transfers (which seem to stick to everything except the PC board) and many other direct transfer methods. Also, this method works quite well with double-sided PC board designs, because the component holes are drilled "before" the artwork is put down on the board. Thus, you can draw your pads on the top and bottom, connect the pads, and end up with the top and bottom artwork perfectly aligned.

I consider this manual technique to be an excellent option for hobbyists or audiophiles who do not anticipate the need to fabricate a large quantity of PC boards. To achieve a high-quality result, this method is very time-consuming. However, the materials are inexpensive, the technique is easy and relatively foolproof, and the time involved with a hobby is not normally held at a premium. After all, it

An illustration of how a positive transparency is sandwiched between the photosensitized PC board material and the glass plate on top.

isn't far removed from painting pin stripes on model cars or assembling railway stations for model railroad sets.

Making Printed Circuit Boards by Computer/Photographic Methods

If you are a serious and prolific hobbyist or audiophile, you will probably develop the need to fabricate PC boards in a more rapid and repeatable method than is possible manually. The best method I have discovered for accomplishing this is to utilize one of the many CAD programs available for PC board layout design in conjunction with a positive-acting photographic procedure.

Computerized PC board layout programs provide a method of starting with any circuit schematic and adapting it to an accompanying professional-quality PC board artwork design. The computer layout program makes the entire process incredibly easy and virtually foolproof, with a host of options for the highest levels of sophistication. Personally, I use the *Electronics Workbench* professional versions of *Layout* and *Ultiboard* for my designs.

Depending on the CAD circuit board design program you choose, the procedures and techniques will vary, but the end product will be a professional PC board design displayed on your computer monitor, waiting for you to do something with it. Once this design phase has been accomplished, the following "general" procedure can be followed for fabricating the PC board:

1. Generate the PC board artwork using a CAD layout program. Specify "pads with holes for prototypes." This will greatly aid in drilling the PC board after the etching process is completed. Specify a "bottom view, reflected image" printout. This will convert the artwork to a bottom view, as though you were observing the PC board while holding it upside down.
2. Print the artwork onto a clear transparency. Transparencies are available at almost any department store or office supply store. They are typically used with inkjet printers to make colored transparencies for overhead projectors. In this case, the inkjet transparency will become a photopositive of the PC board artwork. Print the artwork onto the transparency as dark (opaque) as possible. Some inkjet printers have special settings for extradark printing; use such an option if possible. If you don't have such an option on your particular printing system, simply make two or three transparencies of the artwork, and sandwich them together (be careful to ensure that the artwork transparencies are perfectly aligned with each other). Another option is to allow time for the ink to thoroughly dry on a printed transparency, and then run the same transparency back through the printing process, printing a second "coat" of ink on top of the first printing. This process will work with some inkjet printers employing very accurate paper placement mechanisms; other printers will end up blurring the artwork image.
3. Allow considerable time for the ink to dry on the transparency (transparencies require much more time than paper). Hold the transparency up to a light and carefully check for voids or semiopaque areas in the artwork. If you see small pinholes or semiclear areas, touch them up with a permanent ink marker.
4. For the photoexposing process, you will need a source of ultraviolet light. Obtain four 18-in-long fluorescent lights (and fixtures) with a power rating of 15 W each. These types of fluorescent lights are sold at hardware and department stores for the purpose of adding extra lighting to an area of the home, such as over a sink or in a hallway. Mount the fluorescent lights side-by-side (not end-to-end) on a piece of plywood or

particleboard, measuring about 12-in wide by 24-in long. Drill four holes in the four corners of the plywood, and insert four pieces of threaded rod that will be used for adjustable legs. When completed, this "exposure light" will be oriented so that the fluorescent tubes are on the underside of the plywood base, and the four legs will be adjusted (with appropriate nuts and washers) so that the fluorescent tubes will be about 2½ in above the exposure area when the entire unit is placed on a flat surface. In other words, you end up with four fluorescent lights, side-by-side, upside down, and illuminating the surface about 2½ in below them. There are other methods of adequately exposing photoresist, but this is the best method I've found.

5. You will need to obtain a clear pane of glass, and it should be a little larger dimensionally than the largest PC board that you ever anticipate exposing. Additionally, you will need some presensitized "positive-acting" PC board blanks, a supply of photodeveloper solution for positive-acting photoresist, and a supply of etchant solution. Other than the pane of glass, all of these supplies are readily available from a variety of electronic supply companies.

6. Cut the presensitized PC board blank to the required size of your artwork.

7. Find a darkened work area for the exposing process. This area can be light enough for you to easily see, but it should be void of strong light sources or sunlight. Lay the PC board blank down on a flat, clean work surface. Peel off the protective film from the presensitized surface. Lay your artwork transparency on top of the presensitized surface. (Make sure you haven't accidentally gotten the transparency upside down. I've made that mistake more than once!) Lay the clear glass pane on top of the transparency, sandwiching the transparency between the glass and the presensitized surface. Ensure that the transparency is laying perfectly flat. Set the exposure light (from step 4) above the PC board setup. If you've adjusted the four threaded-rod legs correctly, the fluorescent tubes should be about 2 to 2½ in above the glass pane. Turn on the fluorescent lights and expose the PC board for 8 minutes.

8. Immediately after exposure, submerse the PC board into the developing solution. Within about 30 seconds, you should begin to see the exposed photoresist washing off into the developer solution, leaving a clear pattern of the PC board artwork. As soon as the exposed areas are clean of photoresist (showing the bright copper foil), remove the PC board from the developing solution and wash in cold running water.

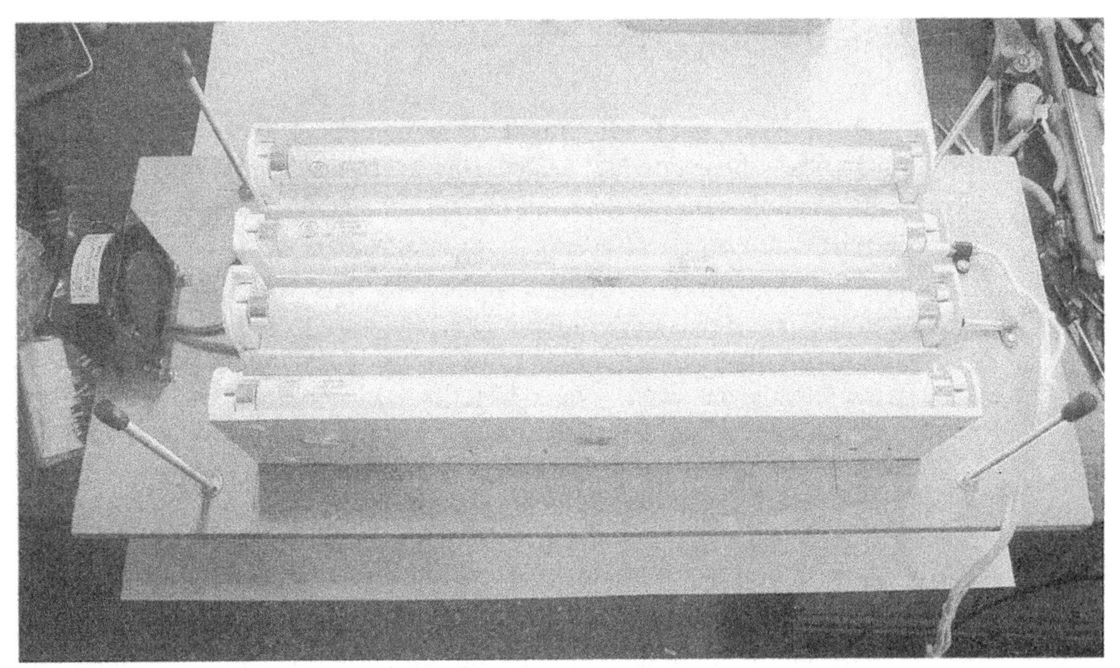

An upside-down view of the exposure light as described in the text. As can be seen in the left side of the photo, I installed a small muffin fan for cooling purposes, but it shouldn't be needed for intermittent use.

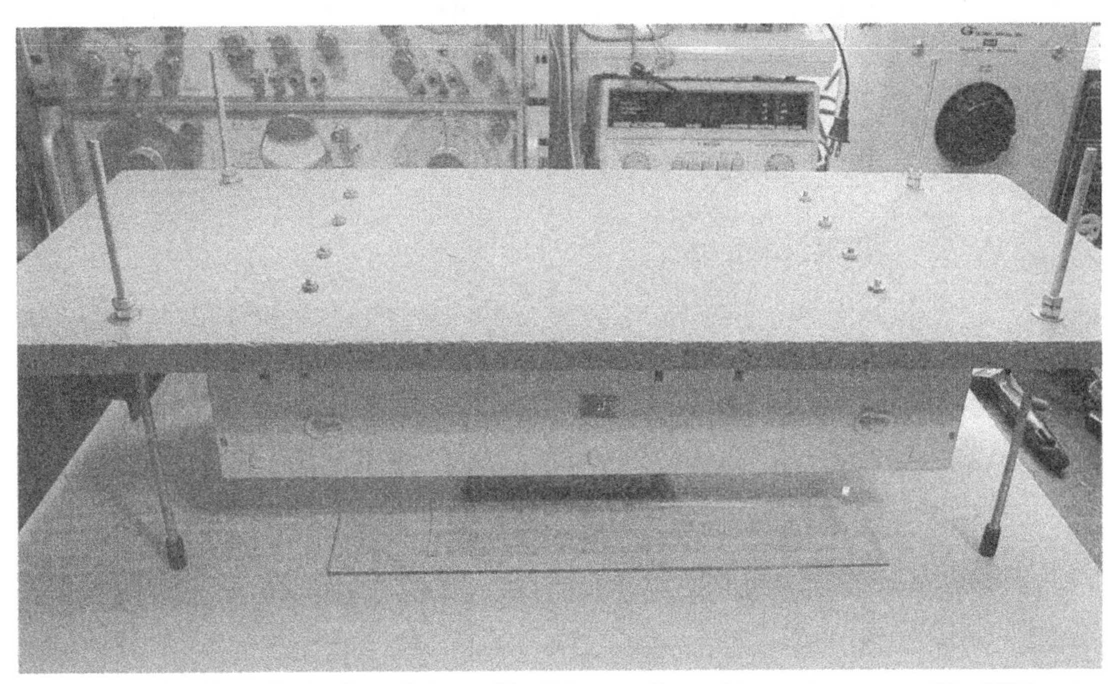

The exposure light from the previous photograph as it is normally used to expose a presensitized PC board.

9. Carefully pat the PC board dry with a paper towel (the photoresist can easily be scratched) and closely examine the photoresist artwork. Areas containing pits or voids can be touched up with a little fingernail polish; any excess photoresist can be scratched off with a small solder pick or similar tool. When you are satisfied that the artwork is ready, immerse the PC board in a glass or plastic tray containing the etchant solution (this will typically be ferric chloride). When all of the exposed copper foil has been removed by the acid, remove the PC board from the etchant solution and thoroughly wash in cold running water.

10. The photoresist can now be removed with fine steel wool, fingernail polish remover, or a variety of other solvents. The PC board should now be ready for drilling the component holes. If you anticipate constructing a lot of PC boards, you should invest in a small drill press for this task. A hand-held drill will function well, but your arm will get mighty tired before the job is finished. The hole size should be slightly oversized in respect to the component lead diameters.

If you would like to utilize the artwork included in this textbook to construct some of the relevant projects, you can copy the artwork onto a clear acetate sheet with a copying machine. (Remember to set the copy machine for the darkest copy possible.) This creates the photopositive. Then, starting with step 4 as previously listed, continue through the same procedure.

Heatsinks

I devoted considerable text to the subject of *thermal dynamics* in my 1999 book entitled *High-Power Audio Amplifier Construction Manual.* I don't have room in this text to go into a highly detailed discussion of thermal dynamics, but I wanted to provide some helpful guidelines regarding heatsinks.

If you're lucky enough to obtain a large heatsink with a manufacturer's part number on it, the specifications are usually available. However, it is more likely that a large heatsink won't provide you with a clue to the manufacturer or the thermal specifications. Hobbyists often collect a large junkbox of old heatsinks, but they don't have any idea of how to determine their suitability for a particular audio power amplifier project. There is a better method than just "thinking that it

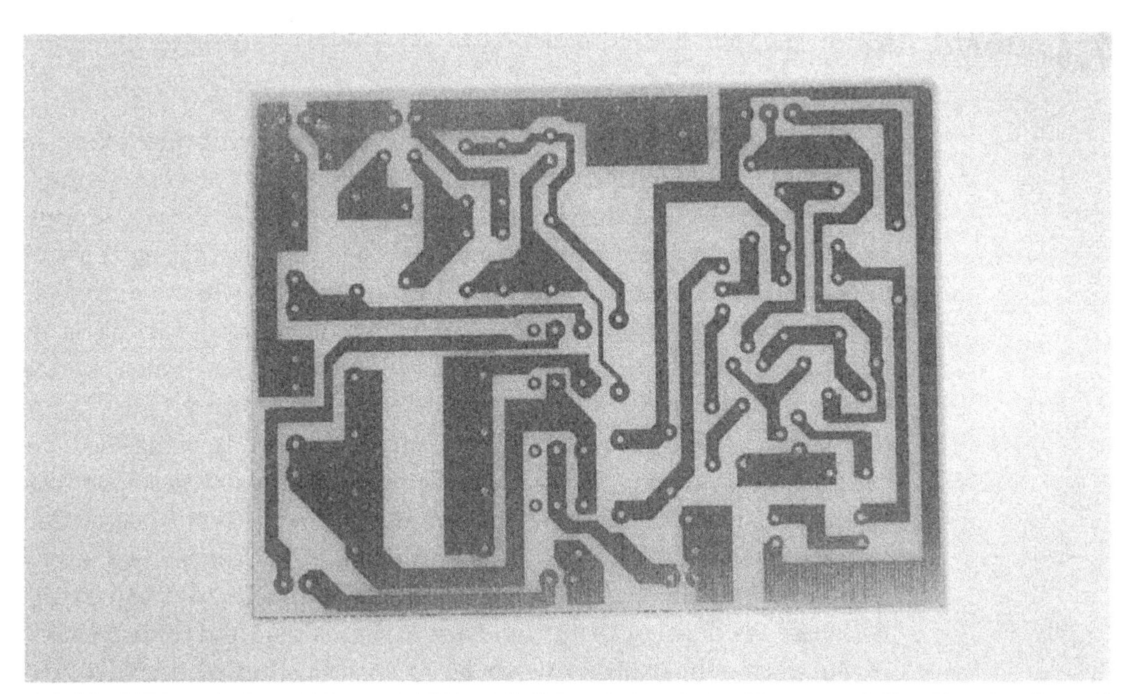

A finished PC board after etching and drilling. I left the photoresist on the copper foil so the artwork would show up better in the photograph.

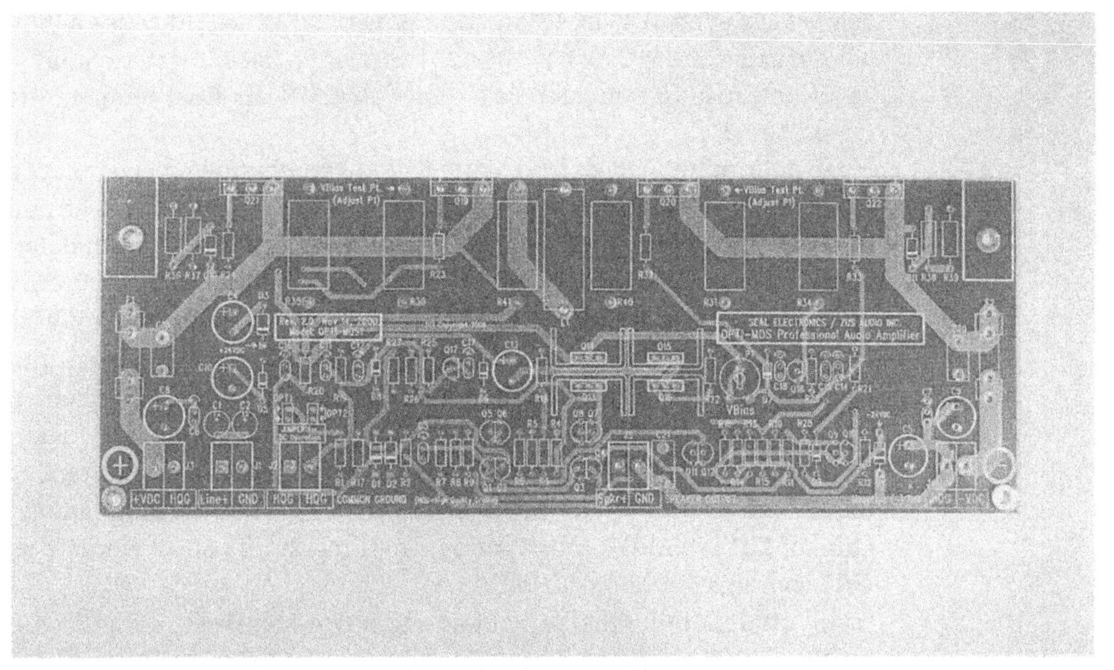

A commercial double-sided PC board, with silkscreen, plated-through holes and fabricated from 3-oz copper foil. This PC board design makes the Fig. 6-21 audio power amplifier more compact and easier to assemble, but the end performance is the same as provided by the single-sided version artwork that is supplied in this book.

looks big enough." I am not providing the following procedure to replace good analytical methods of determining precise thermal requirements from a design/engineering perspective. Rather, the following procedure is a rule-of-thumb method of determining if a surplus or junkbox heatsink will function well with an intended project.

The most important thermal parameter of a large heatsink is its *thermal resistance* (R_{THSA}). This is the thermal resistance of the heatsink to the ambient air. It describes the heatsink's temperature rise, usually in degrees centigrade, relative to power, in watts, that it is called upon to dissipate. For example, a thermal resistance specification of *1°C/W* means the heatsink temperature will "rise 1-degree Celsius for every 1 W it is required to dissipate." Likewise, a thermal resistance specification of *0.4°C/W* means the heatsink temperature will "rise 0.4 degree Celsius for every 1 W it is required to dissipate."

Suppose a heatsink with an R_{THSA} specification of 0.5°C/W was required to dissipate 50 W of power. How hot would it get? Ambient air is normally considered to be about 25°C, so the heatsink would start out at that temperature. Since its temperature would rise 0.5-degrees for every 1 W of dissipation power, 50 W would cause a temperature rise of 25°C ($0.5 \times 50 = 25$). Since the heatsink started out at 25°C, and rose in temperature by another 25°C, its final temperature would be 50°C.

To determine a rough idea of the R_{THSA} of an unknown heatsink, you can force the heatsink to dissipate a known quantity of power and measure its temperature rise. From these two variables, you can determine the approximate R_{THSA}. For example, suppose you temporarily mounted a power resistor to an unknown heatsink, connected a variac to the power resistor, and adjusted the voltage across the resistor until it was dissipating 50 W continuously. When the heatsink reached its maximum temperature (this would take considerable time, due to normal thermal delays), suppose the temperature was 45°C. Subtracting the 25° ambient temperature, you would be left with a rise in temperature of 20° for 50 W of dissipation. Dividing 20° by 50 W would provide your answer of 0.4°C/W.

If you determine the R_{THSA} of an unknown heatsink, you still need to know the anticipated dissipation of the output stage of an amplifier to determine the suitability of the heatsink for the project. As a rule of thumb, assume the efficiency of the amplifier to be 66% (roughly two-

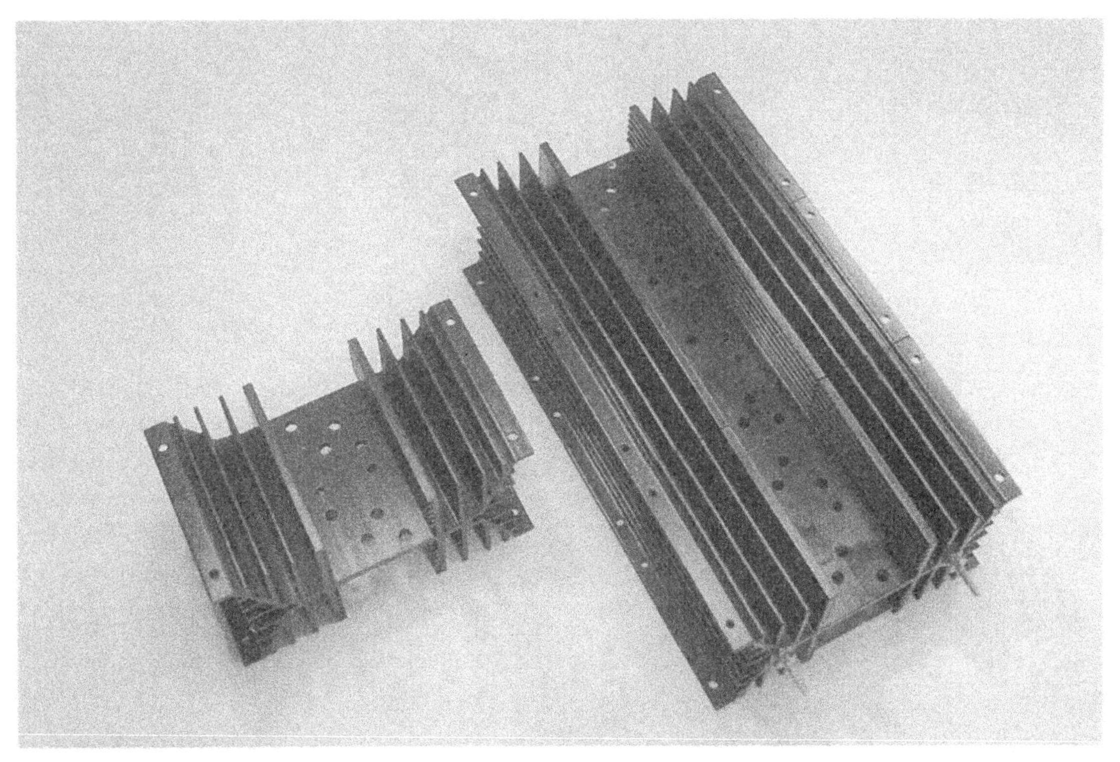

An example of how three heatsinks of the type illustrated in the left side of the photo can be joined together with two-threaded rods to create a larger heatsink for higher-power applications.

thirds) for all class B output stage designs. (I can't provide a good rule of thumb for class A designs, due to their wide variance in efficiency levels.) Thus, if you were constructing a class B amplifier project capable of delivering 200-W RMS into a speaker load, you would anticipate about 100 W of wasted heat energy that would have to be dissipated by the heatsink.

If you were hoping to use a heatsink with an R_{THSA} of 0.4°C/W to dissipate 100 W of power, the temperature rise of the heatsink would be 40°C ($100 \times 0.4 = 40$). Adding the 40° rise to the 25° ambient temperature would give a final temperature of 65°C. Is this too hot?

Manufacturers typically install thermal switches on the main heatsinks of power amplifiers to promote thermal shutdown if the heatsink temperature exceeds about 75° to 100°C. Therefore, a final temperature of 65°C would seem pretty good. However, this still wouldn't represent a "maximum" temperature, because the efficiency levels of

Several examples of common "flat-faced" heatsinks incorporated into larger audio power supplies.

power amplifiers change depending on the speaker load impedance, as well as a variety of factors involving the current-limiting protection in the amplifier electronics. However, if you utilize the aforementioned procedure and can come out with a final temperature of between 75° and 100°C, your heatsink choice should be pretty well suited to the application. In fact, the rule-of-thumb estimates that I've provided are a little conservative, so your final project should run cooler than most commercial units of comparable size and power ratings.

Many manufacturers would consider my aforementioned rule-of-thumb method overly conservative, and from a practical perspective, they would be correct. As stated in previous chapters, music programs contain a very wide dynamic range. Thus, if you were playing a typical program through a 100-W RMS amplifier, and you turned the volume up to where you began to hear considerable distortion from clipping, you wouldn't even be close to an "average" 100-W output to the speaker system. Many commercial power amplifier manufacturers incorporate main heatsinks in their power amplifiers

A small wind tunnel for forced air cooling of semiconductor output devices. A common size 4 ½ rectangular cooling fan can be mounted to either end to create the forced air flow.

that are only capable of dissipating 50%, or in some cases, as low as 25%, of their maximum rated output power. Their reasoning is that the amplifier should never be called on to dissipate greater levels of power due to the dynamic range of the music. Consequently, if you attempted to sinewave test such an amplifier at its maximum RMS output for more than 5 or 10 minutes, it would certainly go into thermal shutdown.

As can be readily seen, there is considerable leeway in deciding upon the required heatsink size (in terms of R_{THSA}). If you intend on using an audio power amplifier for sustained, heavy-duty operation, such as a professional application, wherein the music could be highly compressed and/or overdriven, you would probably want to follow my aforementioned rule of thumb without any derating. However, for a domestic hi-fi application, I wouldn't feel uncomfortable with derating the R_{THSA} of the main heatsink by about 50% (providing a good thermal shutdown system was properly incorporated).

High-capacity wind tunnels for high-power audio amplifiers. Each unit is rated for a continuous dissipation of 1-KW. The four heatsink sections making up each wind tunnel are electrically isolated from each other, allowing the power supply rails to be connected directly to the heatsink sections. This technique eliminates the need for output device insulators, and also reduces the thermal innefficiencies caused by the insulators. Such wind tunnels are typically used for audio power amplifiers with output power ratings of about 2.2-KW RMS."

As a final note on heatsinks in general, I might mention that forced-air cooling will reduce the R_{THSA} parameter by a factor of 2 to 3. Many audiophiles design their projects with a derated heatsink (based on still air convection), but provide a thermostatically controlled fan that will begin to provide forced-air cooling if the heatsink temperature exceeds a preset limit. Thus, they obtain the best of both worlds. They aren't annoyed with fan noise at low-listening levels, but they can still obtain relatively high-power output levels without going to overly large enclosure sizes.

Project Enclosures

I receive a lot of questions about project enclosures. From a functional perspective, low-level and line-level audio systems will be less sus-

An example of how two large heatsinks can be used together by sandwiching a piece of aluminum plate between their flat mounting faces, and mounting the output devices to the lip of the aluminum plate. In this case, the aluminum plate is also used for mounting the PC board. The heatsink below the aluminum plate is identical to the top heatsink.

ceptible to external noise sources if they are enclosed in a conductive metal enclosure of some type. With audio power amplifiers, it really doesn't make a whole lot of difference, providing good grounding and wiring techniques are incorporated. I've seen beautiful audio power amplifier cabinets made from various hardwoods, or incorporated into older, existing wooden enclosures of various types. Typically, in such cases, the power supply and amplifier electronics are mounted to a plate metal base of some type, with a wooden top and sides.

Electronic supply houses typically have a good selection of various enclosure types. Audiophiles usually choose rack-mount enclosures for audio power amplifiers, but some suppliers are now offering older-style metal chassis with screen-mesh covers that are a good choice for some solid-state amplifiers as well as being the traditional style for vacuum-tube amplifiers.

The top view of the internal layout of a typical stereo high-power audio amplifier. Note that the power transformer is kept a considerable distance from the amplifier electronics, and plenty of space is provided for the separation of wiring runs. The smaller PC board assembly to the right of the power transformer is the soft-start and protection circuitry. Each reservoir capacitor is 22,000 μF, providing a total of 44,000,μF, per power supply rail. The power transformer is a 625-VA model manufactured by Piltron Manufacturing, Inc. The enclosure is manufactured by Par-Metal Products, Inc., and the amplifier modules are manufactured by SEAL Electronics/ZUS Audio Systems, Inc.

Scavenging is a good method of obtaining a wide selection of enclosures. Old CD players, some computer enclosures, and, of course, old scrapped audio systems contain usable enclosures for audio projects. If you live close to a city junkyard or landfill, you should check out the scrap metals they may have available. Aluminum plate material is very expensive when purchased from retailers, but scrap pieces are often available at junkyards for 50 cents a pound. Old road signs are often made of aluminum plate material.

APPENDICES

APPENDIX

A

Audio Electronics Terminology

The following terms, definitions, and abbreviations are commonly used in the field of acoustics and audio electronics. Not all of these terms will be found within the context of this textbook, since I have kept the variety of synonymous terms to a minimum for ease of understanding. However, the chances are that you will run across many of these terms and "buzz words" in other audio electronics books and magazines. I felt it would be helpful to include them here.

A, a (1) Ampere, basic unit of electrical current. (2) Signal gain or loss, in decibels.

absolute phase The maintenance of the exact phase relationship between the input signal and output signal of an audio amplifier (or other audio system).

absolute value The numerical value of a number without regard to sign (neither "+" or "−"). Absolute values are used in certain electronic equations and graphs in situations where polarity is of no concern. For example, when calculating the pulsating DC output of a bridge rectifier, the polarity of the AC input is of no importance, since the bridge rectifier will convert the AC input to a single-polarity DC output. However, the absolute value of the AC input will determine the rectified DC amplitude at the output of the bridge.

AC Alternating current (also used to designate "periodic" voltage).

AC coupling A coupling arrangement that will not pass direct current or a DC component of a signal. The most common AC coupling devices are capacitors and transformers.

AC mains Often a synonym for "AC power" or "mains power." It refers to the public AC power line or supply.

active device (active component) Any electronic device capable of providing power gain (i.e., voltage and current gain). Examples of active components used throughout this textbook are bipolar junction transistors (BJTs), junction field-effect transistors (J-FETs), metal-oxide semiconductor field-effect transistors (MOSFETs), vacuum tubes ("valves" in British terminology), and operational amplifiers (op-amps).

active loading The technique of replacing a passive load (such as a resistor) with an active load (such as a BJT constant-current source) to improve the gain and linearity of a transistor (or other active device) stage.

A_e, A_V Voltage gain.

A_{eol}, A_{vol} Open-loop voltage gain.

Ah Ampere-hour.

airpath The path of cooling air around a heatsink surface.

AM Amplitude modulated. The effect of two or more signals being combined in a nonlinear circuit so that upper and lower sidebands are produced. In audio terminology, amplitude modulation is the result of intermodulation distortion.

amp (1) An abbreviation for ampere. (2) Colloquial abbreviation for "amplifier." Not used in this book to avoid confusion with "amp," as used for the abbreviation of ampere.

ampere Basic unit of electrical current flow (i.e., electron flow).

amplify To increase in size; make bigger.

amplified diode Also known as a "Vbe multiplier." In audio power amplifier terminology, a specialized circuit used to provide linear bias and temperature compensation to an output stage for the purpose of reducing crossover distortion.

amplitude The magnitude of a varying quantity. Can be used in terms of "steady state" (i.e., quiescent) or "instantaneous" units of measurement.

amplitude distortion Also called "large-signal distortion" or "large-signal nonlinearity (LSN)." A type of distortion which begins at, or relates to, a specific amplitude or level. In the context of audio power amplifiers, this usually relates to the "beta droop effect" that occurs in the output stage.

analog An electrical variable capable of continuous variation, or in possession of an infinite number of states. Normally used in contrast to "digital" variables, which can only function within the confines of two states, or a specific number of rigidly designated steps. All audio program material is analog, or converted to analog during reproduction.

ant Antenna.

antisurge A type of fuse, which can withstand momentary surge currents that occur during power-up of electrical equipment. Typically called "slow-blow" or "slo-blo" fuses.

assy Assembly.

asymmetric (1) A circuit topology that is not symmetrical in terms of component physics. (2) A nonsymmetrical quality of a complex waveform, such that the long-term absolute sum of its positive cycles does not equal the long-term absolute sum of its negative cycles.

aud Audio.

audiophile An enthusiast whose hobby, dedication, and appreciation is aimed toward the quality of sound.

audio signal generator An instrument designed to produce pure AC (sine wave) signals for testing purposes. Most audio signal generators have frequency ranges far above and below the audio range, and many have selectable square-wave outputs with the equivalent fundamental frequency range.

AUT Amplifier under test.

autonull A circuit incorporating sufficient feedback to cancel out, or "null," unwanted DC offset voltages. In the context of audio power amplifiers, this usually refers to global negative feedback of sufficient quantity to "null" (zero out) any DC voltage occurring at the input stage. "DC autonulling" is a method of automatically nulling the DC offset voltage at the output of an audio power amplifier, typically using a DC "servo." See "servo."

aux Auxillary.

auxiliary power supply In the context of audio power amplifiers, a power supply (usually of a lower DC voltage than the DC supply rails) used to provide operational power to various auxiliary circuits (i.e., relay circuits, DC offset protection, etc.).

A-weighting A filtering process used in testing. Emphasis is placed on 3.5 kHz (the most sensitive area of human hearing). Used in evaluating distortion and SPL specifications.

AWG American wire gauge.

B, b (1) Magnetic flux density in gauss or tesla. (2) The base terminal of a BJT. (3) Current gain factor of a BJT (beta or h_{FE}) (4) Feedback factor.

back EMF (1) Also called "reverse EMF" and "inductive kickback." The bothersome or potentially damaging effect of a high-voltage spike being produced in an inductor when its electromagnetic field collapses too rapidly. The rapid field collapse is usually the result of an abrupt switch-off of inductor current flow. (2) The characteristic of an inductor that opposes a change in current flow.

Baker clamp diode(s) Diodes installed reverse-biased from the DC supply rails to the output rails of an audio power amplifier. Their purpose is to protect against inductive kickback voltages (i.e., back-EMF). Intrinsic Baker clamp diodes are incorporated into some modern MOSFET devices. Baker clamp diodes are also known as "flyback" diodes, "catching" diodes, or "freewheeling" diodes.

backtalk The undesirable condition of one audio system "leaking," or feeding, an unwanted signal into another audio system. This condition is similar to "crosstalk," but differs in the respect that crosstalk occurs internally within only one audio system.

balance (1) The comparable sonic level adjustment between various audio channels. In stereo systems, the control adjustment determining the ratio of output amplitude between the right and left channels. (2) The matching of components and component variables (i.e., voltage, impedance, current) in an electronic circuit.

band-gap reference References, or specifications, related to specific points within a total bandwidth. For example, a chart illustrating THD at 100 Hz, 1 kHz, and 10 kHz would be a band-gap reference (i.e., there are "gaps" between the frequency points).

bandwidth The range of frequencies that has been specified as performance limits for a component, circuit, or filter. Defined as between the −3 dB (half-power) points at the high-pass and low-pass ends of the frequency response. In high-performance audio power amplifiers, these limits will range from about 3 Hz to at least 50 kHz.

Belcher test An audio purity test designed by Dr. Alan Belcher. It utilizes a pseudorandom noise signal to simulate musical (complex) input signals. Since various aspects of the "simulated music" are known, a comparison of input and output signals reveals an analysis of the way the amplifier processes complex signals of any kind.

beta The current gain variable of a BJT.

bi-amping A system whereby the music signal is filtered through an active crossover network, separated into two frequency bands (high frequency and low frequency), and individual audio power amplifiers are utilized to amplify each frequency band. In this way, passive crossovers are eliminated in the speaker system(s), with the power amplifiers driving the individual speaker components directly.

bias The quiescent current or voltage conditions needed to keep an electronic circuit in its active operational region. A set of voltage and current conditions that establish a desired quiescent circuit operation.

biphase Two phase. Usually associated with the output of a coupling transformer used to drive a "push-pull" output stage.

bipolar Essentially a synonym for "bidirectional."

bipolar junction transistor (BJT) The technical name for the common transistor. See "transistor."

BJT Bipolar junction transistor; the most common type of three-layer transistor.

BJT gain droop (or BJT droop) A large-signal nonlinearity mechanism inherent to all BJTs used as output devices in power amplifiers. It is a result of H_{fe} variations during peak periods of current flow (i.e., a cyclic distortion).

bleeder A resistor used to slowly discharge a filter capacitor after the mains power has been shut off. The use of bleeder resistors in power supply design is primarily a safety precaution.

bootstrapping A technique for lifting a generator circuit above ground by a voltage value derived from its own output signal. In the context of modern audio electronic equipment, the technique of bootstrapping is used more often for the purpose of impedance modification.

breakdown Loss of the nonconductive properties of an insulator. In semiconductor terminology, a destructive conduction of current and loss of control usually associated with exceeding the maximum voltage, current, or temperature ratings of the device.

breakthrough signal A general term to describe any type of undesirable interference signal (from any source) which gets into and contaminates an audio signal path.

bridge mode A connection mode utilizing two audio power amplifiers, with each amplifier being 180 degrees out of phase from the other. A speaker is connected across the "high sides" (i.e., the terminals connected to the output rails) of each amplifier, greatly increasing the power output (to the speaker) over that which could be obtained with a single power amplifier. Amplifiers connected in a bridge mode are said to be "bridged."

bridge rectifier Either a discrete or modularized form of four solid-state diodes, dedicated to the rectification of AC (usually AC mains power).

buffer An impedance matching circuit. Typically, a good buffer will have very high input impedance with very low output impedance.

BW, Bw Bandwidth.

C, c (1) Schematic symbol for a capacitor. (2) Collector terminal of a BJT. (3) Capacitance. (4) Coulomb.

capacitor distortion A term usually applied to a type of distortion created by the use of coupling capacitors in the signal pathway. It is created when the signal frequency is low enough to superimpose the coupling capacitor's nonlinear charge/discharge rate onto the signal waveshape.

cascade In general usage, this term describes the technique of arranging two or more circuit stages in a series fashion, wherein the output of a pervious stage supplies the input to the succeeding stage.

cascode A method wherein a common-base transistor stage is used as a collector load for a common-emitter transistor; a technique often used to increase gain, linearity, and reduce "Early effect" of a common-emitter stage. Although the method is defined using BJTs as an example, the same technique can be accomplished with FETs and vacuum tubes.

catching diodes See "Baker clamp diodes."

cathode (1) The negative end of a conducting diode or LED. (2) Positive end of a zener diode in active regulation (i.e., while conducting in the zener region). (3) The electron emitter of a vacuum tube.

CB Common base.

C_{dg} Drain-to-gate capacitance.

C_{ds} Drain-to-source capacitance.

CE Common emitter.

center tap The midpoint of a transformer winding. Can also apply to the midpoint of various types of balanced circuits.

CFP Complementary feedback pair.

C_{gs} Gate-to-source capacitance.

channel separation A stereo (or multichannel) audio specification, usually expressed in negative −dB, defining the signal isolation from one channel to another. The opposite of crosstalk.

choke A single-coil inductor.

C_{iss} Gate-to-source capacitance with input short-circuited in a common source configuration.

C/L Closed loop.

class A A type of output stage biased so that all of the output devices remain in their active region at all times.

class A-B A type of output stage biased so that each output device (there must be at least two) will conduct, or remain active, in excess of 180 degrees of the AC cycle. Essentially the same as a class B, but "overbiased" to some extent.

class B A type of output stage biased so that each output device (there must be at least two) will only conduct for one-half of the AC cycle (approximately 180 degrees). The output devices will always be driven in a complementary mode, facilitating a summation through the load that will be an almost identical reproduction of the full 360-degree AC signal. Class B is by far the most common type of output configuration used in audio power amplifiers.

class C Used only in RF (radio frequency) amplifiers.

class D The class designation applied to pulse-width modulated (PWM) audio amplifiers.

class G The class designation applied to modified class A-B or class B output stages that incorporate multiple power supply rail voltages. During periods of low power demand, the output stage will utilize the lower-voltage power supply rails. During periods of increased power demand, the output stage will automatically switch (via commutating diodes) to a higher-voltage power supply rail. The primary advantage is increased efficiency.

class H The class designation applied to modified class A-B or class B output stages that incorporate variable-voltage power supply rails.

The power supply rails automatically increase in voltage level as output power demand increases. As with class G, the primary advantage is increased efficiency.

class S A method of using a class A stage with very limited current capability, interfaced to a class B stage to make the load appear as a higher impedance, which is within the class A amplifier's drive capability.

closed-loop gain The gain factor of an amplification system incorporating a negative feedback loop.

clip The condition that occurs when an amplifier stage is driven to the point where it cannot remain in its active region. When clipping occurs, the output signal will flatten out at the peaks, as though they had been "clipped" off with a pair a scissors.

CM Common mode.

C_{ob} Output capacitance.

C_{obo} Open-circuit output capacitance in a common base configuration.

C_{obs} Short-circuit output capacitance in a common base configuration.

coil An inductor or choke.

coloration A metaphorical term used to describe some aberration, or distortion, occurring in the original audio material. Audiophiles usually associate this term with the insertion of artificially introduced harmonics.

common emitter A basic BJT circuit configuration; the input is applied to the base, the output is taken from the collector, and the emitter is common to both input and output signals.

common mode A signal that is common in waveform and phase superimposed on multiple conductors. Typically, common mode signals are undesirable "garbage" signals originating from stray EMI and RFI fields.

common mode distortion The type of distortion caused by "common mode stress." See "common mode stress."

common mode rejection The ability of a circuit to cancel out, or reject, common mode signals. Usually expressed as a ratio in decibels. Can also be expressed in terms of a literal ratio (i.e., 10,000 to 1, etc.), in which case it is referred to as the "common mode rejection ratio."

common mode stress A somewhat conjectural term describing a condition that may occur at the input stage of audio power amplifier designs incorporating high negative feedback. This would only take place if the common mode voltage of the input signal and negative feedback signal exceeded the V_{th} (threshold voltage) of the input transistors.

comparator A circuit or device that monitors two or more variables and provides an algebraic summation of their values, or provides a logic output if the variables exceed a designated limit or condition.

compensation In the context of audio processing systems, it is the tailoring of the amplifier's open-loop gain and phase characteristics so that the amplifier is dependably stable when the global feedback loop is closed.

complementary Equal but opposite (conjugate devices). Matched set of semiconductor devices (i.e., BJTs, MOSFETs, JFETs) of opposite polarity.

complementary feedback pair An audio power amplifier output configuration (sometimes called a Sziklai pair, after the inventor). For half of the complementary topology, a low power NPN BJT is connected to a high power PNP BJT. The collector of the NPN is connected to the base of the PNP; the emitter of the NPN is connected to the collector of the PNP. BJT polarities are reversed for the opposite side of the complementary configuration.

compression A type of "dynamic gain reduction." As signal level is increased, gain is proportionally decreased, maintaining a constant output signal level. In the context of audio electronics, compression circuits are typically inactive until maximum output levels are reached. Consequently, the compression circuit(s) activate and dynamically control the amplifier gain as a method of eliminating "clipping."

conduction angle The polar angle of conduction for a specified output device. For example, the conduction angle for a complementary BJT operating in class B mode would be approximately 180 degrees (i.e., the BJT would conduct for approximately half of the full waveform, or 180 degrees).

consonant Musically pleasing.

constant current source An active circuit configuration that provides a regulated preset electrical current. Typically used in differential amplifier circuits.

convection In the context of high-power electronic devices, the type of heatsink cooling afforded by natural air movement.

C_{oss} Short-circuit output drain-to-source capacitance in a common source configuration.

cross-conduction A destructive condition in audio power amplifier output stages wherein two complementary devices are conducting substantial currents at the same instant in time.

crossover network Speaker system frequency divider and routing network. Typically a passive network that routes the high frequencies to the tweeter, midrange frequencies to the midrange speaker, low frequencies to the bass driver, etc.

crossover area The conduction area of a complementary output stage wherein both complementary devices are close to their cutoff points. The transitional area of conduction in complementary devices (i.e., the point at which one device turns on as the other device is turning off).

crossover distortion Distortion occurring in the crossover area caused by nonlinearities inherent when one output device turns off as the other output device is turning on. Crossover distortion is greatly reduced by applying a slight forward bias to the output devices.

crosstalk Leakage signals from one channel to another in a multichannel audio system.

crowbar protection A risky, but simple, method used by some audio power amplifier manufacturers to provide DC speaker protection. A triac is used to monitor the amplifier's output. If sustained DC occurs, the triac will fire and short the amplifier's output rail to ground.

C_{rss} Short-circuit reverse transfer capacitance; input to output; source short-circuited to gate; common source configuration.

current capability In the context of audio power amplifiers, the maximum amperes available from the output of a power amplifier.

current draw The amount of current utilized from the AC mains.

current dumping An adaptation of the basic class S operating mode. A class A amplifier directly drives the speaker load until the output current approaches a predetermined limit. At this point, class B "dumper" transistors begin to provide (or "dump") additional drive current to the load as required. See "class S."

current feedback A feedback system where electrical current is the sensed variable instead of voltage. In the context of audio power amplifiers, current feedback is almost always used in short-circuit protection circuits.

current mirror Any electronic circuit or device that copies a current at a specified ratio. In audio electronic circuitry, current mirrors are most often incorporated as collector loads for a pair of differential transistors. In this application, current mirrors force a current balance through both legs of the differential pair, greatly reducing harmonic distortion within the stage.

current source See "Constant current source."

D,d (1) Drain lead of an FET. (2) Dissipation factor; reciprocal of storage factor Q. (3) Thickness of the dielectric material in a capacitor (measured in centimeters).

damping control Virtually extinct in modern audio power amplifiers, damping control provided a method for the user to vary the damping factor by means of variable feedback systems. Damping control was often found in older vacuum tube audio amplifiers.

damping factor The ratio of the output impedance of a power amplifier output stage and the impedance of a speaker system.

Darlington pair A method of cascading two BJTs so that their beta values multiply. For example, by connecting two transistors with beta specifications of 100 in a Darlington configuration, the resultant beta becomes 10,000.

dB Decibel (one-tenth of a bel); the logarithmic ratio between two levels of power or voltage.

DC, dc Direct current.

DC autonulling See "autonull."

DC balance Usually applied to differential BJT or JFET circuits. A quiescent circuit condition when two or more critical voltages are equal.

DC blocking Essentially the same thing as AC coupling. See "AC coupling."

DC offset In the context of audio electronic circuitry, the small, steady-state DC present at an output stage. This voltage should be extremely small or virtually nonexistent.

DC servo See "servo."

DC speaker protection A protection circuit that monitors the output of an audio power amplifier and disconnects the speaker system if any appreciable DC voltage is detected at the output.

decade Any interval of 10.

decibel One tenth of a bel. A bel is a base 10 logarithm of the ratio of two amounts of power, with one power value being the reference.

derate To use an electronic component below its maximum ratings.

differential amplifier In its simplest form, a two-transistor circuit that shares a common emitter load or current source. If the output is taken between the transistor collectors, a differential amplifier will provide high common mode rejection. (Differential amplifiers are also constructed from JFETs and MOSFETs.)

direct coupled Circuits or stages that do not utilize capacitors or transformers for the transference of signals.

discrete A general term used to describe circuitry that does not contain integrated circuits. In other words, circuits constructed from individual resistors, diodes, transistors, etc.

dissipation The amount of waste energy disposed of in the form of heat.

distortion In audio technology, the term refers to any deviation, modification, or corruption to the original audio program material, with the exception of amplification.

DPDT Double-pole, double-throw (i.e., relative to switches and relays).

DPST Double-pole, single-throw (i.e., relative to switches and relays).

drain One of the three main terminals of a JFET or MOSFET.

driver (1) One of multiple acoustic transducers (speakers) within a speaker system. (2) One of the multiple high-current active devices located in an audio power amplifier.

dummy load (1) Any type of artificial load connected to the output of an electronic device and used for testing purposes. (2) A testing load that looks like a speaker system to a power amplifier, but does not produce any sound. Most dummy loads are simple 8-ohm or 4-ohm noninductive power resistors. Various reactive components may be incorporated into a dummy load to more closely simulate the real-world reactive characteristics of a speaker system.

DVM Digital voltmeter.

dynamic headroom An attempt to define the short-term, or "burst," power capabilities of an audio power amplifier. The IHF defines dynamic headroom as the maximum RMS power capability for 20 milliseconds after a 500-millisecond recovery period. See "headroom."

dynamic power rating The short-term (under 1 second) maximum output power rating of an audio power amplifier.

dynamic range The difference between the maximum output level and the noise floor. For all practical purposes, it equals the sum of the signal-to-noise ratio and the maximum headroom.

dynamic speaker A type of acoustic transducer (speaker) utilizing a speaker cone, moving voice coil, and stationary magnetic field. The most common type of speaker.

E, e, emf (1) Electromotive force, measured in volts. (2) Emitter lead of a BJT.

Early effect The reduction of the effective base width of a BJT when the width of the collector-base junction is increased by increasing the collector-base voltage. Early effect results in a change of a BJT's signal-handling parameters as internal voltage swings increase (resulting in increased harmonic distortion). High-quality audio amplifiers avoid Early effect as much as possible by using topologies requiring the minimum internal voltages swings (e.g., cascode stages).

earth ground (1) The ultimate reference for all electrical operations. (2) The green or green-yellow conductor in most power cables, intended for connection to earth ground.

eddy currents A circulating current induced in a conducting material by a varying magnetic field. Eddy currents circulating in the core material of a power transformer represents a loss factor.

EF Emitter follower.

electrostatic coupling Usually an undesirable phenomenon. Refers to the induction of noise or signals into neighboring conductors or circuitry via an electric field.

electrostatic shield (1) A shield incorporated into power transformers to prevent capacitive leakage between primary and secondary windings. (2) Conductive shielding used to protect static sensitive (MOS or CMOS) devices from destructive electrostatic discharges. (3) A safety shield used to protect humans from flashover current.

electrostatic speaker A type of loudspeaker that operates on the principle of a variable "electrostatic" field.

EMI Electromagnetic interference.

equal loudness contours A graph with multiple curves plotting frequency (Hz), sound pressure level (SPL), and loudness (phon) to graphically illustrate the nonlinear response of human hearing.

equalizer See "graphic equalizer."

ERP Effective radiated power.

eV Electron-volt.

F, f (1) Frequency, in hertz. (2) Farad; basic unit of capacitance. (3) Temperature, measured in degrees Fahrenheit.

feedback A percentage of an output signal that is applied "back" to the input. Feedback can be either "negative" or "positive." Negative feedback (NFB) is utilized in audio systems for correction and stabilization. It can be applied "globally," "locally," or "nestedly." Positive feedback is usually undesirable (except in oscillators) and is often referred to as "regenerative feedback."

feedforward An alternative to negative feedback for error-correction purposes. The error-correction signal is derived in parallel with the amplification process and subtracted from the output.

FET Field-effect transistor.

fidelity Faithfulness in reproduction.

floating An electrical circuit or variable that is not referenced to ground.

flushout The functional name for a resistor placed between the base and emitter junctions of a BJT to aid in the removal of charge carriers. The purpose is to improve the transistor's transient response.

flyback diodes See "Baker clamp diodes."

FM Frequency modulated.

frequency distortion Refers to nonuniformities of the frequency response within the frequency range.

frequency range The range of frequencies between the typical −3 dB (upper and lower frequency) rolloff points.

frequency response Often used as a synonym for "frequency range" (See "frequency range"). The term applied to relative level or gain variations in respect to frequency.

f_T Transition frequency of a BJT.

fundamental frequency (1) The lowest frequency component of a complex electrical signal. (2) A harmonically pure sinusoidal waveshape. (3) The harmonically pure test signal applied to an audio system for distortion analysis. Synonymous terms are *first harmonic, fundamental,* and *fundamental component.*

G, g (1) Giga, prefix for 1,000,000,000. (2) Gate lead of an FET. (3) Conductance, measured in siemens or mhos.

gain An increase in voltage, current, or power. Sometimes used in conjunction with a "negative" value, in which case it is synonymous with "attenuation."

germanium The element from which the first commercial transistors were manufactured. Due to the severe disadvantages of germanium BJTs in respect to leakage and temperature sensitivity, virtually all germanium BJTs have been replaced by silicon BJTs.

g_{fs} Forward transfer admittance.

GHz Gigahertz.

global feedback In the context of audio circuits, the negative feedback taken from the output stage and applied back to the input stage. In contrast, "local feedback" is negative feedback confined to a single stage.

g_m Mutual conductance.

graphic equalizer A grouping of selective tone controls, or filters, with the filter adjustment controls arranged on the front control panel in such a manner to form a visual "graphic" representation of the broadband filter settings.

grid The control element of a vacuum tube.

ground Sometimes used as a synonym for "earth ground." (See "earth ground.") More often, it is used to denote the chassis common. Occasionally used to erroneously denote circuit common.

ground lift Any method of isolating one common point from another. For example, the isolation of earth ground from circuit common, etc.

ground loop A condition when two or more system components share a common electrical ground line and unwanted (spurious) voltages are unintentionally induced. Often, the ground circuit formed acts like a shorted turn in the presence of power line magnetic fields, causing a substantial current to flow (even though the ground conductor's DC resistance is very low). Ground loops are a key reason for undesirable "hum" conditions in audio amplifiers.

group delay A time delay imposed on a specific group of signals.

H (1) Magnetic field strength (amperes per meter). (2) Henry; basic unit of inductance.

harmonic An integer product of a fundamental frequency. The term "subharmonic" is descriptive of frequencies that are integer fractions of a fundamental frequency.

harmonic distortion Any distortion, or coloration, caused by the artificial generation of harmonics. Usually associated with subtle nonlinearities rather than profound types of distortion, such as clipping.

headroom The area, or level, of linear operation between 0 dB and clip.

heat exchanger (1) A synonym for heatsink. (2) A device used for cooling incorporating an internal chamber of circulating water, or some other form of cooling liquid.

heatsink A thermally conductive material used to transfer wasted heat away from a heat-sensitive device or area.

HF High frequency.

H_{FE}, h_{FE} Beta; current gain parameter of a BJT. (Note: Uppercase letters denote DC gain parameters, while lowercase denotes small-signal AC gain figures.)

H-field The magnetic component of an electromagnetic wave.

HT High tension (i.e., high voltage and current).

HV High voltage.

hybrid A manufactured package containing both integrated circuit technology and discrete components. Sometimes, the term is used to describe a circuit using a combination of active devices from different electronic families (e.g., a vacuum tube/transistor circuit).

Hz Hertz.

I Electrical current, measured in amperes.

I_B BJT base current.

IC Integrated circuit.

I_C BJT collector current.

I_{cbo} BJT base leakage current; measured with the collector open.

I_{ceo} BJT emitter-to-collector leakage current with the base open.

I_D FET drain current.

$I_{D(ON)}$ On-state drain current.

I_{DSS} FET leakage current from drain to source with the gate lead open.

IF Intermediate frequency.

I_{GSS} Gate current with the drain to source short-circuited.

impedance Total opposition to current flow. Impedance is the summative effect of frequency-related opposition (reactance) and DC opposition (resistance).

impedance matching The technique of optimizing the output impedance of one device with the input impedance of another device to facilitate the most efficient voltage, current, or power transfer.

IN, in Input.

inductance A characteristic of all coils and transformers. It is the property that opposes a change in current flow by either storing or releasing energy contained within its magnetic field.

infrasonic A synonym of "subsonic" and "infrabass." Frequencies below the range of human hearing.

input impedance The impedance that exists between the input terminals of a circuit or device when the source (i.e., the output of a previous stage or device) is disconnected.

inrush current Synonymous with "surge current." The transient high current that occurs when the AC mains power is initially applied to an electronic system. Inrush current is a result of the rapid charging of power supply filter capacitors.

instability In the context of audio electronic circuitry, it is the undesirable tendency to break into either periodic or continuous oscillation. Instability is often used to describe any kind of intermittent problem or abnormal operation.

insulated gate bipolar transistor (IGBT) A relatively new electronic device, the IGBT is a literal cross between a MOSFET and a BJT. They exhibit the low drive requirements of a MOSFET together with the low saturation voltage of a BJT.

intermodulation distortion (IM or IMD) Distortion caused by "beat frequencies" created when complex AC signals pass through nonlinearities in an audio circuit or device. Since harmonic distortion is also a result of circuit nonlinearities, intermodulation distortion will usually be somewhat proportional to harmonic distortion.

interstage crosstalk A general term for any kind of signal leakage between stages of audio circuitry.

inverter (1) Any circuit or device that inverts. (2) A converter system that provides AC mains power from 13.8 VDC (automobile battery) or similar conversion function.

inverting A 180-degree phase inversion.

I/O Input-output.

I/P Input.

J Joule; basic unit of energy.

JFET Junction field-effect transistor.

jitter Signal distortion and/or intermittent signal loss due to timing errors. Most often applied to digital audio systems.

Johnson noise See "thermal noise."

joule Unit of energy. 1 watt = 1 joule per second.

junction The combination point of two different semiconductor materials (i.e., P and N material). In BJTs, all dissipation heat originates from the junction areas.

K (1) Cathode. (2) Kelvin; basic unit of temperature on the Kelvin scale. (3) Coupling coefficient.

k (1) Kilo; prefix for 1,000. (2) Dielectric constant.

KHz Kilohertz.

KV Kilovolt.

KW Kilowatt.

KWh Kilowatt-hour.

L Inductance; measured in henries.

large signal A relative classification for circuit analysis. In audio terminology, it usually relates to voltage or current swings above 20% of the maximum capability of the device or circuit under analysis.

lateral MOSFET A planar type of MOSFET (in contrast to the V-groove double-diffused MOSFETs commonly referred to as "V-MOSFETs," "DMOS," "HEXFETs," "Vertical DMOS," or "D-MOSFETs). Lateral MOSFETs were designed specifically for audio applications and have significant advantages over vertical MOSFETs for audio power amplifier applications.

leakage inductance The residual inductance in a transformer winding when all other windings of the same transformer are shorted.

LF Low frequency.

limiter In the context of audio power amplifiers, a high-quality compressor circuit designed to keep the output of a power amplifier from going into "hard" clipping. The high-frequency components of a "clipped" waveform are extremely destructive to HF drivers (i.e., tweeters).

Lin circuit The classic three-stage solid-state transformerless design topology that is still the basic foundation of about 99% of all audio power amplifiers currently being manufactured.

linearity The straightness of a graphical line representative of the transfer function (input Vs output) of an active device. The phrase "high linearity" is casually synonymous with "accurate reproduction."

line level In the context of audio terminology, the semistandardized output signal level of preamplifiers and audio program sources (i.e., tape decks, FM receivers, CD players, etc.). It is typically about 1 to 2 V RMS. Virtually all audio power amplifiers are designed to accept line level signals as inputs.

listening level An individual assessment of the optimum SPL for a specific room or audio system, or both.

load A very general term applied to any device, component, circuit, or system that receives a useful output. In the context of audio circuits, it represents the speaker system(s) or dummy load(s). See "dummy load."

load line A graphical analysis, usually in reference to an active device, representing resistance (or impedance), voltage, and current excursions. Load lines are typically drawn to determine the optimum quiescent settings for a BJT, JFET, or MOSFET single-stage circuit.

local feedback Feedback applied to a single stage or component.

long-tailed pair British synonym for a differential amplifier. See "differential amplifier."

loudness control A type of "bass boost" filter, usually incorporated into the volume control of an audio preamplifier, to provide increased bass volume at lower listening levels. Such filters provide automatic compensation for the loss of human hearing sensitivity of lower frequencies at reduced listening levels.

LP Low pass.

LSI Large-scale integration.

m (1) Milli; prefix for a thousandth (1/1,000). (2) Meter; basic unit of length.

M, meg (1) Mega; prefix for a million. (2) Mutual inductance.

mic Microphone.

microphony The phenomenon of mechanical vibration causing electrical noise or distortion. May have some relevance to vacuum tube amplifiers, but is essentially nonexistent in solid-state amplifiers.

midband The audio frequency range where the human ear is most sensitive. About 500 Hz to 5 kHz, sometimes defined as "centered" at 3 kHz.

Miller capacitance The intrinsic capacitance appearing between the high-level terminal and control terminal of an active semiconductor device. In other words, between the collector and base leads in a BJT and between the drain and gate leads in a FET. Internal Miller capacitance will vary depending on the voltage applied to the high-level terminal and the gain characteristics of the device.

Miller integrator In the context of audio power amplifiers, an essential element of the Lin topology. A Miller integrator is constructed by placing an external capacitor between the base and collector leads of a transistor (or gate and drain leads of an FET).

mom Momentary.

monoblock (or "monobloc") A single-channel (monaural) audio power amplifier constructed in a dedicated chassis with a dedicated power supply.

MOSFET Metal-oxide semiconductor field-effect transistor.

music power A term used for defining the short term output power of an audio power amplifier.

muting The act of disabling the output from an amplifier or other audio system. Automatic muting systems (also called "turn-on delay" circuits) are commonly used with audio power amplifiers to mute the speakers during "power on/off" cycles. This eliminates the annoying (and potentially destructive) "turn-on" thumps associated with power-up amplifier stabilization.

mV, mv Millivolt.

n (1) Nano; prefix for 1/1,000,000,000. (2) Turns ratio.

NC (1) Normally closed. (2) No connection.

neg Negative.

nested feedback A method of injecting negative feedback from the output stage of an amplifier to each preceding stage, creating multiple NFB loops "inside" the global NFB loop. The technique has seldom been used in conjunction with discrete circuitry, but is commonly found in operational amplifier design.

neu Neutral.

NF Noise figure.

NO Normally open.

nom Nominal.

noninverting An electronic circuit, stage, or system that maintains the same phase at the output that is received at the input.

nonlinearity The opposite of "linearity." See "linearity."

Nyquist oscillation Oscillation caused by excessive phase shift in a negative feedback loop resulting in the negative feedback turning to "positive" feedback and sustaining oscillation.

O/L Open loop.

offset voltage Slight variance of an established DC voltage. A general term that can apply to minor voltage variances within any electronic circuit or system.

on resistance The DC resistance between an FET's source and drain leads when fully turned on by an applied voltage between the gate and source leads.

open-loop gain The inherent gain factor of an amplification system without any type of negative feedback applied.

osc Oscillator.

out Output.

out-of-phase (1) Analogous to a phase reversal between two signals (when two signals are 180 degrees apart in phase). (2) A general term to describe any phase differential between two or more AC voltages.

output impedance The impedance presented by a source to a load.

output transformer In the context of audio power amplifiers, the transformer that couples the amplifier output stage to the speaker system. Seldom used in modern solid-state amplifiers, due to the nonlinear "reactive" response. Virtually always used in vacuum tube amplifiers.

overall feedback Synonym for "global feedback." See "global feedback."

overload level An ill-defined term used to loosely describe the level of an "overdriven" power amplifier where the sonic quality is no longer acceptable, somewhere slightly below the "hard-clipping" level.

p Pico; prefix for a trillionth.

P Power, measured in watts.

P, pri Primary of a transformer.

P1 In audio power amplifier terminology, the low-frequency open-loop response pole.

P2 In audio power amplifier terminology, the high-frequency open-loop response pole.

parallel mode Two or more audio power amplifiers connected together, so that the inputs and outputs are paralleled. A risky method of driving very low impedance speaker systems.

parasitics (1) Undesirable stray inductance and capacitance. (2) Undesirable instability variations riding on top of, or mixed into, an output signal.

passive components Electronic components that are not capable of providing power gain (i.e., resistors, capacitors, inductors, diodes, etc.). In contrast, BJTs, FETs, vacuum tubes, and operational amplifiers are "active" components.

PC (1) Personal computer. (2) Printed circuit.

PCB Printed circuit board.

P_D Maximum or total power dissipation (sometimes denoted "P_T").

peak The highest instantaneous level of an AC signal or waveform.

peaking out An AC signal or waveform on the verge of clipping.

peak limiting An electronic method of limiting destructively high peak output levels.

pF Picofarad.

pf Power factor.

phase The term used to describe the relative "timing" position of an AC waveform. One complete cycle of any AC waveform is considered to equal 360 degrees, similar to a full circle. Consequently, the phase relationship of any AC waveform can be compared to the timing of another AC waveform, of the same frequency, in terms of degrees.

phase distortion A general term applied to any type of phase problem that can result in audible or spatial differences between the original program material and the reproduced sound of an audio system.

phase-linear crossover network A specialized form of speaker crossover network that maintains in-phase relationships of the frequency bands applied to the individual speaker elements.

phase linearity A graphical analysis of the absolute phase deviation of an audio power amplifier. Phase shift is compared between the input and output signals throughout the audio bandwidth.

phase response Similar to phase linearity, but defined as a "plus" or "minus" degree phase shift throughout the audio bandwidth (i.e., not a graphical analysis).

phase splitter An electronic circuit used to extract two AC signals, one the inverted complement of the other, from a single AC signal.

pi-mode class See "sliding bias."

pink noise Artificially manufactured noise that averages out to equal energy per octave, or other specified increment. As such, it provides a good pseudorandom imitation of all music.

PK, pk Peak.

polarity In the context of electrical and electronic circuits, the "charge" orientation in respect to some common point of reference (either "positive" or "negative"). Can also refer to the orientation of one or more analog (audio) signals at a specific stage of processing.

polycarbonate A type of capacitor dielectric. Considered "good" quality.

polyester AKA Mylar. A type of capacitor dielectric. Considered "good" quality.

polypropylene A type of capacitor dielectric. Considered "high" quality.

polystyrene A type of capacitor dielectric. Considered "high" quality.

pos Positive.

pot Colloquial abbreviation for potentiometer. See "potentiometer."

potentiometer A three-terminal variable-resistive device. Often used in power amplifier circuitry for precise adjustment of bias voltages and/or stage balance (the small type of potentiometers normally soldered into electronic circuits are called *trimpots*). Larger types of potentiometers are commonly used for operator controls (i.e., volume, tone, balance, etc.). By connecting the "tap" terminal to either "end" terminal, a potentiometer becomes a variable resistor, called a *rheostat*.

power bridging The act of connecting two or more power amplifiers together in bridge mode. See "bridge mode."

power density The power output capability of an audio power amplifier expressed in terms of the physical size of the amplifier. For example, watts per cubic foot, watts per standard rack unit, etc.

power factor A term describing AC mains power utilization. It is defined as the ratio of the actual consumed power to the "apparent" consumed power. The best theoretical power factor is 1, indicating that the load is entirely resistive.

predriver In the context of audio power amplifiers, a term often applied to various transistors (or other active devices) used to "drive" the output devices. A rather ambiguous term, since its usage has never actually been standardized.

pri Primary.

program The original recorded or reproduced music and/or vocal signal. For example, a "compact disc" contains the "program" the user desires to listen to.

program asymmetry The phenomenon of complex AC signals (i.e., music) becoming asymmetrical in terms of its "average" sonic pressure (i.e., for short periods of time, complex musical programs can actually increase or decrease room pressure). When converted to electrical signals, program asymmetry results in DC shifts, both at the inputs and outputs of audio equipment.

PSRR Power supply rejection ratio.

PSU, PS Power supply unit.

pulse width modulation (PWM) A technique wherein an audio signal is encoded into a corresponding square wave duty cycle. The fundamental frequency of the square wave is ultrasonic, and when applied to a speaker, the speaker coil will convert the duty cycle into a "power wave." The power wave will be (hopefully) a good reproduction of the original audio signal. Often, pulse width modulated amplifiers are referred to as "digital amplifiers." The advantage of pulse width modulated amplifiers is extremely high efficiency. Pulse width modulation is also a technique incorporated into many types of "switching" power supplies, as a method of regulation.

push-pull A condition where output devices alternately control approximately one-half of the full output signal.

PWM Pulse width modulation.

Q (1) A figure of merit for an energy-storing device, tuned circuit, or resonant system; equal to the reactance divided by the resistance. (2) When specifying inductor quality, the ratio between the inductance and the coil DC resistance. (3) Quantity.

quasi-complementary An older method of output stage architecture, utilizing multiple transistors of the same "polarity" (i.e., all NPN, for example) to simulate true complementary pairs. Of each quasi-complementary pair, one NPN output transistor operates in a common-collector configuration, while the other NPN transistor operates in a common-emitter fashion. Although popular in the late 60s and early 70s, the advent of high-quality and relatively inexpensive "true" complementary BJT pairs replaced the need for quasi-complementary designs. A few manufacturers still use quasi-complementary architecture today, usually to reduce costs or to take advantage of a more reliable (or rugged) nature of a specific device type.

quiescent The steady-state condition of an electronic circuit without any signal applied.

R Resistance, in ohms.

radiation pattern As pertaining to acoustical physics, the dispersion pattern of sound waves. Regarding audio electronics, it most often applies to the way sound is radiated from a speaker system.

Rb, rb Base resistance of a BJT.

Rc, rc Collector resistance of a BJT.

rcvr Receiver.

$r_{ds(on)}$ Small signal; drain to source on-state resistance.

$r_{DS(ON)}$ DC drain-to-source on-state resistance.

Re, re Emitter resistance of a BJT.

recombinational noise Random noise generated in the normal operation of all semiconductor devices, caused by the recombination of free and spatially separated holes and electrons.

rect Rectifier.

rectifier Any device dedicated to the function of converting AC to DC. Can be in the form of solid-state devices (diodes) or vacuum tubes. The term is sometimes used in a general way to mean "bridge rectifier." See "bridge rectifier."

ref Reference.

regulation The process of holding various quantities (i.e., voltage, current) constant in a system.

regulator A circuit or integrated circuit dedicated to maintaining regulation of an electrical variable. See "regulation."

reservoir The adjective often applied to the large "filter" capacitors utilized in most high-current DC power supplies. The connotation alludes to a storage area, or "reservoir," of electrical energy.

regulated power supply Any type of power supply with special circuitry added to maintain its output under a variance of loading conditions.

reservoir capacitors Another name for "filter capacitors." The large smoothing capacitors used to convert the "pulsating DC" output of power supply rectifiers to relatively constant DC.

residue In the context of electrical signals, a general term that applies to the unwanted content (i.e., distortion or noise) within a signal, or output, voltage. It is most often used in conjunction with audio distortion analyzers, in reference to the leftover "residual" signal.

resolution The ability to delineate, detail, or distinguish between nearly equal values of quantity. For example, a hypothetical control knob graduated into 100 equal increments has a minimum resolution of 1 increment. Resolution is often used to denote the smallest possible increment of any device or control.

resonant power supply A type of "switching" power supply incorporating a resonant "tank circuit" to reduce HF and RF emissions and improve efficiency. Resonant power supplies (or some variant of them) are used in some professional audio power amplifiers.

resonance (1) The condition existing in a circuit when the inductive reactance equals the capacitive reactance. (2) In the context of audio amplifiers, the undesirable condition wherein the inductive and capacitive elements of the amplifier circuit, in combination with the circuit gain, produce damped or sustained oscillation. (3) The natural frequency of mechanical vibration in a speaker system (or any other type of audio transducer).

RF Radiofrequency.

RFI Radiofrequency interference.

RF filtering Filters designed to remove any RF interference from audio circuitry (especially sensitive preamp and high-gain circuits). Special RF filter/shut-down circuits are sometimes incorporated into high-power audio amplifiers to automatically shut down the amplifier if RF oscillation is detected.

RF oscillation In the context of audio electronic circuitry, sustained oscillations at radio frequencies. This condition is very destructive to output stages and some speaker systems. It is notorious because it cannot be heard and is often difficult to easily detect.

ringing The condition of producing a damped oscillation.

ripple (1) The AC residual of a rectified and filtered DC power supply. (2) A general term for any repetitive and cyclic variation occurring in a processed waveform.

ripple current The cyclic AC "current" variations in power supply filter capacitors resulting from the pulsating DC output of the rectifier circuit.

ripple voltage The AC voltage component riding on the DC voltage of power supplies. The fundamental frequency is twice that of the mains supply (assuming a full-wave rectifier is incorporated in the power supply design). Ripple voltage is a primary cause of "hum" in audio power amplifiers.

$R_{j\text{-}a}$ Thermal resistance of a semiconductor; core to ambient.

$R_{j\text{-}mb}$ Thermal resistance of a semiconductor; core to mounting base.

R_L General abbreviation for a load.

RLY Relay.

rolloff In the context of audio systems, the frequency point at which the output voltage decreases to a specified amplitude (usually by −3 dB, denoting the "half-power" point). The frequency spectrum between the points of HF rolloff and LF rolloff determines the "bandwidth." In general usage, rolloff is applied to any signal voltage that begins to decrease as frequency varies.

$R_{mb\text{-}sink}$ Thermal resistance of a semiconductor; mounting base to heatsink.

RMS, rms Root mean square.

rmt Remote.

R_{ON} On resistance of a MOSFET.

R_S, r_S Source impedance.

R_{TH} Thermal resistance.

R_{THCS} Thermal resistance from a semiconductor case to heatsink.

R_{THNS} Thermal resistance of a semiconductor insulator.

R_{THJC} Thermal resistance of an internal semiconductor junction to its outside case.

R_{THJA} Thermal resistance of a semiconductor junction to the ambient air.

R_{THSA} Thermal resistance of a heatsink to the ambient air.

S, s (1) Siemens; basic unit of conduction (replaced the older term "mho"). (2) Source lead of an FET. (3) Seconds (time or arc). (4) Area of one plate of a capacitor (measured in square centimeters).

safe operating area (SOA) A BJT parameter curve illustrating all of the possible "safe" voltage/current zones wherein there is no danger of the BJT going into secondary breakdown.

saturation (1) A condition in BJTs when they are driven beyond the "fully-on" limits. When saturated, BJTs require longer "turn-off" time periods. (2) When the core material of an inductor reaches its limit of permeability. (3) In the context of acoustics, the point at which an increase in acoustic energy does not produce a proportional increase in loudness.

SB Sideband.

schematic A line drawing representation of an electrical circuit.

sec Secondary.

secondary breakdown A destructive failure of a BJT, not resulting from exceeding any maximum voltage or current parameters, but due to certain combinations of voltage and current interacting with the physics of BJT design.

sensitivity In the context of audio electronic equipment, the input signal level required to produce the maximum rated output signal or power.

separation A figure of merit, describing the signal voltage isolation between two or more channels. The antonym of crosstalk. See "crosstalk."

servo A type of negative feedback designed to "autonull" (i.e., cancel out) DC offsets at the output of audio power amplifiers. DC servo systems are designed to apply negative feedback to continuously correct for DC conditions only, with little or no feedback applied to the signal (i.e., little or no AC feedback).

shield The general term applied to the "outside" conductive layer of many types of cabling. The purpose of shielding is to isolate the internal conductors from the effects of RFI, EMI, and intraceable signal leakage. Cable shielding is commonly implemented in the form of wire mesh or conductive foil.

shoot through See "cross-conduction."

shunt Synonymous to "parallel." Applied to circuit, component, feedback, and control schemes. "Shunting" is the act of connecting various components or circuits in parallel.

siemens Unit of conduction, synonymous with "mho" (an older term). The reciprocal of ohms.

sig Signal.

signal-to-noise ratio (SNR or S/N) The definitive term describing the random noise performance characteristics of audio equipment. It represents the RMS value of the random noise components averaged across the audio bandwidth and compared as a ratio (usually in terms of negative decibels) to the 0-dBr level of the audio system.

single ended A general term, applied in contrast to "dual" or "dual ended." Often applied to power supplies, output stages, or any type of circuit that is asymmetrical (i.e., nonbalanced, noncomplementary, nonbridged, etc.).

single pole (1) A single path or single contact pair of a switch or relay, respectively. (2) The basic rolloff response of a passive RL or RC network (i.e., −6 dB per octave).

skin effect The tendency of high frequencies to flow through the outside surface, or "skin," of a conductor.

slew-induced distortion (SID) Distortion resulting from an inadequate transitional speed capability of an audio system.

slew rate (or) slew limit The maximum rate of change of a signal or output voltage (typically measured in volts per microsecond). The term is usually applied to operational amplifiers and audio power amplifiers to define how rapidly the output will respond to instantaneous input changes.

sliding bias The term used to describe the type of dynamic bias applied to the output stage of audio power amplifiers designed to operate in class A during low-level operation and "slide" into class A-B operation during high-level operation.

small signal (1) A general term used to denote low-level output of an audio power amplifier. Typically, "small signal" is at or less than

20% of its rated power output. (2) The signal level within the preamplifier stages.

SNR, S/R Signal-to-noise ratio.

snubbing capacitors Capacitors placed across the diode bridge terminals in a power supply to reduce the RF noise emitted by the diode switching.

sone A unit of loudness. A 1-kHz tone that is 40 dB above a listener's threshold of hearing produces a loudness of 1 sone.

sonic accuracy The relative identical nature of a reproduced program with the original program. Synonym for fidelity.

sound pressure level (SPL) A value, expressed in decibels, representing the pressure of sound waves relative to a reference pressure. Reference pressures in common use are 0.0002 μbar and 1 μbar.

source follower A FET equivalent of a BJT emitter follower.

soft start A type of power supply (used in many high-power audio amplifiers) that will power up slowly as a means of eliminating AC power line surges.

spatial perception The ability of human hearing response to sense the point of origin of a sound in space. In the general sense, it applies to three-dimensional sound attributes.

SPDT Single-pole, double-throw (pertaining to switches and relays).

speaker level Signal voltages at the correct level to drive speaker systems (can range anywhere from 2 to over 200 V RMS).

spk, spkr Speaker or loudspeaker.

SPST Single-pole, single-throw (pertaining to switches and relays).

sq Square.

stability The capability of an electronic circuit or device to maintain a stable set of quiescent parameters without degenerating into self-sustaining oscillations or dynamic shifts.

star ground A nodal point for two or more ground connections. So named, due to the use of a "star" lockwasher often used to assure good conductivity and mechanical reliance.

steady-state (1) Synonymous with "quiescent." See "quiescent." (2) A steady signal level or a continuous periodic waveform.

stereo Audio signals in the form of two simultaneous channels, representing left and right listening perspectives.

stiff (1) In the context of audio power amplifiers, any amplifier with a high damping factor specification (i.e., very low output impedance). (2) A power supply with very low output impedance. (3) A low-compliance speaker.

stray capacitance Incalculable circuit or system capacitance resulting from component spacing, PC board track proximity, and wiring paths. Synonymous with parasitic capacitance. See "parasitic."

stray field Usually applies to an undesired EMI field, but is sometimes used to denote ESD or RFI fields.

subharmonic A frequency that is an integer fraction of a fundamental frequency.

subjectivism The antithesis of rational and scientific evaluation. Human perception without qualification.

subsonic A synonym for "infrasonic" or "infrabass." See "infrasonic."

substrate The base, or "foundational," material on which solid-state devices are constructed.

surge A temporary "swelling" (increase) in either voltage or current.

supply rails (or DC supply rails) The DC power supply busses internally located within audio electronic equipment.

sw Switch.

swamp (1) A slang term meaning to cover up one signal with another signal of higher amplitude. (2) A type of class D amplifier developed by Infinity Systems. SWAMP was an acronym for "SW-itch AMP-lifier."

sweep To span a variable range. Usually applied to frequency, denoting the variance of a frequency between two specified points.

swing The maximum output excursion (almost always in terms of voltage) of a circuit, amplifier, or operational amplifier.

Sziklai A complementary active device configuration, synonymous to the "complementary feedback pair." See "complementary feedback pair."

T, t (1) Tesla; basic unit of magnetic field strength. (2) Time, usually in seconds. (3) Temperature, usually in degrees centigrade unless otherwise specified. (4) Transformer.

T_A, T_{amb} Ambient temperature.

tandem mode (1) Synonymous with "parallel mode." See "parallel mode." (2) When two or more power amplifiers are equally driven by a single source.

THD Total harmonic distortion.

thermaled out A slang expression to describe an audio power amplifier that has automatically shut itself off due to excessive heating.

thermal noise Electrical noise produced by the random motion of free electrons in conductors and semiconductors. The effect increases with temperature. Also known as "Johnson noise."

thermal runaway A self-perpetuating thermal degeneration condition characteristic to BJT devices. It occurs when a quiescent bias enables a collector current to flow that is large enough to cause heat build-up in the BJT collector junction. Due to the positive temperature coefficient of BJTs (i.e., relative to current flow), the heat build-up instigates an increase in collector current, creating more heat, causing more current to flow, etc. If the BJT quiescent bias is not reduced to compensate for this "runaway" condition, the BJT will destroy itself. Some types of MOSFETs are susceptible to this condition also.

thermistor A type of semiconductor resistor, designed to change resistance value in accordance with temperature changes. Thermistors are available with either positive or negative temperature coefficients.

timbre The "signature" of a musical instrument. The unique blend and phasings of harmonics that create a unique sound peculiar to a specific instrument.

T_J Junction temperature of a semiconductor device.

t_{OFF} Turn-off time.

t_{ON} Turn-on time.

tolerance The maximum allowable deviation from published specification. The term is usually applied to acceptable variances of discrete components.

tone burst A type of test pattern (i.e., a specific number of cycles within a square wave envelope) used to test transient response in audio equipment.

tone generator A synonym for "audio signal generator." See "audio signal generator."

topology The "architecture" (i.e., the physical structure, or layout) of a populated printed circuit board.

toroidal A type of inductor or transformer constructed on a round, doughnut-shaped, core. Although more expensive, toroidal transformers are more efficient, weigh less, and produce less EMI than standard E+I (i.e., laminated) transformers.

total harmonic distortion (THD) A measurement of distortion in which undesired harmonics are generated due to circuit nonlinearity. Since the primary method of THD extraction will include whatever noise characteristics are inherent to the amplifier under test, the technically accurate way of expressing THD measurements is "% THD + N" (the "N" stands for noise).

trace (1) A "line" of copper track on a PC board. (2) The luminous line on an oscilloscope.

transconductance The phenomenon of controlling a current with a voltage. Transconductance amplifiers convert a signal voltage to a corresponding (or proportional) current signal. Traditionally, transconductance has been thought of as the "amplification factor" of vacuum tubes and FETs, however, the term is also appropriate for BJTs.

transducer Any device capable of converting one form of energy to another. In the context of audio systems, the most notable transducers are speakers (i.e., they convert electrical energy to sonic energy).

transfer function The mathematical relationship between the output and the input of an electrical circuit or signal path.

transformerless A direct-coupled audio power amplifier (i.e., devoid of any type of output transformer, hence "transformerless").

transient Any brief and abrupt change in signal properties or circuit operation. "Spikes" or "glitches" are slang synonyms.

transimpedance The process of converting a current to a proportional voltage. Analogous to "transresistance," but used in the context of AC voltages and currents.

transistor The common name for "bipolar junction transistor," or BJT. A three-layer, three-lead, two-junction active solid-state device. The most common active "building block" of solid-state circuitry.

transition frequency The high-frequency parameter of a BJT. Essentially, it is the high-frequency point at which the transistor's gain qualities drop to −3 dB.

transresistance The process of converting a current to a proportional voltage. Analogous to "transimpedance," but used in the context of DC voltages and currents.

tri-amp Colloquial name for any audio system using active crossovers to divide the audio signal into three frequency groupings (treble, midrange, bass). The three audio channels are then amplified by three dedicated audio power amplifiers and applied to a speaker system without any internal crossover network (the crossover was accomplished "prior" to the amplification stage).

t_r Rise time.

t_{rr} Reverse recovery time.

trimcap A small, PC-board-size adjustable capacitor.

t_{STG} Storage temperature.

turn-on delay circuit See "muting."

twisted pair A type of audio cable. As the name suggests, the conductors are twisted around themselves, providing improved immunity to RF and other transmitted interferences (both transmitted and received). Commonly used in conjunction with balanced sources and destinations.

ultrasonic Frequencies above the human hearing range.

unbalanced In the context of audio systems, any signal without a complementary counterpart (i.e., a "single-ended" signal).

unstabilized (unstabilised) A British term meaning "nonregulated." The term is typically applied to "raw" (nonregulated) power supplies.

V, v Voltage; electromotive force.

VA Voltampere.

variac An adjustable autotransformer, normally used for testing purposes. The turns ratio is established to provide a range of zero to about 110% of the AC mains voltage. Variacs *do not* provide isolation, so appropriate caution should be exercised when using them.

VAS Voltage amplifier stage.

V_{be} Base-to-emitter voltage of a BJT.

$V_{BE(SAT)}$ Base-to-emitter saturation voltage of a BJT.

V_{CBO} Collector-to-base voltage of a BJT with the emitter open (sometimes denoted "BV_{CBO}").

V_{CE} Collector-to-emitter voltage of a BJT.

V_{CEO} Collector-to-emitter voltage of a BJT with the base open (sometimes denoted "BV_{CEO}").

$V_{CEO(SUS)}$ Maximum collector-to-emitter voltage with the base open.

V_{DS} Drain-to-source voltage of an FET.

V_{EBO} Emitter-to-base voltage of a BJT with the collector open (sometimes denoted "BV_{EBO}").

V_{GS} Gate-to-source voltage of an FET.

$V_{GS(th)}$ Gate-to-source threshold voltage.

V_{in} Input voltage.

volt The basic unit of electromotive force (EMF). The force of 1 volt is required to move 1 ampere of current through 1 ohm of resistance.

VOM Volt-ohmmeter.

Vout, Vo Output voltage.

VTVM Vacuum tube voltmeter.

V/μs Volts per microsecond.

W Watt; basic unit of power.

watt The basic unit of "power" (energy transfer). One Joule per second is equal to 1 watt.

weighting The technique of filtering out selective bands of noise components before calculating the signal-to-noise (SNR) performance of a piece of audio equipment.

Winding resistance The DC resistance of the wire in an inductor or transformer coil.

X Reactance (measured in ohms); the opposition to alternating current exhibited by reactive components.

Y Admittance (measured in siemens or mhos); the reciprocal of impedance.

Y_{fs} Forward transadmittance of an FET.

Z Impedance (measured in ohms); the reactive and resistive opposition to the flow of alternating current.

Z_{IN} Input impedance.

Zobel network The RC circuit placed across the output of almost all audio power amplifiers. Its function is to help stabilize the power amplifier under a variety of load situations.

Z_{OUT} Output impedance.

Sources of Information and Materials

Parts and Materials Suppliers

SEAL Electronics provides audio power amplifier kits, PC boards, power supply components, electronic components, and other audio electronic construction supplies.

SEAL Electronics
P.O. Box 268
Weeksbury, KY 41667
(606) 452-4135
e-mail—sealelec@eastky.net
website—www.sealelectronics.com

Parts Express and MCM Electronics carry a complete line of passive and semiconductor components for audio electronics projects, together with an excellent selection of other audio equipment needs.

Parts Express
725 Pleasant Valley Drive
Springboro, OH 45066
1-800-338-0531

MCM Electronics
650 Congress Park Drive
Centerville, OH 45459-4072
1-800-543-4330

Jameco Electronics
1355 Shoreway Rd.
Belmont, CA 94002-4100
1-800-831-4242

Digi-Key Corporation
701 Brooks Ave. South
Thief River Falls, MN 56701-0677
1-800-344-4539

Mouser Electronics
958 North Main St.
Mansfield, TX 76063-4827
1-800-346-6873

Par-Metal is an excellent source for audio equipment enclosures, providing decorative and rugged enclosures for almost any audio equipment need.

Par-Metal Products, Inc.
29 Ewing Ave.
N. Arlington, NJ 07031
(201) 955-0800

Circuit Specialists is an excellent source for almost any PC board fabrication need, including presensitized PC boards, chemicals, and equipment.

Circuit Specialists
220 S. Country Club Dr.
Building 2
Mesa, AZ 85210
1-800-528-1417

Electronics Workbench is the industry leader in computerized circuit design and PC board layout software.

Electronics Workbench
908 Niagara Falls Boulevard
Suite # 068
North Tonawanda, NY 14120-2060
1-800-263-5552

Electronic Surplus Dealers

B.G. Micro
555 N. 5th Street, Suite 125
Garland, TX 75040
1-800-276-2206

All Electronics Corporation
14928 Oxnard Street
Van Nuys, CA 91411
1-800-826-5432

Fair Radio Sales
1016 E. Eureka St.
P.O. Box 1105
Lima, OH 45802
(419) 227-6573

Recommended for Further Reading

High-Power Audio Amplifier Construction Manual
1999, by G. Randy Slone
Published by McGraw-Hill
ISBN# 0-07-134119-6

Audio Power Amplifier Design Handbook—Second Edition
2000, by Douglas Self
Published by Newnes Publications
(an imprint of Butterworth
Heinemann, Ltd.)
ISBN# 0-7506-4527-X

High Performance Audio Power Amplifiers
1996, by Ben Duncan
Published by Newnes Publications
(an imprint of Butterworth
Heinemann, Ltd.)
ISBN# 0-7506-2629-1

Valve & Transistor Audio Amplifiers
1997, by John Linsley Hood
Published by Newnes Publications
(an imprint of Butterworth
Heinemann, Ltd.)
ISBN# 0-7506-3356-5

TAB Electronics Guide to Understanding Electricity and Electronics–Second Edition
2000, by G. Randy Slone
Published by McGraw-Hill
ISBN# 0-07-136057-3

Recommended Periodicals

audioXpress
P.O. Box 876
Peterborough, NH 03458
(603) 924-7292

Poptronics
Gernsback Publications, Inc.
275-G Marcus Blvd.
Hauppauge, NY 11788
(631) 592-6720

Nuts & Volts
T & L Publications, Inc.
430 Princeland Court
Corona, CA 92879
(909) 371-8497

APPENDIX C

PC BOARD ARTWORK

This appendix contains all of the "full-size" artwork diagrams for the applicable projects in this textbook. We concluded that it would be more convenient to place such diagrams in a separate appendix, to facilitate easier copying for fabrication processes, and also as an aid to the aesthetics of the textbook design.

G. Randy Slone and the McGraw-Hill Editorial Department

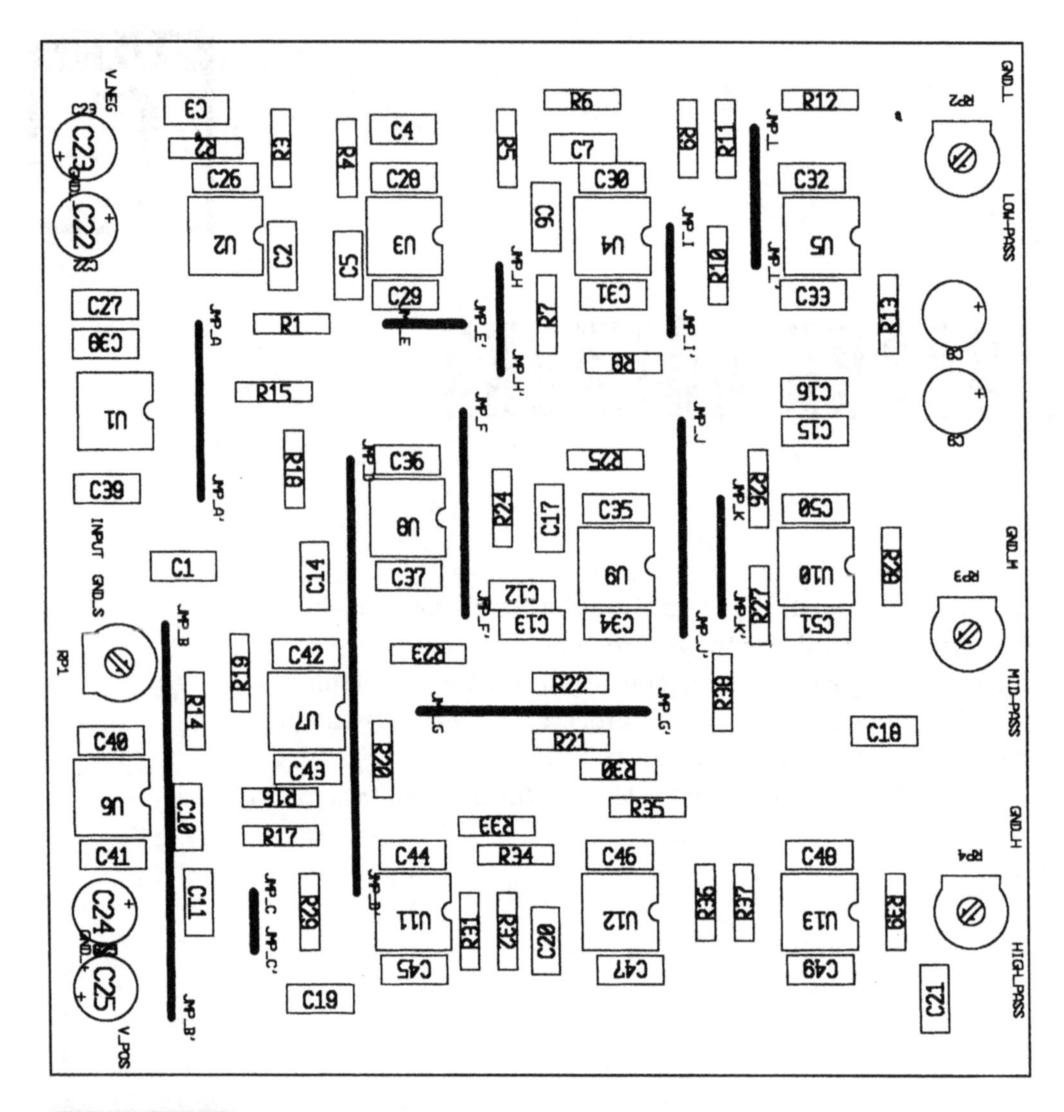

FIGURE C.1

Component layout of the Fig. 4-12 phase-linear filter. (Repeat of Figure 4-14.)

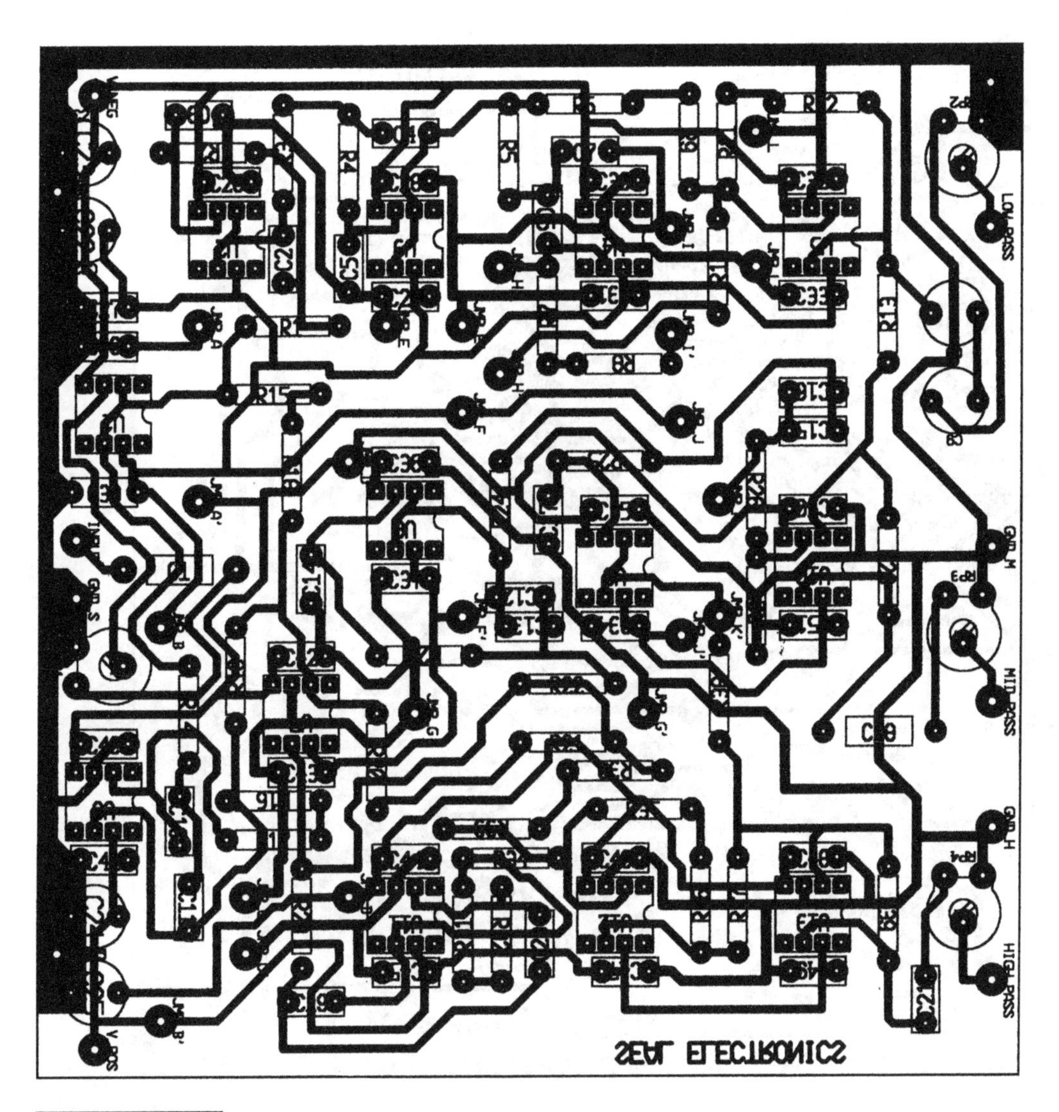

FIGURE C.2

Top view layout of the Fig. 4-12 phase-linear filter. (Repeat of Figure 4-15.)

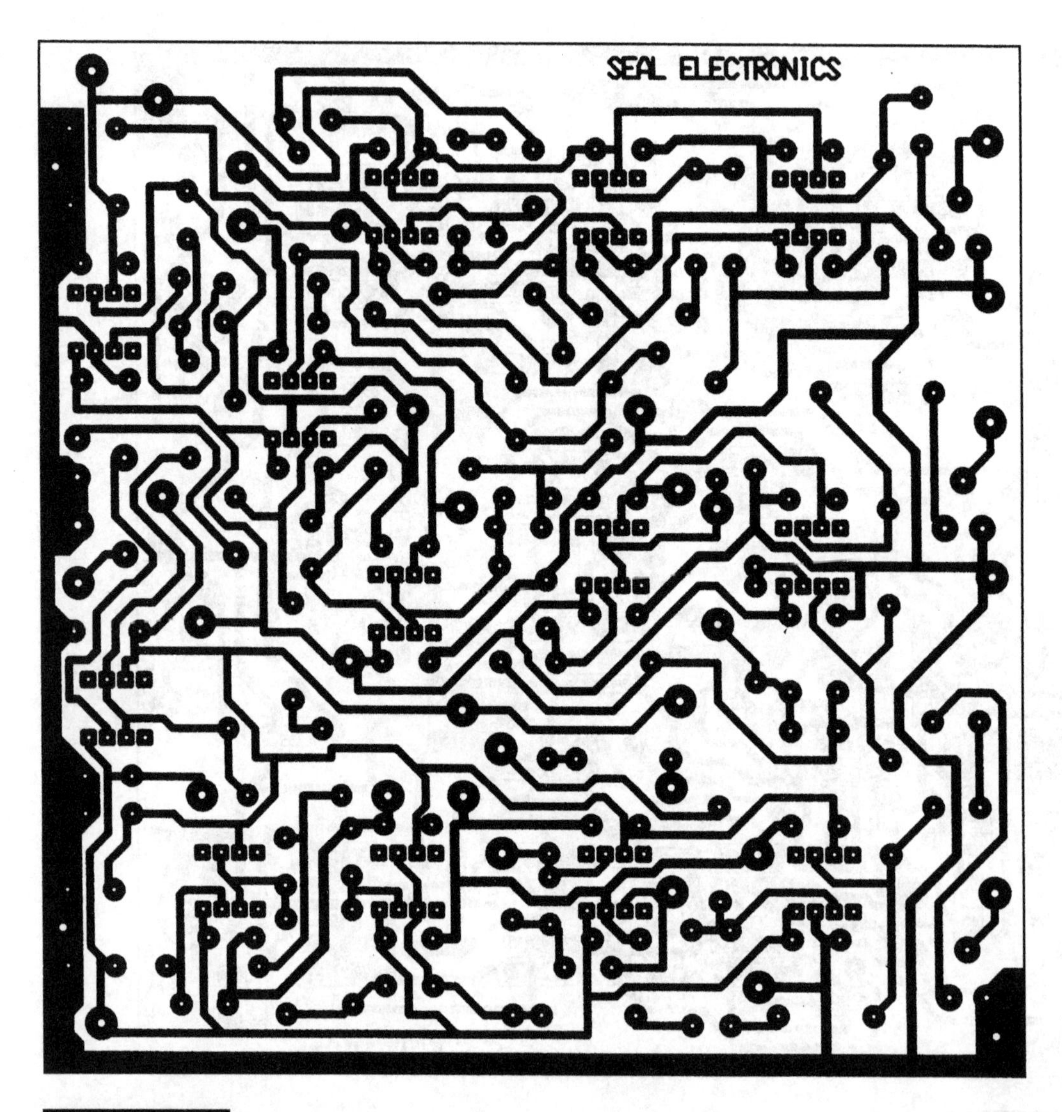

FIGURE C.3

Bottom view "reflected" artwork for the Fig. 4-12 phase-linear filter. (Repeat of Figure 4-16.)

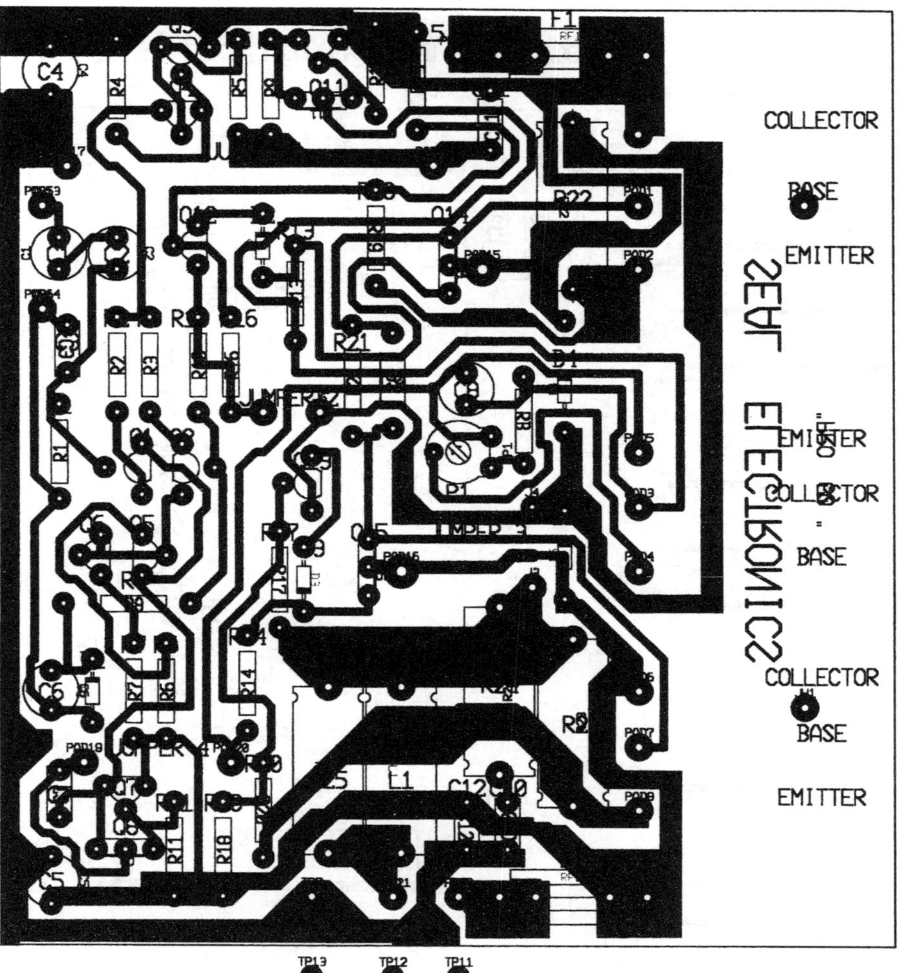

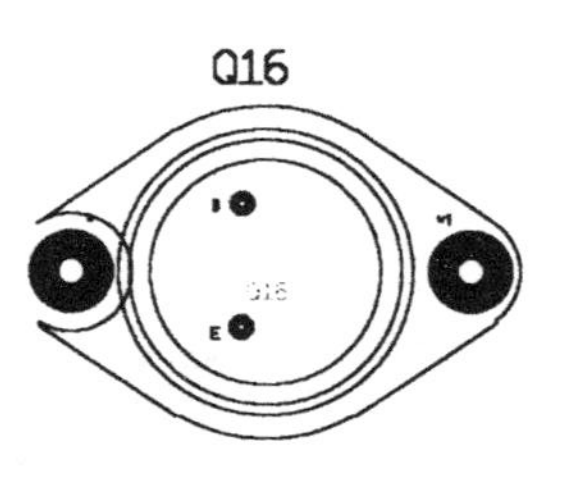

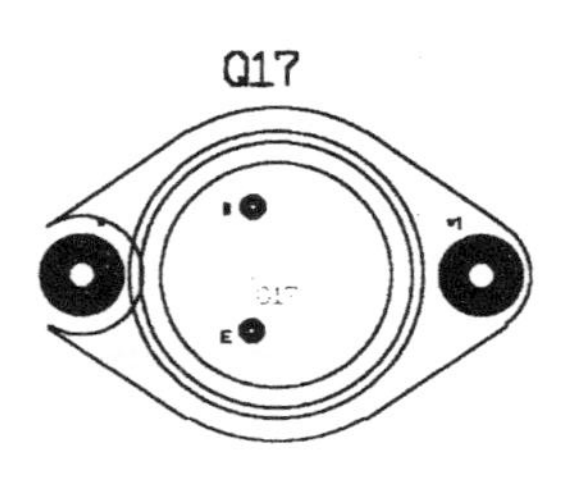

FIGURE C.4

Top view of the Fig. 6-2 amplifier. (*Note*: This view is 88 percent of full size.) (Repeat of Figure 6-3.)

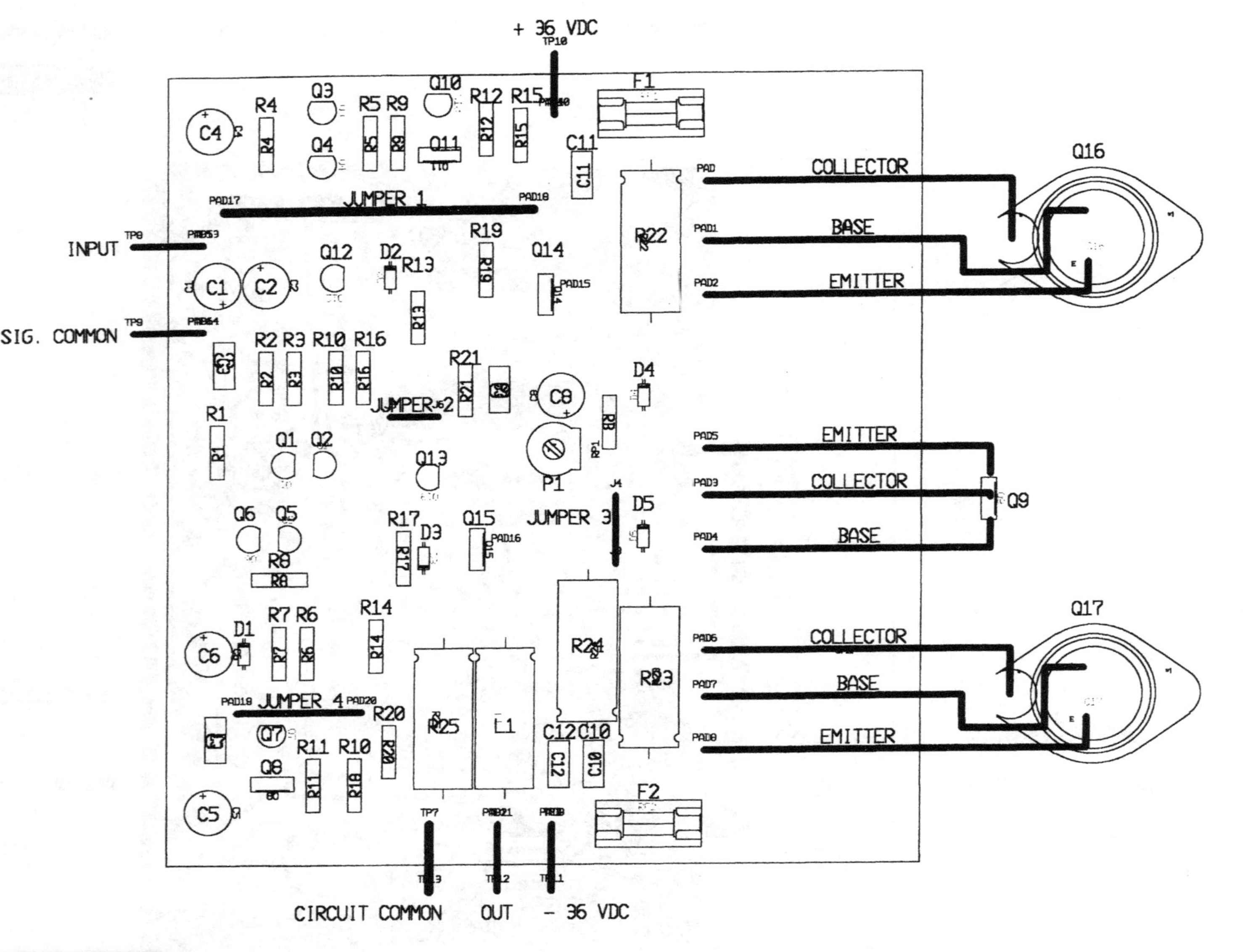
+ 36 VDC
INPUT
SIG. COMMON
JUMPER 1
JUMPER 2
JUMPER 3
JUMPER 4
COLLECTOR
BASE
EMITTER
Q16
Q9
Q17
CIRCUIT COMMON
OUT
- 36 VDC

FIGURE C.5

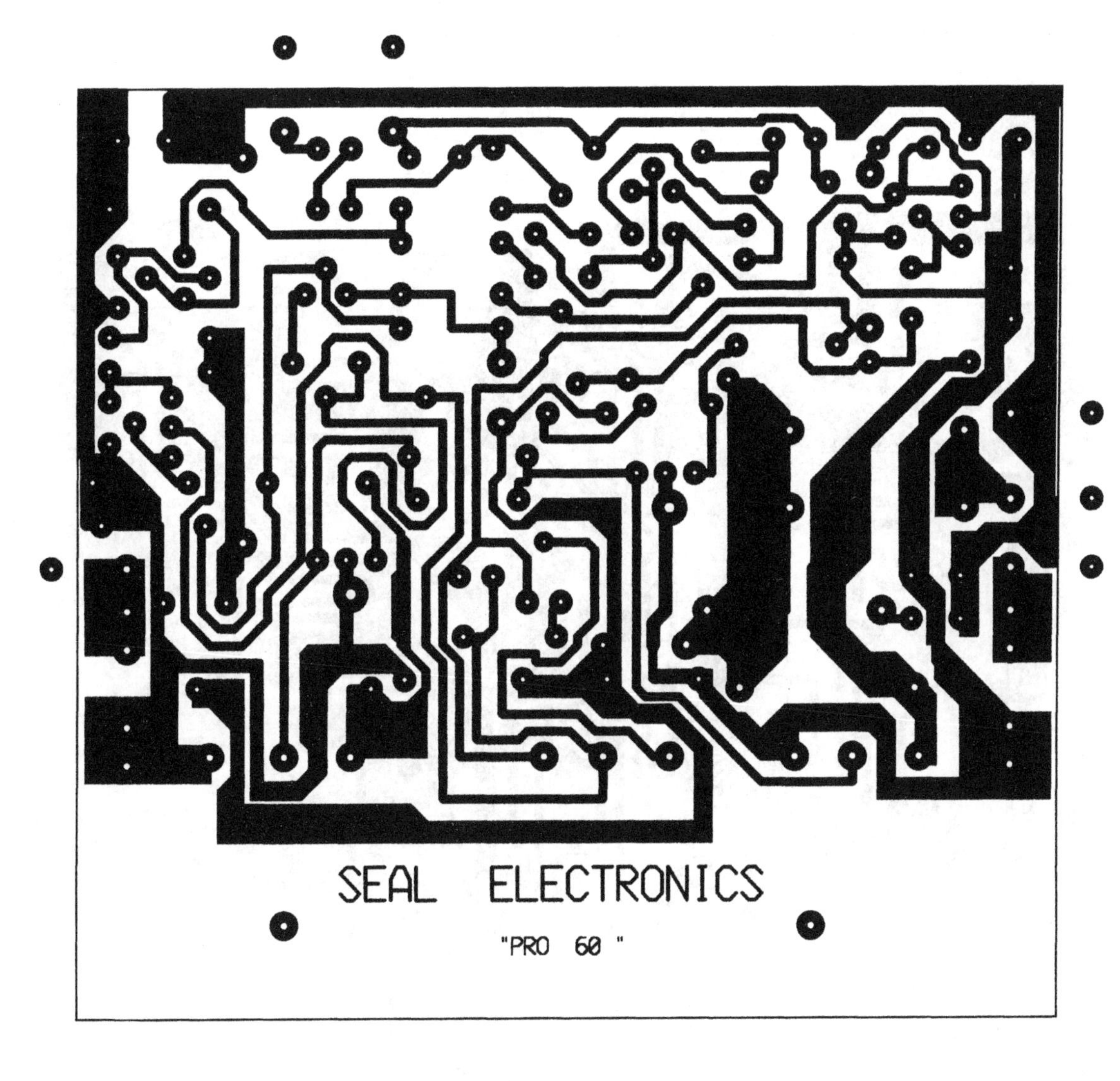

FIGURE C.6

Reflected artwork for the Fig. 6-2 amplifier. (*Note*: This view is 90 percent of full size.) (Repeat of Figure 6-5.)

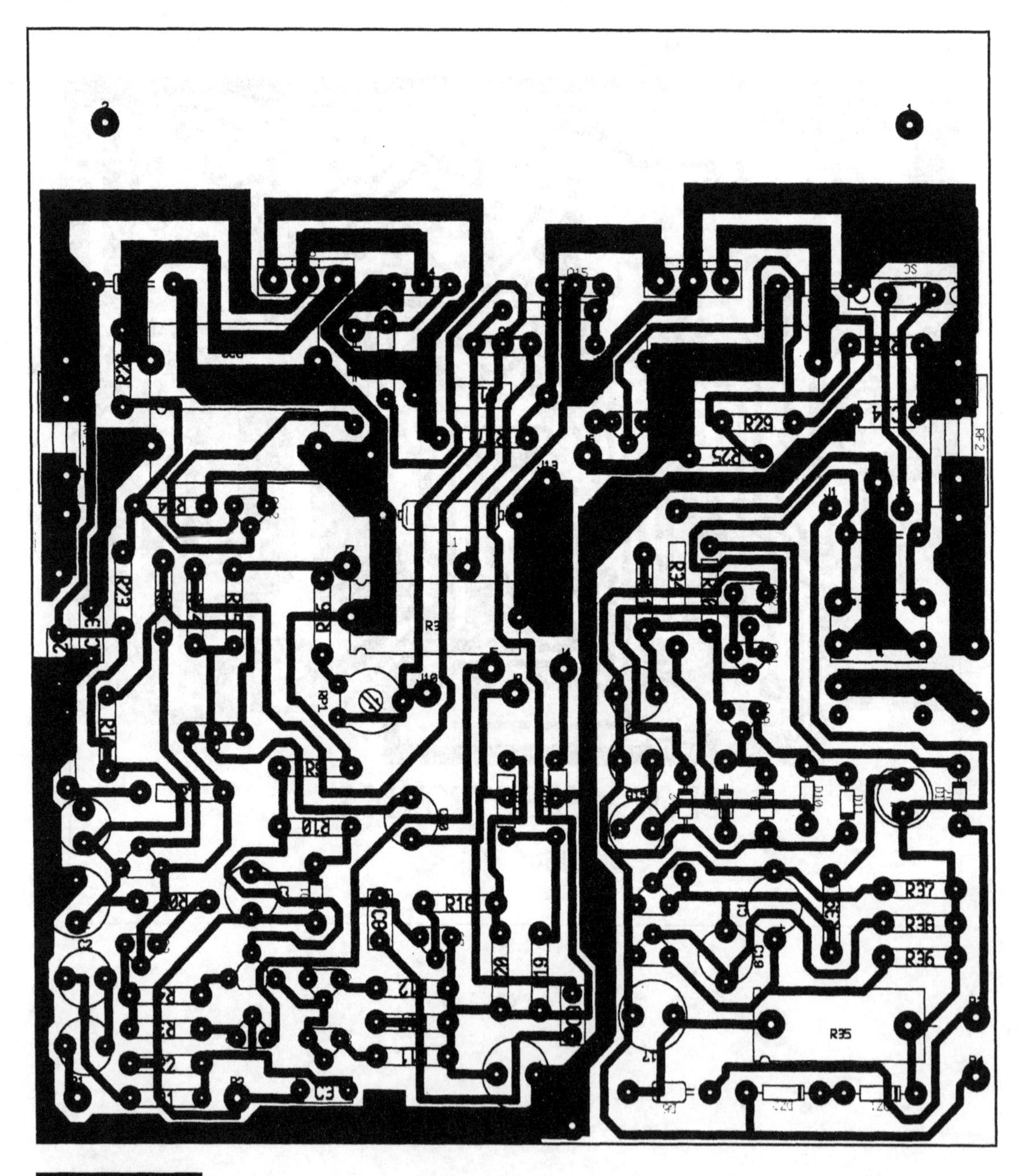

FIGURE C.7

Top view of the Fig. 6-6 amplifier. (*Note*: This view is 97 percent of full size.) (Repeat of Figure 6-7.)

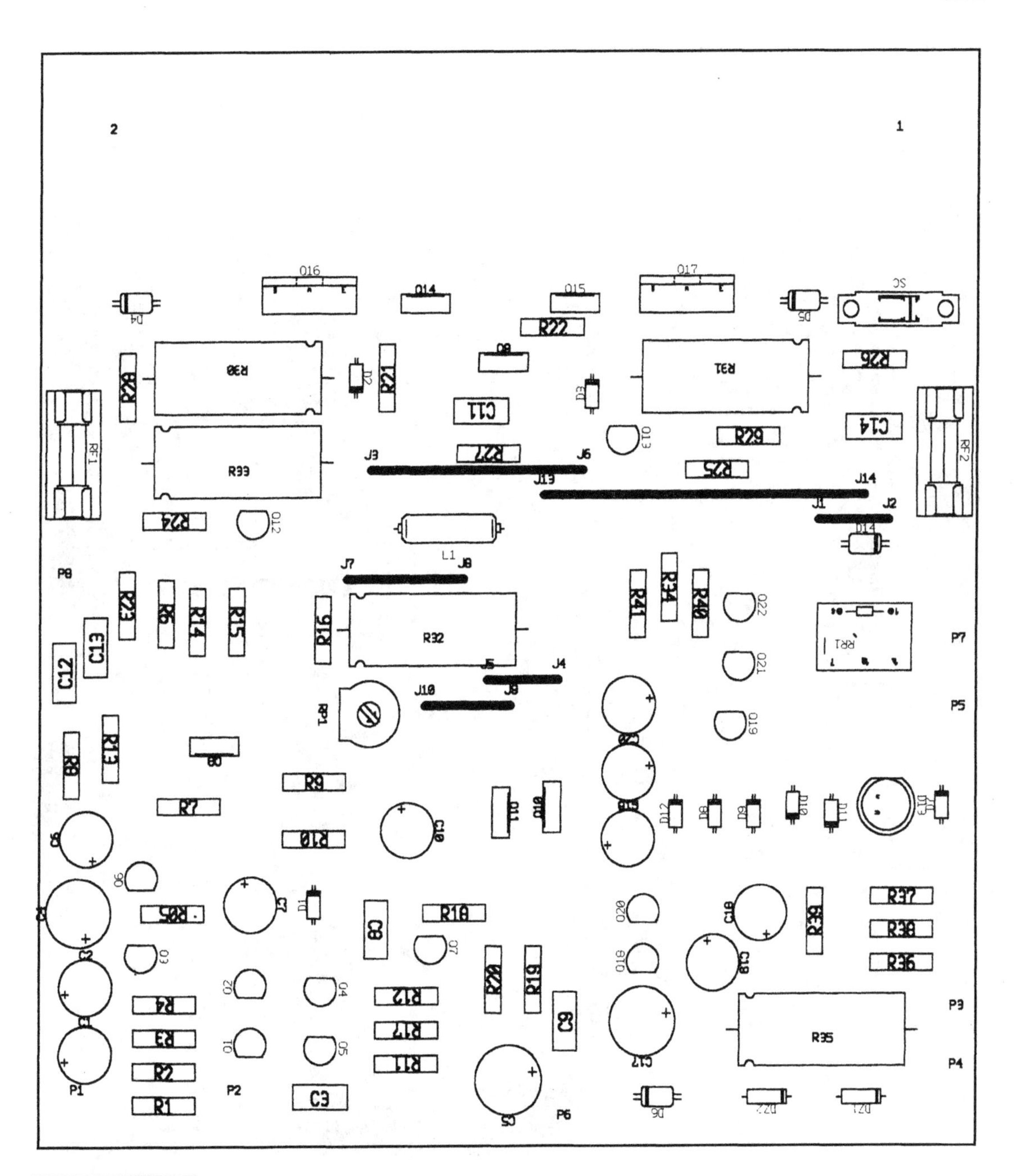

FIGURE C.8

Top layout view of the Fig. 6-6 amplifier. (*Note*: This view is 97 percent of full size.) (Repeat of Figure 6-8.)

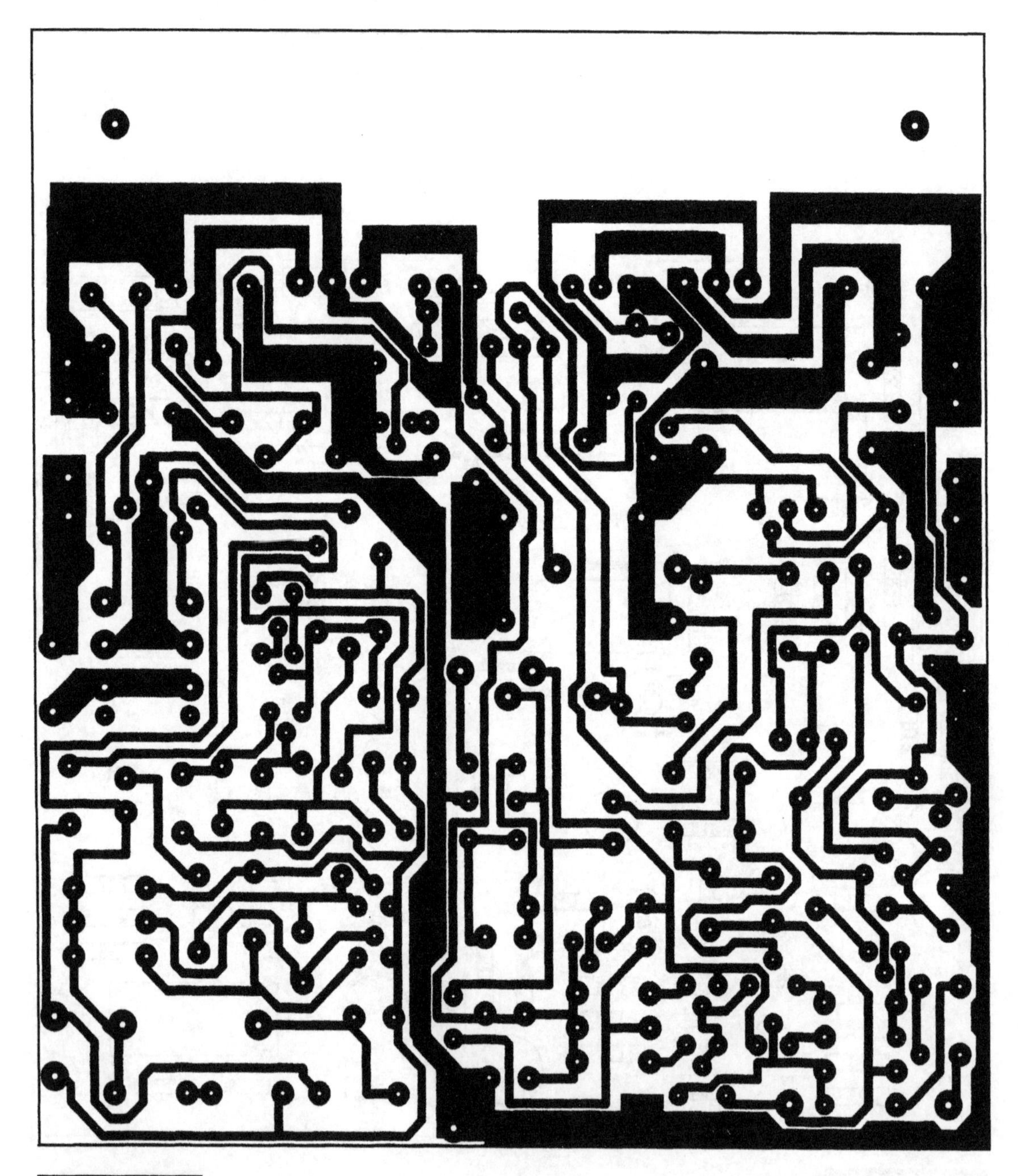

FIGURE C.9

Reflected artwork for the Fig. 6-6 amplifier. (*Note*: This view is 98 percent of full size.) (Repeat of Figure 6-9.)

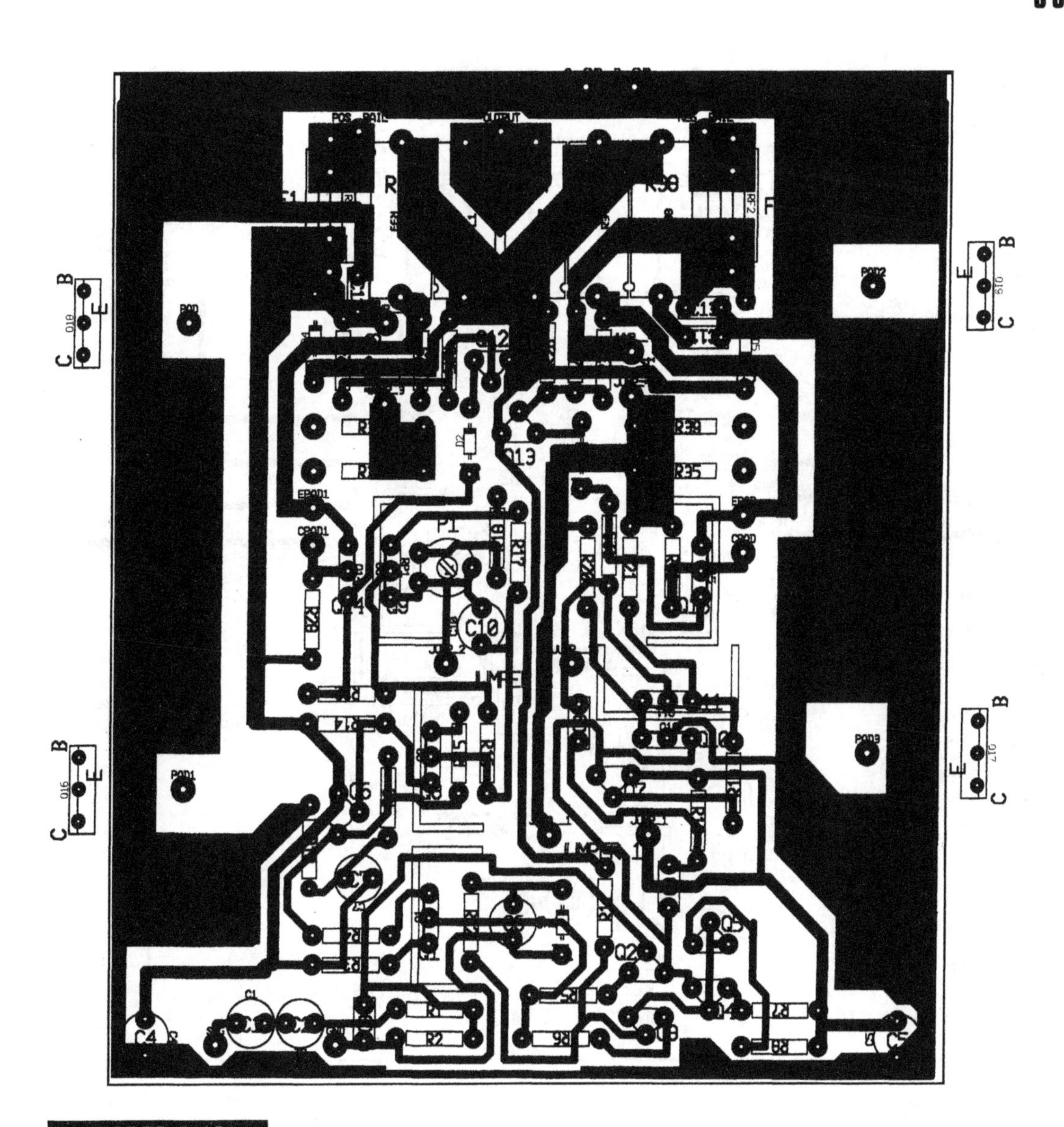

FIGURE C.10

Top layout view of the Fig. 6-11 amplifier. (*Note*: This view is 80 percent of full size.) (Repeat of Figure 6-12.)

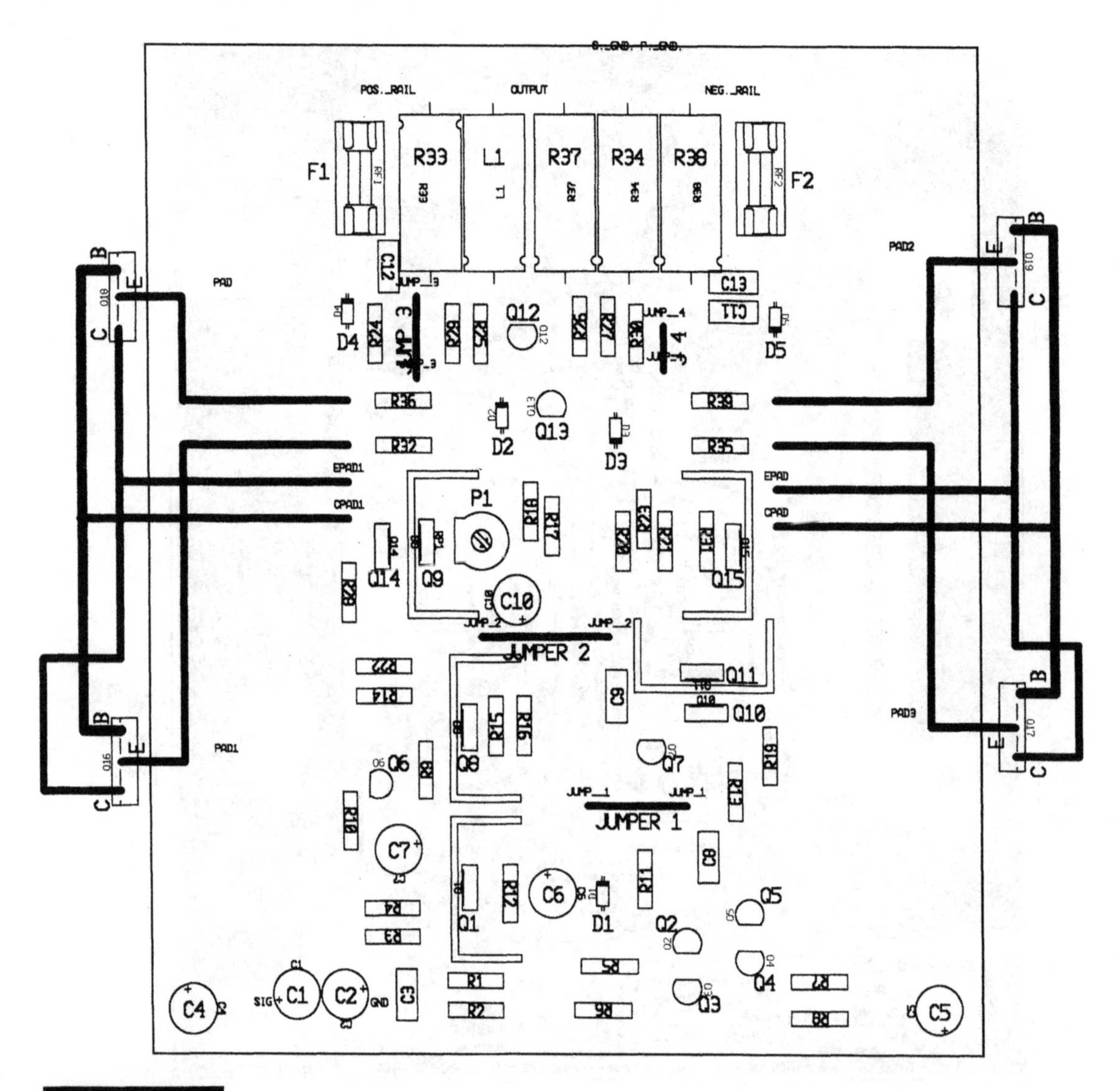

FIGURE C.11

Top view of the Fig. 6-11 amplifier. (*Note*: This view is 78 percent of full size.) (Repeat of Figure 6-13.)

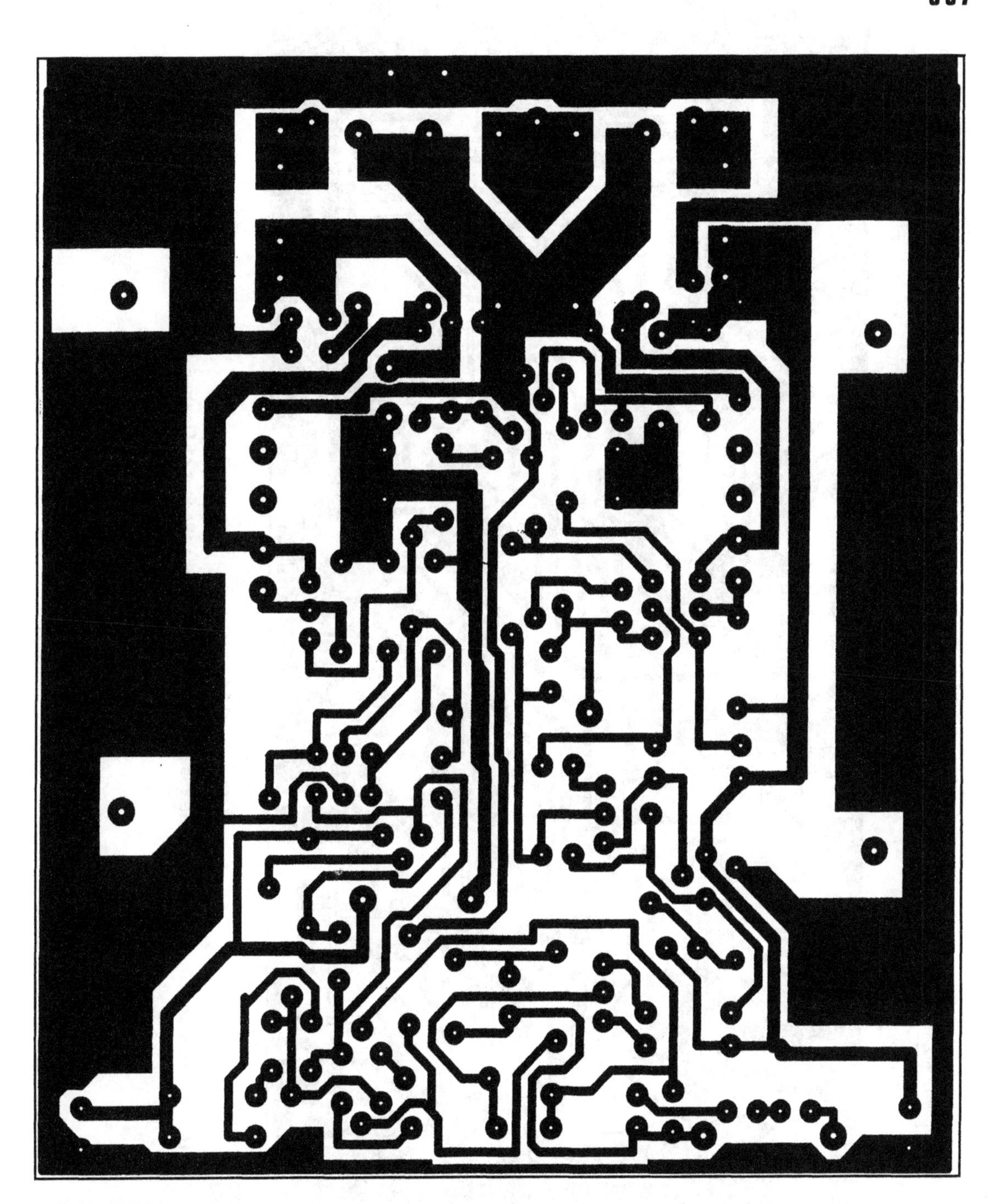

FIGURE C.12

Reflected artwork for the Fig. 6-11 amplifier. (*Note*: This view is 97 percent of full size.) (Repeat of Figure 6-14.)

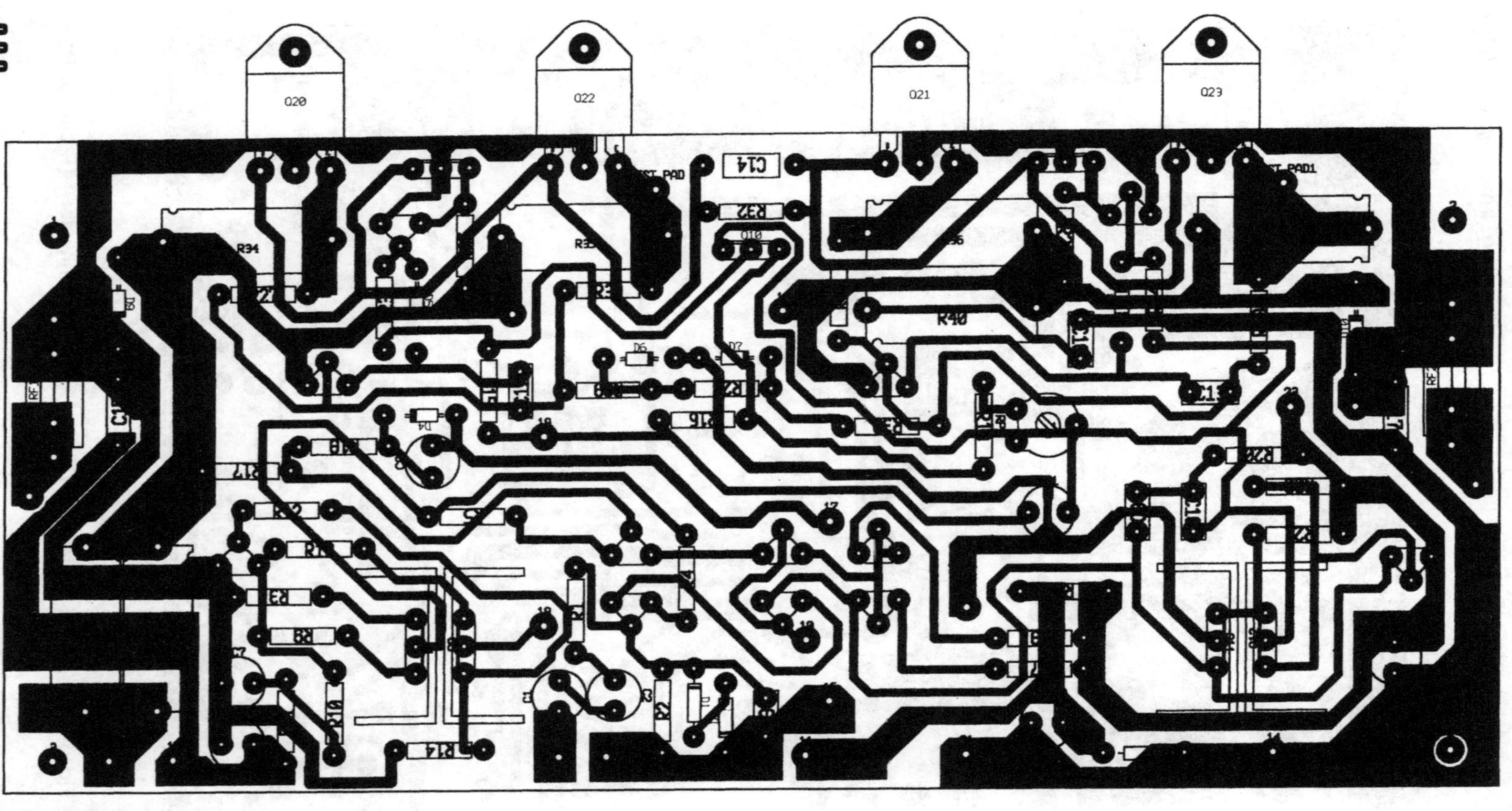

FIGURE C.13

Top layout view of the Fig. 6-15 amplifier. (*Note*: This view is 87 percent of full size.) (Repeat of Figure 6-16.)

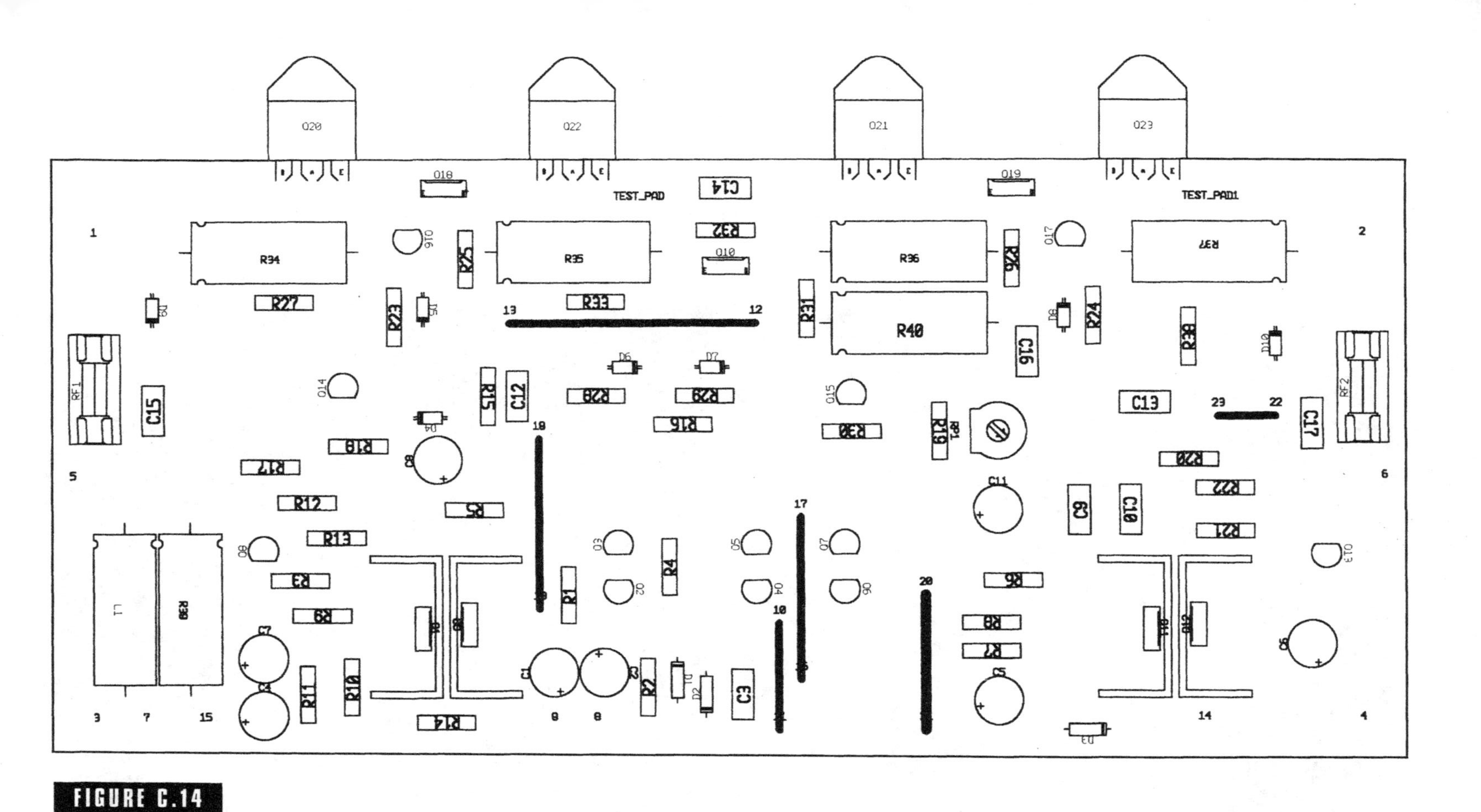

FIGURE C.14

Top view of the Fig. 6-15 amplifier. (*Note*: This view is 87 percent of full size.) (Repeat of Figure 6-17.)

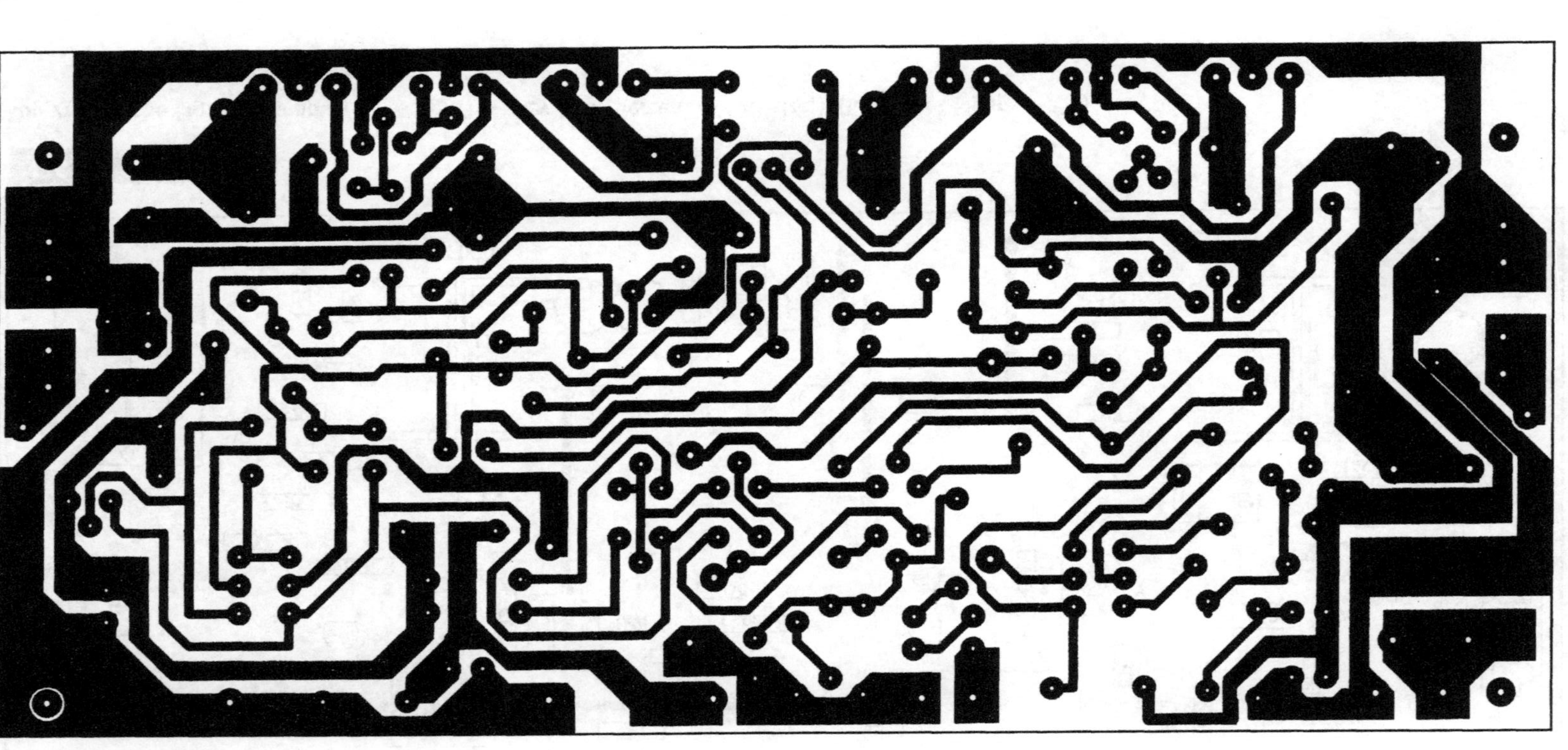

FIGURE C.15

Reflected artwork for the Fig. 6-15 amplifier. (*Note*: This view is 87 percent of full size.) (Repeat of Figure 6-18.)

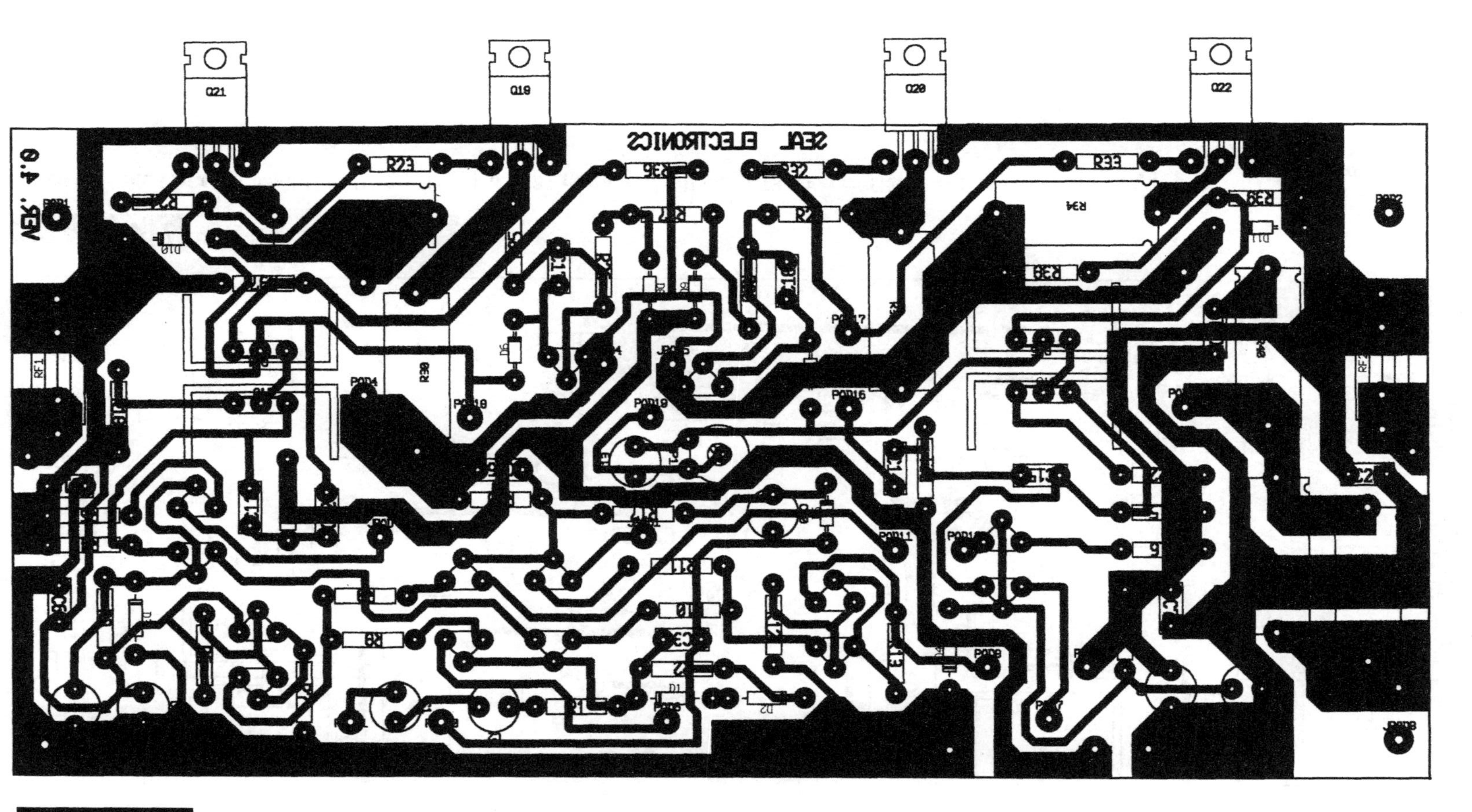

FIGURE C.16

Top layout view of the Fig. 6-21 amplifier. (*Note*: This view is 87 percent of full size.) (Repeat of Figure 6-22.)

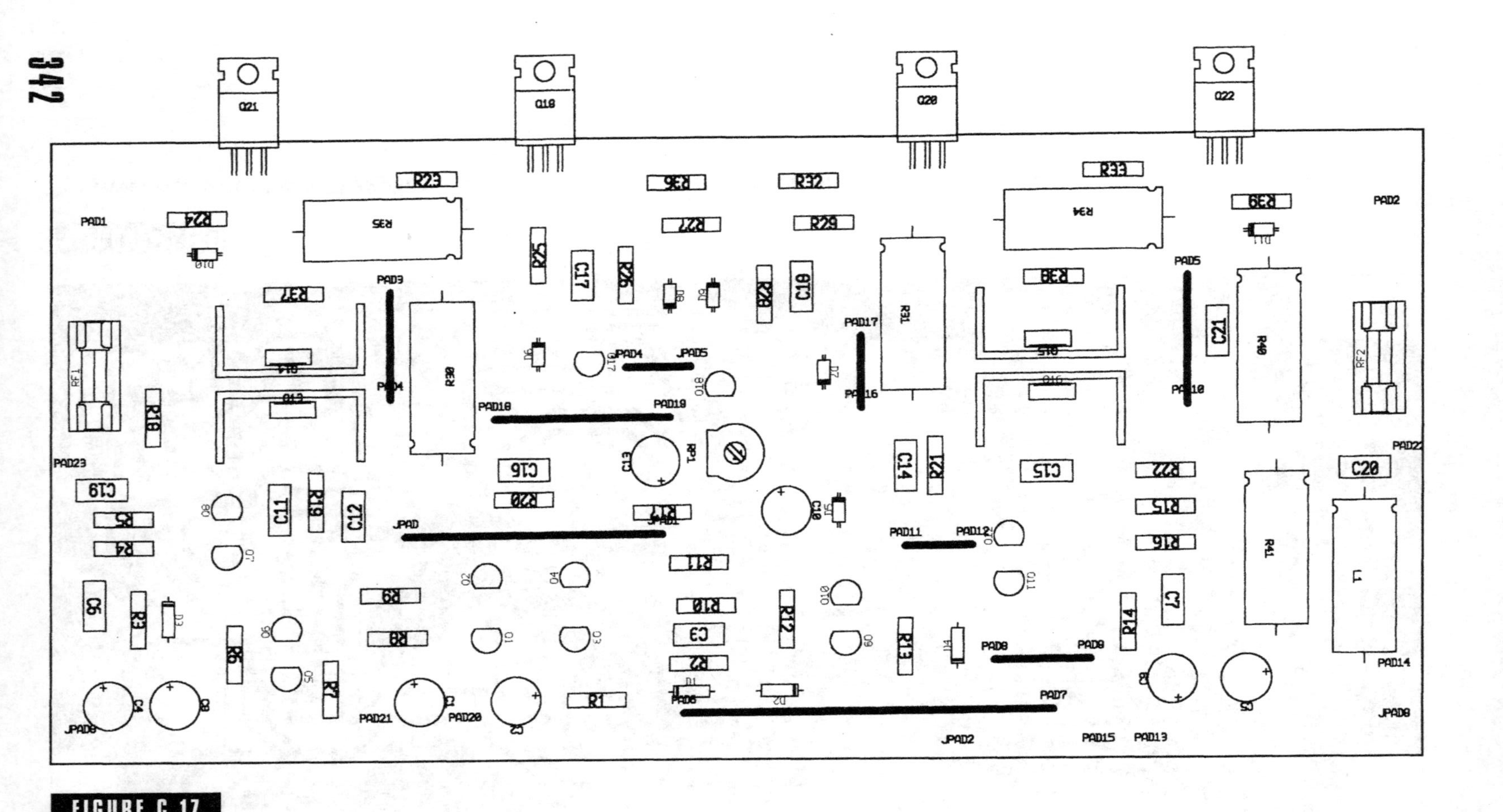

FIGURE C.17

Top view of the Fig. 6-21 amplifier. (*Note*: This view is 87 percent of full size.) (Repeat of Figure 6-23.)

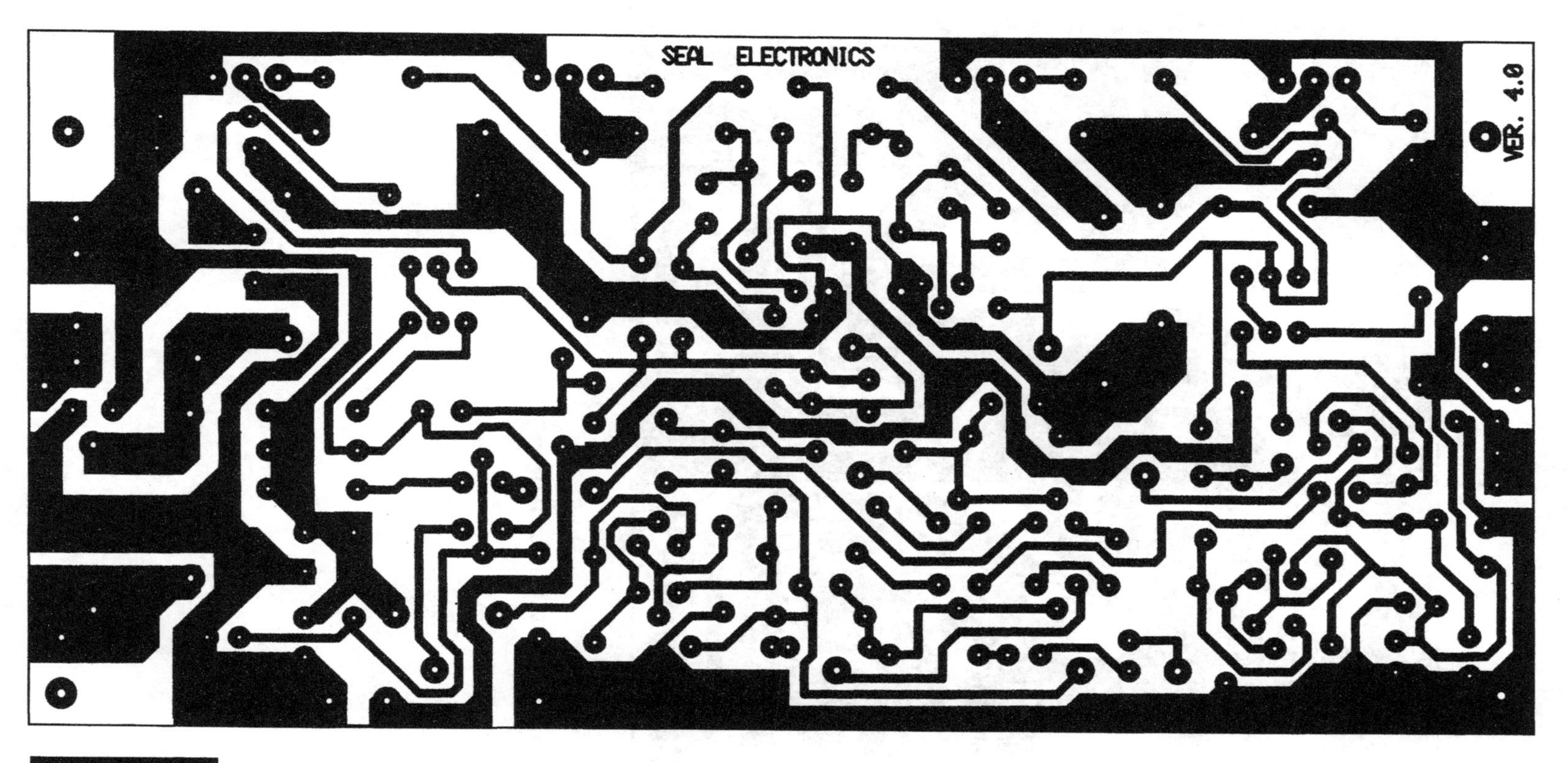

FIGURE C.18

Reflected artwork for the Fig. 6-21 amplifier. (*Note*: This view is 87 percent of full size.) (Repeat of Figure 6-24.)

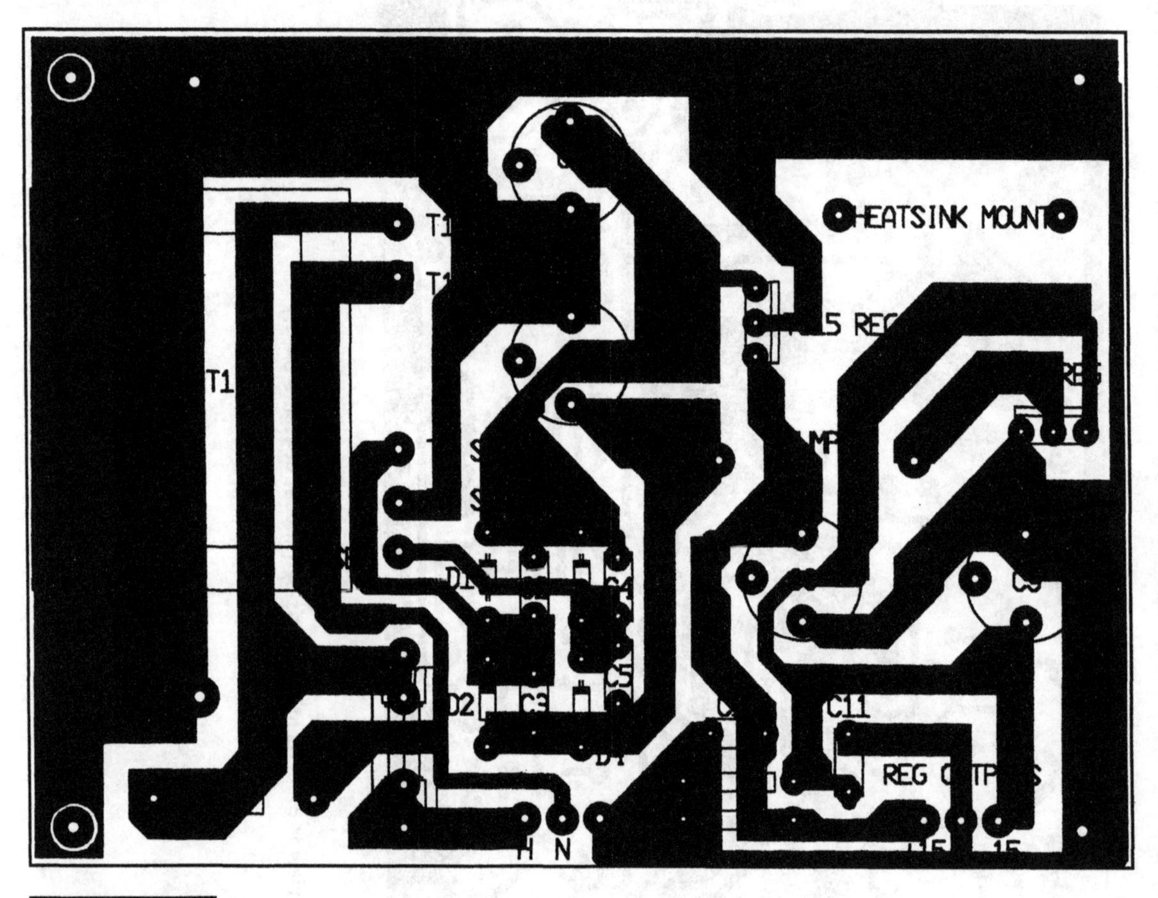

FIGURE C.19

Top view silkscreen of the Fig. 7-8 power supply. (Repeat of Figure 7-9.)

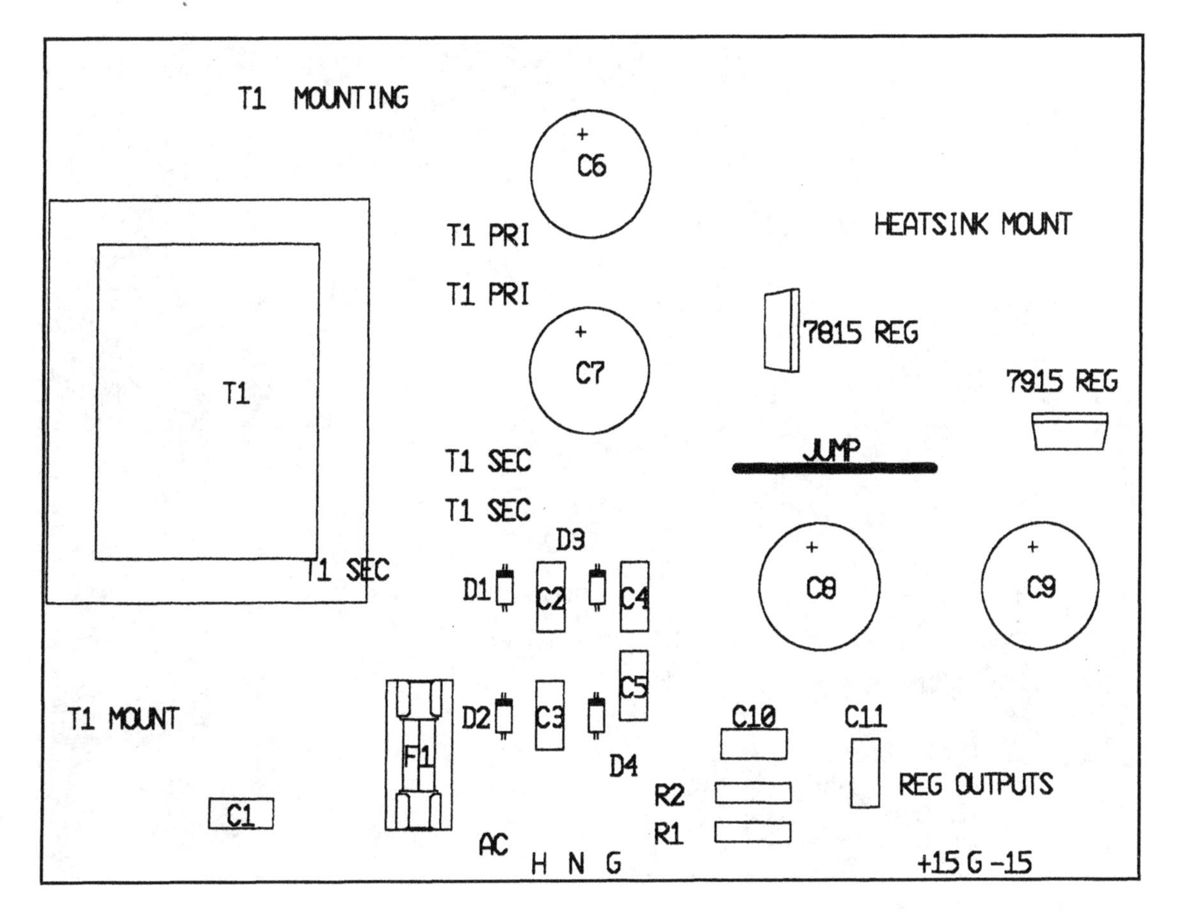

FIGURE C.20

Top view component layout of the Fig. 7-8 power supply. (Repeat of Figure 7-10.)

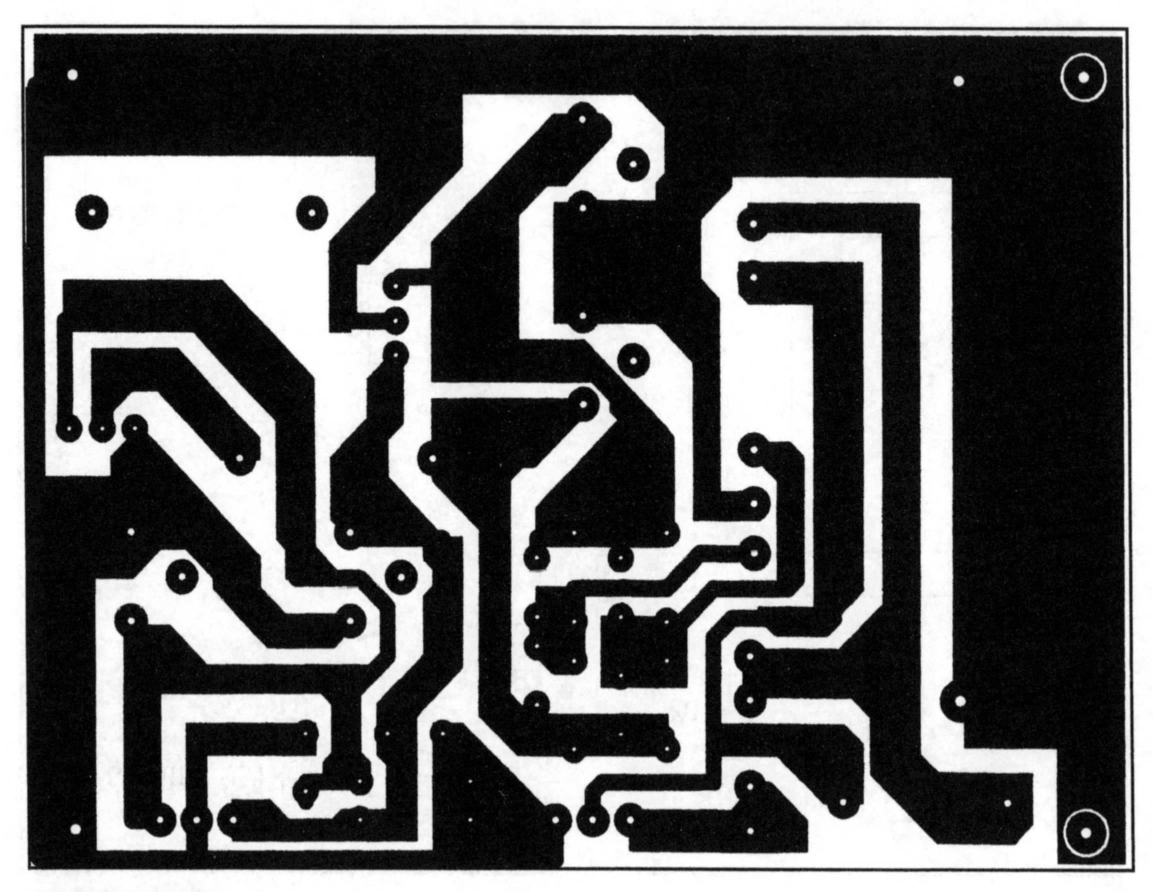

FIGURE C.21

Bottom view reflected artwork of the Fig. 7-8 power supply. (Repeat of Figure 7-11.)

INDEX

ABOUT THE AUTHOR

G. Randy Slone is an electronics engineer, a consultant, and author of five books, including *High-Power Audio Amplifier Construction Manual* and *The TAB Guide to Understanding Electricity and Electronics.* As a process control engineer, his consulting clients have included DuPont, Champion International, and Ralston Purina. A former college instructor, Slone is the owner/operator of SEAL Electronics, and the current senior design engineer for ZUS Audio Inc. He spends much of his time working in his state-of-the-art home electronics laboratory.

CPSIA information can be obtained
at www.ICGtesting.com
Printed in the USA
LVHW060613060720
659790LV00023B/321